Chemistry at the Frontier with Physics and Computer Science

Theory and Computation

Chemistry at the Frontier with Physics and Computer Science

Theory and Computation

Sergio Rampino
Scuola Normale Superiore
Faculty of Science
Pisa, Italy

ELSEVIER

Elsevier
Radarweg 29, PO Box 211, 1000 AE Amsterdam, Netherlands
The Boulevard, Langford Lane, Kidlington, Oxford OX5 1GB, United Kingdom
50 Hampshire Street, 5th Floor, Cambridge, MA 02139, United States

Notices

Knowledge and best practice in this field are constantly changing. As new research and experience broaden our understanding, changes in research methods, professional practices, or medical treatment may become necessary.

Practitioners and researchers must always rely on their own experience and knowledge in evaluating and using any information, methods, compounds, or experiments described herein. In using such information or methods they should be mindful of their own safety and the safety of others, including parties for whom they have a professional responsibility.

To the fullest extent of the law, neither the Publisher nor the authors, contributors, or editors, assume any liability for any injury and/or damage to persons or property as a matter of products liability, negligence or otherwise, or from any use or operation of any methods, products, instructions, or ideas contained in the material herein.

ISBN: 978-0-323-90865-8

For information on all Elsevier publications
visit our website at https://www.elsevier.com/books-and-journals

Publisher: Susan Dennis
Acquisitions Editor: Charles Bath
Editorial Project Manager: Czarina Mae Osuyos
Production Project Manager: Sruthi Satheesh
Designer: Victoria Pearson

Typeset by VTeX
Transferred to Digital Printing 2022

Working together
to grow libraries in
developing countries

www.elsevier.com • www.bookaid.org

To Cécile, Julie and Lilas

Contents

Part III
Electronic structure and chemical bonding

Biography

Sergio Rampino

Sergio Rampino was born in Mesagne (Apulia, Italy) in 1984. He graduated with honors in Chemistry (2007) and Italian Language and Literature (2012) at the University of Perugia, where he also obtained his PhD in Chemistry (2011). In 2017 he was appointed lecturer in Theoretical and Computational Chemistry at the Scuola Normale Superiore in Pisa, where he presently teaches both undergraduate and PhD students. His research, partly carried out at several European research and computing centers, has focused on several topics of general, physical and inorganic chemistry ranging from the quantum dynamics of elementary reactions to relativistic density-functional theory, the analysis of chemical bonding, and the use of virtual-reality technology for chemistry. In 2016 he was awarded the 'Eolo Scrocco' prize by the Division of Theoretical and Computational Chemistry of the Italian Chemical Society.

Preface

Ogni petra àza parete.
('every stone makes the wall a little higher', old apulian saying)

When I entered the classroom to attend my first university lesson, this was a general chemistry one and I was an eighteen-year-old student with a background in arts and humanities and with no idea of what to expect for the next five years. It took me several months to become acquainted with a new paradigm of conceiving and relating to reality, and a couple of years of classes and laboratories of organic, physical, inorganic and biological chemistry before my compass could find the North in a highly fascinating, albeit less popular, facet of chemistry at the crossroads with physics and computer science.

It is in this subdiscipline of chemistry, labeled by the adjectives 'theoretical and computational', that I have pursued the academic path since the start of my PhD project in 2007. In about fifteen years of research I have been lucky enough to work on such diverse and fascinating subjects as the dynamics of chemical reactions, the analysis of chemical bonding, the effects of relativity on chemistry, and the integration of immersive virtual reality in chemical applications. I met extraordinary people and had the chance of visiting research and computing centers in wonderful places such as Edinburgh, Bristol, Toulouse, Paris, Zurich, Barcelona and Cracow, the fond memories of which shall forever be inextricably linked with the topics of my research.

In July 2020, I received an invitation by the Publisher to author a new book in the field of theoretical and physical chemistry, and I had no hesitation in accepting as I regarded it as an opportunity to sum up the work of all these years and offer a personal perspective, with concrete examples from my own research activity, on the discipline on which I had, over time, specialized. I am extremely grateful to the acquisitions editor Anneka Hess for that contact. During the following months, I received precious help from project managers Devlin Person and Veronica III Santos, and from the permissions coordinator Narmatha Mohan. I am grateful to them for their support during the various stages of the book production. I would also like to thank Vicky Pearson Esser for her excellent job in designing the book cover, and the production manager Sruthi Satheesh, together with any other person who was involved in the production of the book.

The organization of this book partly reflects the structure and content of the courses that I held at the Scuola Normale Superiore in Pisa, Italy, from 2017 to 2022, and of a series of outreach lectures that I delivered for a younger audience within the initiative 'La Normale a Scuola' organized by the Scuola Normale Superiore as a support to high-school students during the Covid-19 global health emergency. I am grateful to all the students who attended my courses over these years for their precious feedback. In turn, the presentation of the theoretical aspects of the book, especially those related to the fundamentals of quantum mechanics (Chapter 2) and electronic-structure theory (Chapter 12 and part of Chapter 13), draws on the introductory courses in theoretical chemistry held by Prof. Francesco Tarantelli at the University of Perugia in 2005–2007, while the description of the Verlet algorithm in Section 7.1 draws on useful lecture notes by Prof. Hannes Jónsson.

I am very grateful to my academic mentors Prof. Antonio Laganà and Prof. Vincenzo Barone, for the opportunity of working with them and for the many, many stimulating discussions, and to all the people who have collaborated with me over the years, many of whom figure as coauthors of the articles cited in this book. In particular, I am grateful to Stefano Crocchianti, Dimitrios Skouteris, and Loriano Storchi for initial guidance in, respectively, scientific programming, quantum reactive scattering, and parallel computing, and to Prof. Antonino Polimeno and Prof. Mirco Zerbetto for the opportunity of working on interesting new topics opening new perspectives in my research. I am also grateful to Dr. Kenneth Lawley, Prof. Kim Baldridge, Prof. Stefano Evangelisti, Prof. Gabriel Balint-Kurti, Prof. Ernesto García, Prof. Irene Burghardt, Dr. Mariusz Sterzel, and Prof. Carole Morrison for hosting short-term scientific visits to their groups financed by the European COST and HPC-Europa programmes. Special thanks are finally due to my younger collaborators whose work I had the chance of supervising and to whom I wish the best for their future: Matteo De Santis, Lorenzo Paoloni, Simone Potenti, Surajit Nandi, Bernardo Ballotta, Luca Sagresti, Silvia Alessandrini, and Giovanni Nottoli.

The writing of this book has occurred to me at a time when the Irish flute entered wildly into my life. I am grateful to fellow musicians Stefano Battaglia, Yuri Bernardini, Diego Ceccarelli, Lorenzo Del Grande, Antonio Malacarne, Nico Marraccini, Carlo Rogo, and Giovanni Stea for the hundreds of jigs and reels and for the memorable *craic*, and to maestro Carlo Ipata for several fruitful and lively flute lessons. It has also occurred in not so easy times of precarity, sometimes paradoxically made harder by the same transdisciplinary vocation that animates this book, and that hardly reconciles with the high degree of specialization that characterizes the current academic world. For their support and their presence, I am grateful to my family, Cécile Pirat, Julie and Lilas Rampino, to my family of origin, Gabriele Rampino, Irene and Lucia Ammaturo, Adelaide, Guido, Giuliana and Daniele Rampino, to Drs. Paolo Catanzaro and Arianna Luperini, and to Arianna Federici.

To the loving memory of Jon Mikel Azpiroz.

Sergio Rampino
Scuola Normale Superiore
March 2022

Chapter 1

Introduction and scope

1.1 Introduction and scope

This book is an introduction to theoretical and computational chemistry that aims at framing chemistry between its neighboring disciplines physics and computer science. Modern chemistry may be defined as the branch of science that aims at modeling and understanding the properties and behavior of matter at the atomic and molecular level. Theoretical and computational chemistry, as opposed to experimental chemistry, is a subdiscipline of chemistry pursuing the same objective 'comfortably' from behind a desk, with the aid of pen and paper as well as of modern information and communication technology, rather than in a laboratory. In other words, it uses the laws of physics to predict the behavior of molecular systems in terms of their constituting particles, the nuclei and the electrons.

This well explains the first adjective ('theoretical')[1] of the dyad that characterizes this discipline and that appears as a subtitle to this book. The second adjective ('computational')[2] owes its raison d'être to the fact that the mathematical apparatus involved by the theory is so complex that the use of a computer is mandatory in order to efficiently perform the required number of operations and to handle the involved (often huge) amount of data.

Theory and computation are also the two extremes of a journey with intermediate stages that leads from physics to chemistry through mathematics and computer science, and that fits well into the following hierarchical scheme:

- Theory
- Model
- Method
- Implementation
- Computation

These five stages, which reflect the daily activity of a theoretical and computational chemist, may be described as follows. The physical theory required for an accurate modeling of atoms and molecules turns out to be too complex and onerous to be used as is, so that approximations are often introduced through

---

[1] The word 'theory' curiously derives from a Greek root meaning *observation*, which is the main activity of the 'experiment'. Theory is indeed constructed by rationalizing the outcome of observation, and then used to make predictions on new observations: one may succinctly say that it follows observation in order to precede it.

[2] The word 'computation' has a Latin root meaning to count, to calculate.

Chemistry at the Frontier with Physics and Computer Science
https://doi.org/10.1016/B978-0-32-390865-8.00008-8

models that simplify the physical and mathematical treatment. Models are put into action through methods, which in turn require an effective implementation on a computer. This leads to the last stage, computation, where calculations are performed and their outcome is elaborated and transformed into chemical insight.

The purpose of this book is to illustrate – and accompany the reader through – this journey by conjugating the big picture to the most practical implications or, in other words, by framing chemistry in the overall context where it relates to physics and computer science while at the same time offering practical examples of how to solve a chemical problem from a theoretical and computational perspective, going from theory to models, methods, implementation, and computation. For that purpose, the two central concepts in chemistry of chemical reaction and chemical bonding will be taken as leading themes to illustrate the treatment of the two basic particles that constitute matter at the atomic and molecular level: the nuclei and the electrons, respectively. Accordingly, after reviewing the physics of molecular systems and introducing the Born–Oppenheimer approximation, which allows for the separation of the nuclear and electronic motions, the concepts of chemical reaction and chemical bonding are explored with reference to the underlying nuclear dynamics and electronic structure.

Constant attention is given throughout the book to practical aspects connected to the numerical resolution of the relevant equations and the related implementation in computer programs, which have played a key role in the development of the discipline itself. For a choice of three physical problems (wavepacket dynamics, Hartree–Fock equations, electron-cloud redistribution) a detailed account from theory to implementation is given through the series of chapters *From theory to computing*, while the application of the discussed methodologies to relevant chemical problems from the author's research are illustrated through the chapter series *Applications*. Computer-code examples will be given in the Fortran 95 programming language, and should be fully accessible to readers with intermediate-level experience in computer programming. On the other hand, the relation between chemistry and computer science is nowadays rapidly evolving due to the constant progress in information and communication technology, so that the role of computer science in chemistry goes well beyond the traditional aspect of scientific computing. The final part of the book deals with the most recent perspectives due to the connection of chemistry with computer science.

In more detail, the book is structured as follows:

In Part I, *Physics and chemistry*, the physics of molecular systems is reviewed with emphasis on the paradigm change from classical to quantum mechanics (Chapter 2). Chapter 3 explores the relation between physics and chemistry and provides an introduction to the concepts of chemical reactions and chemical bonding that will be developed in Parts II and III, respectively, with a focus on the perspective of a chemist as opposed to that of a physicist. Chapter 4 provides a brief historical account on the birth and evolution of theoretical and computational chemistry with reference to the topics treated in this book.

Part II, *Nuclear dynamics and chemical reactions*, is devoted to the treatment of the nuclear motion and the illustration of the modeling of chemical-reaction dynamics. In particular, Chapter 5 revisits chemical reactions from the physical perspective of collision theory. Chapter 6 is devoted to a cornerstone concept in chemistry, that of the potential-energy surface. Chapter 7 reviews the main theoretical approaches for the treatment of the nuclear motion. Chapter 8, the first of the series *From theory to computing*, focuses on the implementation of a quantum method for modeling the evolution of a simple reaction. In Chapter 9, the relation between the microscopic view of single collisional processes and the macroscopic one of chemical kinetics is addressed. Chapter 10 illustrates the study of an astrochemical reaction through the previously discussed methodologies. Finally, Chapter 11 provides an overview of the main methods tackling nuclear dynamics in more complex contexts.

In Part III, *Electronic structure and chemical bonding*, we move on to the other side of the Born–Oppenheimer approximation and deal with the treatment of the electrons and their role in chemical bonding. After illustrating the fundamental electronic-structure theory (Chapter 12), the implementation of a basic electronic-structure program is discussed in Chapter 13 (second of the series *From theory to computing*). Then the chemical concepts of atoms and bonds are revisited in terms of the previously introduced mathematical and physical formalism (Chapter 14), and the implementation of a computer program for analyzing the electron-charge redistribution occurring upon chemical bonding is discussed (Chapter 15, third of the series *From theory to computing*). Chapter 16, through a study of chemical bonding in a diverse class of metal–carbonyl complexes, illustrates how the chemical concepts of σ-donation and π-backdonation can be modeled through the previously discussed methodologies rooted in the quantum mechanics of the electrons. Chapter 17, the last chapter of Part III, deals with the frontier topic of the effects of relativity in chemistry, with a focus on the chemistry of the so-called superheavy elements.

Part IV, *Chemistry and computer science*, the final part of the book, explores the connections between chemistry and computer science. In particular, Chapter 18 focuses on the traditional aspect of scientific computing. After reviewing the basics of computer programming, the topic of parallel computing (in its high-performance and high-throughput flavors) is discussed, and a concrete example of parallelization of a relativistic electronic-structure program is given. Chapter 19 shows how state-of-the-art virtual-reality technology can enhance scientific research and education in chemistry through the description of two applications directly related to the modeling of chemical reactions and the analysis of chemical bonding discussed in Parts II and III of the book, respectively. In Chapter 20, emerging paradigms based on artificial intelligence and machine learning for a data-driven approach to chemistry are discussed. The concluding chapter, Chapter 21, deals with challenges and implications of the most recent perspective of chemistry as an open science.

1.2 Notation and conventions

For the reader's convenience a summary of the notation and conventions adopted in the book is given in the following. A selection of the most relevant symbols used in the book is given below. Symbols are grouped together on the basis of the domain to which they are relevant, and listed in an order that largely reflects their appearance in the book. A few symbols are used with different meaning in different parts of the book so as to adhere to conventional notations specific to some topics (e.g., the scattering matrix in nuclear dynamics and the overlap matrix in electronic structure are both often notated with S). However, this should have no effect on the overall readability as it should always be clear from the context what the repeated symbols refer to. In some parts of the book, where declared, use is made of atomic units, whereby the mass of the electron m_e, the charge of the electron $|e|$, Planck's constant $\hbar = \frac{h}{2\pi}$, and the Bohr radius $a_0 = \frac{\hbar^2}{k_c m_e e^2}$ are all set to unity (from the expression of a_0, it follows that also the Coulomb constant k_c is equal to 1 in atomic units). Lowercase and uppercase bold characters are generally adopted for vectors and matrices, respectively. Curly brackets are used to indicate sets of variables. The i subscript denotes variables related to the ith electron, the a subscript denotes variables related to the ath nucleus. When more than one subscript is necessary, i, j, k, l are used for electrons or molecular orbitals, a and b are used for nuclei, and p, q, r, s are used for atomic orbitals (basis functions). The symbol δ_{ij} indicates the Kronecker delta ($\delta_{ij} = 0$ if $i \neq j$, $\delta_{ij} = 1$ if $i = j$). Some integrals are expressed in Dirac (or bra-ket) notation, whereby $\langle \varphi | \psi \rangle = \int \varphi^*(x) \psi(x) \, dx$ and $\langle \varphi | \hat{f} | \psi \rangle = \int \varphi^*(x) \hat{f} \psi(x) \, dx$.

General theory

$P(x)$	probability distribution
$p(x)$	probability density
ξ	collective coordinate
p	momentum vector
$\hat{p}$	momentum operator
r	position vector
$\hat{r}$	position operator

Molecular quantum mechanics

$\{r_i\}$	set of electronic coordinates
$\{r_a\}$	set of nuclear coordinates
Ψ	molecular wavefunction
Γ	nuclear wavefunction

ψ	electronic wavefunction
Z_a	atomic number of nucleus a
m_a	mass of nucleus a
m_e	mass of electron
e	charge of electron
$\hbar$	Planck's constant
k_c	Coulomb's constant
E	total energy
$\hat{U}$	time-evolution operator
$\hat{H}$	Hamiltonian operator
$\hat{T}$	kinetic-energy operator
$\hat{V}$	potential-energy operator
$\hat{T}_N$	nuclear–kinetic energy operator
$\hat{H}_e$	electronic Hamiltonian
ε_n	energy of electronic state n

Nuclear dynamics

b	impact parameter
σ	cross section
r_e	diatomic equilibrium distance
k_e	diatomic force constant at equilibrium distance
D_e	diatomic dissociation energy
n	bond-order coordinate
β	bond order–exponential mapping parameter
r_{BC}, r_{AB}	interatomic distances
Φ	$\widehat{ABC}$ angle
R, r, θ	reactant Jacobi coordinates
R', r', θ'	product Jacobi coordinates
γ_v	vibrational wavefunction
ε_v	energy of vibrational state v
χ	wavepacket
S	scattering matrix
$k(T)$	thermal rate constant (function of temperature T)
$k(E)$	microcanonical rate constant (function of energy E)
$\rho(E)$	density of states
v_R	reaction rate

Electronic structure

φ	molecular orbital
χ	atomic orbital (basis function)
ε_i	energy of orbital i

μ	label for N-tuple of molecular orbitals
Θ_μ	Hartree product associated with N-tuple μ
Φ_μ	Slater determinant associated with N-tuple μ
$\hat{h}$	one-electron part of the electronic Hamiltonian
$\hat{j}^{(j)}$	Coulomb operator for orbital j
$\hat{J}$	total Coulomb operator
$\hat{k}^{(j)}$	exchange operator for orbital j
$\hat{K}$	total exchange operator
ρ	electron density
S	overlap matrix
C	molecular-orbital coefficient matrix
D	density matrix
F	Fock matrix
h	one-electron part of the Fock matrix
J	Coulomb matrix
K	exchange matrix

The original figures of this book were produced using the following software: Gnuplot, Pymol, and the PSTricks LaTeX package.

Part I

Physics and chemistry

Chapter 2

The physics of molecular systems

2.1 Classical and quantum mechanics

As already mentioned in Chapter 1, we shall be concerned with the modeling of the structure and behavior of matter at the atomic and molecular level. At this level of detail, we shall consider the nuclei and the electrons as the 'elementary' particles of matter, and we shall therefore focus on the laws of physics that govern their behavior.

Now, how do we imagine nuclei and electrons? Most likely, in the first instance, as particles – more or less point-like – possessing a mass and a charge, and localized in a given point of the three-dimensional physical space. This is certainly how nuclei and electrons appear in popular textbook representations, such as the atom model of Rutherford or Bohr, with the orbiting electrons arranged around a nucleus and caught as in a photograph, or the Lewis diagrams of electron-pair sharing in chemical bonding, with nonbonded electrons depicted as dots near the respective nuclei. Both these representations are rooted in a classical view of physics: within such a framework, modeling and predicting the behavior of a molecular system ultimately means being able to simulate the evolution in time of the position of its constituting particles. In classical mechanics, this can be easily achieved through Newton's laws of motion, which are iconically represented by the mathematical expression for Newton's second law[1]

$$f = ma \, . \tag{2.1}$$

According to Newton's laws, given a particle of mass m in a given point of space, the evolution of its position in time can be easily determined once the forces acting on the particle are known. Thinking of nuclei and electrons as charged point particles interacting through a Coulomb electrostatic potential and generalizing Eq. (2.1) to the multiparticle case, one would readily obtain the tools for simulating the evolution of the resulting molecular system. However, while classical mechanics finds widespread usage in theoretical and computational chemistry for dealing with 'higher-level', coarse-grained aspects of chemistry (see, for instance, Chapters 7 and 11 of this book), it unfortunately fails in

---

[1] As mentioned in Chapter 1, Section 1.2, vectors will generally be indicated with lowercase bold characters throughout the book. Uppercase bold characters will be used for matrices.

the correct description of molecular systems containing the lightest elements (hydrogen, notably) and is completely inappropriate for treating the motion of the electrons.

In fact, as early 20th-century experiments suggested, predictions on the behavior of very light particles based on classical mechanics provide results that are not in agreement with the experimental observations. It was then necessary – and this is a bright example of what the philosopher of science Thomas Kuhn theorized as a scientific revolution [1] – to develop a new theory capable of describing, on an equal footing, the behavior of both macroscopic and microscopic physical systems. This involved the abandonment of some basic premises of classical mechanics, however obvious they may appear, and a paradigm shift towards a new, rather counterintuitive theory, the quantum theory, that we shall briefly review in the following while referring the reader to dedicated textbooks (Refs. [2] and [3], for instance) for further details.[2]

In classical mechanics, in line with the perception of reality that we have everyday through our senses, we assume that objects have properties (position in space, velocity, energy, etc.) that can be described with precise numerical values, and whose existence is objective, i.e., independent of the observer. As already mentioned, the evolution of these values in time can be predicted by using the classical laws of motion. Now, the most fundamental and radical change of view in switching from classical to quantum mechanics is that the unique and objective value of a physical quantity is replaced with the much vaguer concept of a probability distribution: a mathematical function that, for a given value of a variable, returns the probability that the value, after a measurement, will be observed. In other words, in quantum mechanics an object by itself does not possess a precise value for a given quantity, including its position in space. A precise value for a given physical quantity comes to exist only when an observer makes a measurement of that quantity, and the likelihood of the result will be described by the associated probability distribution. This means that measurements of the same property in identical conditions (think of rewinding time and repeating the measure, or performing the measure on clones of the same system) would typically give different results.

The quantum theory, accordingly, involves an inherent uncertainty on the value of physical properties, which replaces the uniqueness and objectivity assumed in classical physics. The measurement of a physical quantity can indeed be conveniently thought of as the throw of a die. This has, in fact, an associated probability distribution $P(x)$ that describes how likely, after the die is rolled, a result x will be obtained. A few examples will help in clarifying this analogy and in becoming familiar with these concepts. The probability distribution associated with the throw of one die is nonzero only in a finite countable set: it

---

[2] It is worth noting here that the early decades of the 20th century were also the stage for another important scientific revolution that gave birth to Einstein's relativistic theory. Relativistic effects in chemistry are certainly smaller than quantum effects, and the conjugation of relativity with quantum mechanics is still an open challenge. Chapter 17 of this book is devoted to some aspects of the relation between relativity and chemistry.

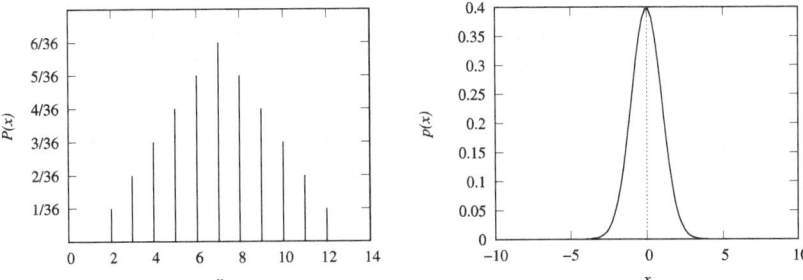

FIGURE 2.1 Probability distribution for the throw of two dice (left-hand side). Probability density described by a Gaussian function (right-hand side).

has indeed a constant value $1/6$ for values of x ranging from 1 to 6. For a throw of two dice whose result is the sum of the results of the throws of each die, the probability distribution looks like that plotted in the left-hand side of Fig. 2.1. It will be a discrete distribution having nonzero values for integer x between 2 and 12, and will be symmetric around $x = 7$. Probability distributions may in general be continuous, and are in this case said to be probability densities. We will notate a probability density as $p(x)$.[3]

In quantum mechanics, the position of an object is described by a three-dimensional probability density function of the spatial coordinates x, y, and z. In general, the position of the particles making up a physical system will be described by a probability density that is a function of the spatial coordinates of all particles that we will collectively notate as ξ. In contrast to the position, the other physical quantities associated with a physical system turn out to be generally described by discrete probability distributions as a result of the action of confining forces on the systems, and are said to be 'quantized' (hence the name of the theory).

Given a probability distribution (or density), an important parameter is the associated average value. This may be regarded as the most representative value of that distribution, the value that is expected to be observed most often if many measurements are performed (it is, in fact, also called the 'expectation value'). In the case of a discrete probability distribution $P(x)$, the average value is de-

---

[3] The probability density is the probability that the result of a measurement will be in an infinitely small interval x divided by the width of this interval. The probability itself is thus given by the integral

$$P(a, b) = \int_a^b p(x)\, dx\,, \tag{2.2}$$

and $p(x)$ is simply the derivative of the probability, $p(x) = dP(x)/dx$. Typically, the probability is normalized to unity:

$$\int_{-\infty}^{\infty} p(x)\, dx = 1\,. \tag{2.3}$$

fined by a weighed sum of the allowed values, each multiplied by its probability:

$$\langle x \rangle = \sum_i P(x_i)x_i .$$

(2.4)

Analogously, the average value of a continuous distribution represented by the probability density $p(x)$ is defined as the integral:

$$\langle x \rangle = \int_{-\infty}^{\infty} p(x)x \, dx .$$

(2.5)

Now, the quantum-mechanical probability distributions of all physical properties of a system are interrelated and a special role in such interrelation is played by the probability density for the position, or more precisely by a mathematical function (in general, a complex one) whose square modulus gives the probability density for the position at a given time t. Such a function is called the system wavefunction and is generally indicated as $\psi(\xi, t)$ (through the book we will also use other Greek letters to differentiate wavefunctions associated with different physical systems). In detail, the probability of finding the system in the multidimensional volume V of the space of the coordinates ξ at a given time t, is proportional to the square modulus of the wavefunction:

$$P(V, t) = \int_V |\psi(\xi, t)|^2 \, d\xi .$$

(2.6)

Due to its relation with $P(V, t)$, the wavefunction $\psi(\xi, t)$ is also called the 'probability amplitude' and is usually normalized[4] such that

$$\int_{-\infty}^{\infty} |\psi(\xi, t)|^2 \, d\xi = 1 .$$

(2.7)

Thus, by itself the wavefunction has no physical meaning; at each instant, however, its square modulus is proportional to the probability of the system to be in the spatial configuration ξ.[5]

---

[4] In principle, it is not required that the wavefunction return a normalized probability density. Indeed, the value of the wavefunction at a given point of the space acquires its meaning in relation to the values of the wavefunction in the other points of the space. An unnormalized wavefunction can be normalized by computing the value a of the integral in Eq. (2.6) and multiplying the wavefunction by $1/\sqrt{a}$.

[5] The interpretation – the bridge between the mathematical apparatus and the reality – of quantum mechanics adopted in this book is that deriving from Max Born (1882–1970, Nobel Prize in Physics 1954) and the so-called Danish school [4].

Given a single-particle system with position vector $r \equiv (x, y, z)$, the average value for the position of the particle is[6]

$$\langle r \rangle = \int \psi^*(r) r \psi(r) \, dr \, , \tag{2.8}$$

where we omitted the integration limits and separated the complex conjugate factors of $|\psi(\xi)|^2$ (in analogy with Eq. (2.9) that we shall discuss later).

Now, a generic physical quantity F will have an associated probability distribution or density featuring an average value $\langle F \rangle$. As already mentioned, such a probability distribution or density is related to the probability density for the position or, more precisely, to the wavefunction. This relation is incorporated in the definition of the quantum-mechanical operator $\hat{F}$

$$\langle F \rangle = \int \psi^*(r) \hat{F} \psi(r) \, dr \, . \tag{2.9}$$

In other words, the operator $\hat{F}$, which is peculiar to the physical quantity F, acts on the wavefunction $\psi(r)$ so as to extract, according to Eq. (2.9), the average value associated with the probability distribution of F.

Important quantum-mechanical operators that we will be concerned with in this book are the momentum, $\hat{p}$, and the kinetic-energy, $\hat{T}$, operators:

$$\hat{p} = -i\hbar \nabla \, , \tag{2.10}$$

$$\hat{T} = -\frac{\hbar^2}{2m} \nabla^2 \, , \tag{2.11}$$

where the gradient operator ∇ is

$$\nabla = \begin{pmatrix} \frac{\partial}{\partial x} \\ \frac{\partial}{\partial y} \\ \frac{\partial}{\partial z} \end{pmatrix} . \tag{2.12}$$

By comparison of Eqs. (2.8) with (2.9), it follows that the operator $\hat{r}$ for the position is simply r. It will also be useful to add that the operator for a physical quantity that is a function $f(r)$ of the position (such as, for instance, the Coulomb electrostatic potential) is simply $f(r)$, since the average value for $f(r)$ is worked out from the probability density $p(r)$ as follows

$$\langle f(r) \rangle = \int p(r) f(r) \, dr \, . \tag{2.13}$$

---

[6] With the notation $\langle r \rangle$ we indicate a vector whose components are the average values of the x, y, and z coordinates, i.e., $\langle x \rangle = \int \psi^*(r) x \psi(r) \, dr$, $\langle y \rangle = \int \psi^*(r) y \psi(r) \, dr$, and $\langle z \rangle = \int \psi^*(r) z \psi(r) \, dr$, respectively.

The extension of the above-mentioned operators to the case of a many-particle system follows easily by substituting r and p with the composite vectors of the position and momentum vectors of all particles.

Another important parameter associated with a probability distribution is its variance, which describes how much the distribution deviates from its average value. One may seek particular states of a system for which the average value of a given physical property F is the only allowed value, i.e., states for which the probability distribution or density for F has null variance and is entirely collapsed on the only allowed value $\langle F \rangle$ while being null anywhere else. It can be shown that such states would be described by wavefunctions that are solutions of the following, so-called 'eigenvalue', equation:

$$\hat{F}\psi(\xi, t) = F\psi(\xi, t) \,. \tag{2.14}$$

In general, this equation does not have solutions for any value of F. Solving the eigenvalue equation means both finding the allowed values for F, the so-called 'eigenvalues', and the corresponding $\psi(\xi, t)$, the so-called 'eigenfunctions'. The ensemble of the eigenvalues constitute the so-called spectrum of the operator. As already mentioned, as a result of the confinement of the system in space due to forces acting on it, the spectrum is often a discrete one, with the eigenvalues being labeled by an integer index called the quantum number. If there is more than one eigenfunction associated with the same eigenvalue, than the eigenvalue is said to be degenerate.

2.2   The Schrödinger equation and the molecular Hamiltonian

The fundamental equation of quantum mechanics is the time-dependent Schrödinger equation [5]:

$$i\hbar\frac{\partial}{\partial t}\psi(\xi, t) = \hat{H}\psi(\xi, t) \,, \tag{2.15}$$

where $\hat{H}$ is the Hamiltonian operator (the quantum-mechanical operator of total energy), corresponding to the Hamiltonian of classical mechanics and gathering all the terms that concur in determining the time evolution of the system. This equation may be regarded as the analog of the Newton equation of classical mechanics, in that it relates the time evolution of the position of a system (represented here by its wavefunction) to the forces acting on the constituting particles and encoded in the Hamiltonian. This is the fundamental equation, the only one that the wavefunction has to satisfy together with some initial postulates and the following symmetry requirements: the wavefunction must be antisymmetric (must change sign) with respect to a permutation of a pair of fermions (such as electrons, protons, and neutrons), and symmetric with respect to a permutation of a pair of bosons (such as photons, phonons and α particles).

Since the Schrödinger equation is a first-order differential equation in t, its solution must have the form

$$\psi(\xi, t) = \hat{U}(t, t_0)\psi(\xi, t_0),\tag{2.16}$$

where the operator $\hat{U}(t, t_0)$ is called the time-evolution operator, or wave operator. It is a linear operator and satisfies the initial condition $\hat{U}(t_0, t_0) = 1$. If the Hamiltonian operator is independent of time, the time-evolution operator reads

$$\hat{U}(t, t_0) = e^{\frac{-i\hat{H}(t-t_0)}{\hbar}},\tag{2.17}$$

and Eq. (2.16) can be reformulated as

$$\psi(\xi, t) = e^{\frac{-i\hat{H}(t-t_0)}{\hbar}}\psi(\xi, t_0).\tag{2.18}$$

In this case the Schrödinger equation can be recast in a time-independent form:

$$\hat{H}\psi(\xi) = E\psi(\xi),\tag{2.19}$$

which is the form in which the Schrödinger equation most often appears and with which the reader will probably be more familiar. Solving Eq. (2.19) means finding the 'shape' of the wavefunction $\psi(\xi)$ that responds to the physics encoded in the Hamiltonian $\hat{H}$. Analytical solutions of Eq. (2.19) only exist for very simple model systems that are commonly treated in quantum-mechanics textbooks, such as the particle in a box, the electron in a central field, the harmonic or anharmonic oscillator. For the vast majority of systems of interest in chemistry, numerical solutions to Eq. (2.19) have to be sought.

As already mentioned, the focus of this book is on molecular systems in terms of their constituting particles the nuclei and the electrons. We shall therefore tailor Eq. (2.19) to systems made up of a given set of nuclei and a given number of electrons. To make the formalism simpler, atomic units will be here adopted (see also Chapter 1, Section 1.2). Accordingly, the mass of the electron m_e, charge of the electron $|e|$, Planck's constant $\hbar = \frac{h}{2\pi}$, the Bohr radius $a_0 = \frac{\hbar^2}{k_c m_e e^2}$ are all set to unity (from the expression of a_0, it follows that also the Coulomb constant k_c is equal to 1 in atomic units).

Indicating with Ψ the molecular wavefunction and with r_i and r_a the position vectors of electron i and nucleus a, respectively, the molecular Schrödinger equation reads

$$\hat{H}\Psi(\{r_i\}, \{r_a\}) = E\Psi(\{r_i\}, \{r_a\}),\tag{2.20}$$

where $\{r_i\}$ and $\{r_a\}$ indicate the sets of coordinates of all electrons and nuclei, respectively. For an isolated, nonrelativistic system of electrons and nuclei interacting through a Coulomb electrostatic potential,[7] the Hamiltonian consists of a term $\hat{T}_e$ accounting for the kinetic energy of the electrons, a term $\hat{T}_N$ accounting for the kinetic energy of the nuclei, and a term $V(\{r_i\}, \{r_a\})$ accounting for the nucleus–electron, electron–electron, and nucleus–nucleus interactions:

$$\hat{H} = \hat{T}_e + \hat{T}_N + V(\{r_i\}, \{r_a\}) \,. \tag{2.21}$$

Based on the quantum-mechanical operators introduced in Section 2.1 (see Eqs. (2.10)–(2.12) and the subsequent discussion), the terms $\hat{T}_e$, $\hat{T}_N$ and $V(\{r_i\}, \{r_a\})$ in atomic units read:

$$\hat{T}_e = -\frac{1}{2} \sum_i \nabla_i^2 \,, \tag{2.22}$$

$$\hat{T}_N = -\frac{1}{2} \sum_a \frac{\nabla_a^2}{m_a} \,, \tag{2.23}$$

$$V(\{r_i\}, \{r_a\}) = -\sum_{a,i} \frac{Z_a}{r_{ai}} + \sum_{i>j} \frac{1}{r_{ij}} + \sum_{a>b} \frac{Z_a Z_b}{r_{ab}} \,, \tag{2.24}$$

where ∇_i^2 is the Laplacian with respect to the coordinates of the ith electron, ∇_a^2 is the Laplacian with respect to the coordinates of the ath nucleus, m_a and Z_a are the mass and the charge, respectively, of the ath nucleus, and $r_{ai} = |r_a - r_i|$, $r_{ij} = |r_i - r_j|$, $r_{ab} = |r_a - r_b|$. The notation $i > j$ and $a > b$ in the summations of Eq. (2.24) indicates that each electron–electron and nucleus–nucleus couple, respectively, is counted only once.

2.3 The Born–Oppenheimer approximation

The Schrödinger equation for a molecular system, Eq. (2.20), however simple and concise it may look, is extremely difficult to solve due to the nature of the molecular Hamiltonian and the high dimensionality of the molecular wavefunction. As a matter of fact, as we shall see throughout this book, a number of approximations have to be introduced in order to transform Eq. (2.20) into a problem that can be managed both from a mathematical and a computational point of view. The first of these dates back to the early days of quantum mechanics and is due to physicists Max Born and Robert Oppenheimer [6]. It is known as the Born–Oppenheimer approximation and has since been of paramount importance for the discussion and interpretation of several aspects of chemistry.

The essence of the Born–Oppenheimer approximation is that the motion of the electrons can be treated separately from the motion of the nuclei. The nuclei, which are at least 1836 times heavier than the electrons, can be considered to move infinitely slower than the electrons or, in other words, the electrons can

---

[7] For two interacting charges q and Q at a distance r, the Coulomb electrostatic potential is given by the expression $k_c \frac{qQ}{r}$.

be considered to rearrange almost instantaneously around the moving nuclei. As a consequence of this approximation, a separate problem for the motion of the electrons at a given fixed geometry of the molecule can be unbundled from Eq. (2.20). This is done by dropping $\hat{T}_N$ in the molecular Hamiltonian and writing the Schrödinger equation for a wavefunction $\psi_n(\{r_i\}; \{r_a\})$ depending on the electronic coordinates and only parametrically on the (fixed) nuclear coordinates:

$$\hat{H}_e \psi_n(\{r_i\}; \{r_a\}) = \varepsilon_n(\{r_a\})\psi_n(\{r_i\}; \{r_a\}) \tag{2.25}$$

$$\hat{H}_e = -\frac{1}{2}\sum_i \nabla_i^2 - \sum_{a,i} \frac{Z_a}{r_{ai}} + \sum_{i>j} \frac{1}{r_{ij}} + \sum_{a>b} \frac{Z_a Z_b}{r_{ab}}, \tag{2.26}$$

where $\hat{H}_e$ denotes the resulting electronic Hamiltonian and the solutions – the electronic states – are labeled after the electronic quantum number n. Eqs. (2.25) and (2.26) correspond to a problem of interacting electrons in an external potential due to a set of nuclei fixed in space, which enter the equations only in the nucleus–electron and nucleus–nucleus interaction terms of the Hamiltonian through their atomic numbers and the set of parameters $\{r_a\}$.

In order to recover the overall problem, the total wavefunction may be expanded in the set of the electronic eigenfunctions of Eq. (2.25) with expansion coefficients $\Gamma_m(\{r_a\})$ depending on the nuclear coordinates [7]:

$$\Psi(\{r_i\}, \{r_a\}) = \sum_m \Gamma_m(\{r_a\})\psi_m(\{r_i\}; \{r_a\}). \tag{2.27}$$

Introducing Eq. (2.27) into Eq. (2.20), multiplying by $\psi_n^*(\{r_i\}; \{r_a\})$, and integrating over all electronic coordinates, one obtains a set of coupled equations for the expansion coefficients:

$$\left(\hat{T}_N + \varepsilon_n(\{r_a\})\right)\Gamma_n(\{r_a\}) + \sum_m \hat{\Lambda}_{nm}\Gamma_m(\{r_a\}) = E\Gamma_n(\{r_a\}), \tag{2.28}$$

where $\hat{\Lambda}_{nm}$ are the so-called nonadiabatic operators:

$$\hat{\Lambda}_{nm} = \int \psi_n^*(\{r_i\}; \{r_a\})\left[\hat{T}_N, \psi_m(\{r_i\}; \{r_a\})\right] d\{r_i\}, \tag{2.29}$$

with $\left[\hat{T}_N, \psi_m\right] = \hat{T}_N\psi_m - \psi_m\hat{T}_N$.

If the Born–Oppenheimer approximation is valid, ψ_m will negligibly depend on $\{r_a\}$, so that $\hat{T}_N$ commutes with ψ_m and the nonadiabatic operators vanish. The nuclear motion is thus completely separated from the electronic motion, and for each electronic state it is possible to write a Schrödinger equation for a nuclear wavefunction $\Gamma_n(\{r_a\})$:

$$\left(\hat{T}_N + \varepsilon_n(\{r_a\})\right)\Gamma_n(\{r_a\}) = E_n\Gamma_n(\{r_a\}), \tag{2.30}$$

where the electronic part of the problem has been condensed in the term $\varepsilon_n(\{r_a\})$ that now – as suggested by a comparison of Eq. (2.30) with Eqs. (2.20) and (2.21) – acts as a potential-energy term for the motion of the nuclei. The Born–Oppenheimer approximation thus gives rise to the concept of a 'potential-energy surface' associated with a given electronic state, i.e., the collection of values $\varepsilon_n(\{r_a\})$ for all nuclear geometries. As we will discuss more extensively in Chapter 6, the potential-energy surface is a cornerstone concept in chemistry, in that it constitutes the 'energy landscape' over which the nuclei move and most of the chemistry is played out, and represents a powerful tool for the rationalization and interpretation of chemical processes occurring within the Born–Oppenheimer limit.

By following the Born–Oppenheimer approximation, we thus reach a cross-roads, with the electronic problem (Eq. (2.25)) on one side and the nuclear problem (Eq. (2.30)) on the other side. As already mentioned in Chapter 1, Part II of this book is devoted to the treatment of the nuclear problem (Eq. (2.30)) with a focus on the modeling of the dynamics of chemical reactions, while Part III is devoted to the treatment of the electronic problem (Eq. (2.25)) with focus on the electron-charge redistribution upon chemical bonding (the concluding chapter of Part II also briefly reviews the main methodologies to solve the molecular Schrödinger equation beyond the Born–Oppenheimer approximation).

However, before immersing ourselves in a detailed exploration of the two sides of the Born–Oppenheimer approximation for a theoretical and computational modeling of chemical reactions and chemical bonding, we will address in Chapter 3 the relation between chemical concepts and their physical counterparts and provide in Chapter 4 a brief historical account on the birth and development of theoretical and computational chemistry with reference to the topics covered in this book.

Chapter 3

Chemical concepts and their physical counterpart

3.1 Reductionism, emergentism, or fusionism?

The discovery of the law of conservation of mass in chemical reactions, formulated independently by Lomonosov and Lavoisier in the late 18th century,
marked the end of alchemy and gave birth to chemistry, which connected with
physics since its early days. During the 19th century, in fact, the two disciplines walked side by side in the attempt to model the behavior of matter at the
macroscopic level through thermodynamics, and in the 20th century they came
to a point of union with the advent of quantum mechanics, which provided the
theoretical framework for an accurate description of chemical systems at the
microscopic level.

In spite of over two centuries of partnership, however, the relation between
physics and chemistry is still an open question in a debate that involves scientists together with philosophers and historians of science. Such debate, that
as we shall see involves the confrontation of rather opposing views, has been
attracting increasing interest over the past few decades and, as highlighted by
Spezia [8], has certainly played a key role in giving birth in the 1990s to a new
branch of philosophy of science, the philosophy of chemistry, witnessed by the
foundation of the scientific reviews *Hyle* and *Foundations of Chemistry*, and by
the establishment in the same years of the *International Society for the Philosophy of Chemistry* (http://ispc.sas.upenn.edu/). While an extensive survey on the
relation between the two disciplines is the subject of several recent specialized
works (see, for instance, the books by Hettema [9] and by Gavroglu and Simões
[10], which approach the subject from a philosophical and historical perspective,
respectively) a bright and concise overview on this matter can be found in the
already cited article by Spezia [8], which we shall closely follow in this section
with the purpose of providing the reader with an overview on the subject.

The main debate is centered on the opposition between reductionism and
emergentism, the key point here being whether chemistry can be reduced to
physics or not. In other words, are chemical concepts reducible to the physics
that governs the behavior of the particles constituting a chemical system, or do
they keep – at least to some degree – their autonomy and cannot be entirely
explained or justified on the basis of physics?

In its strictest sense, "reduction [...] is the explanation of a theory or a set of experimental laws established in one area of inquiry, by a theory usually though not invariably formulated in some other domain". [11] On the other hand, one talks of 'emergent properties' when a reduction of these cannot be operated. A reductionist view might seem more congenial to scientists in general, and certainly a look at the already-mentioned connection between physics and chemistry over the past two centuries may point to a progressive explanation of chemistry in terms of physics. However, there are chemical concepts – even some that are apparently rather simple and clear – that seem resistant to a complete reduction to physics and, as we shall see in Chapter 14 of this book, answering a simple question such as 'what is an atom?' on the basis of physics may not be an easy task.

One of the earliest and most influential perspectives on the side of reductionism comes from one of the fathers of modern physics, Paul A. M. Dirac, who in 1929, in a work that for other reasons will be cited again in Chapter 17 when discussing the effects of relativity in chemistry, wrote:

> *The underlying physical laws necessary for the mathematical theory of a large part of physics and the whole of chemistry are thus completely known, and the difficulty is only that the exact application of these laws leads to equations much too complicated to be soluble. [12]*

In other words, according to the above quote from Dirac, chemistry had been entirely mathematized through physics, the only remaining issue being the technical difficulty in dealing with a complicated and onerous formalism. More recently, Richard Bader, another physicist with whom we will become acquainted in Chapter 14, in a work published in *Foundations of Chemistry* in 2013, wrote:

> *Chemistry is physics, the physics of an atom in a molecule or in an extended system such as a crystal. [13]*

Bader devoted a large part of his work to operate a reduction of the chemical concept of atom to the underlying physics, and arrived at the formulation of a consistent theory, the quantum theory of atoms in molecules, that indeed provides a definition of the atoms exclusively rooted in the physics of molecular systems. In a similar spirit, a few decades before, Löwdin gave his mathematical definition of a molecule:

> *A system of electrons and atomic nuclei is said to form a molecule if the Coulombic Hamiltonian H'–with the centre of mass motions removed–has a discrete ground state energy E_0. [14]*

On the other hand, however, both Bader's definition of the atom (see Chapter 14) and Löwdin's definition of the molecules may result far from satisfactory to a chemist's view. Roald Hoffmann, a Nobel Prize winner in chemistry in 1981, in a private correspondence with Bader dated 2007 and reported in Ref. [15], wrote:

I also have philosophical reservations about the reductionist framework in which AIM (atoms in molecules) resides; I believe the most interesting ideas of chemistry are not reducible to physics. I think we have a fundament (sic) *difference of opinion between us on this matter.*

The same view, whereby chemistry seems to keep its own autonomy and its own explicative and predictive power without the necessity of resorting to microphysics, may be traced back to the first half of the 20th century. On an opposite front with respect to Dirac, in fact, stands one of the fathers of modern chemistry, Linus Pauling, who in 1939 wrote:

I do not think that quantum mechanical calculations of molecular structure or crystal structure will ever make the sort of chemical arguments in my book obsolete. [16]

As is apparent from the above discussion, the debate between reductionists and emergentists is far from being closed. The tension between these two currents is actually fertile for the newborn philosophy of chemistry and we shall see the fruits in the future. As a temporary concluding remark, a further interesting position which in such a heated context certainly deserves a note, is the more temperate perspective – neither reductionist nor emergentist, but rather 'fusionist' – put forward by Heisenberg, who in 1972 wrote that:

Physics and Chemistry have become fused in quantum chemistry. [17]

3.2 Chemical reactions

Chemical reactions are probably the most iconic and fascinating manifestation of chemistry, as they are at the core of what, to an 'innocent' eye, may look like magic: the transformation of matter. In a chemical reaction, one or more chemical substances (the reactants) are converted into one or more different substances (the products). In current chemistry, a chemical reaction can be notated with an associated chemical equation that takes the general form:

$$m\mathrm{M} + n\mathrm{N} \rightarrow p\mathrm{P} + q\mathrm{Q} \,, \qquad (3.1)$$

where the uppercase letters M, N, P, and Q indicate different chemical (atomic or molecular) species and the lowercase letters m, n, p, and q are the stoichiometric coefficients, indicating how many m and n molecules (or moles) of M and N, respectively, have to combine together in order to be converted into p

molecules (or moles) of P and q molecules (or moles) of Q. The number of reactant species participating in the process (two in reaction (3.1)) is called the molecularity. Accordingly, reaction (3.1) is called a bimolecular reaction. Note that, of course, a chemical reaction can involve a different number of reactant or product species from that of reaction (3.1).

Modeling a chemical reaction at the macroscopic level, for a chemist often means estimating the so-called reaction rate v_R, i.e., how rapid the conversion from reactants to products is. This, for reaction (3.1), is defined as follows:

$$v_R = -\frac{1}{m}\frac{d[M]}{dt} = -\frac{1}{n}\frac{d[N]}{dt} = \frac{1}{p}\frac{d[P]}{dt} = \frac{1}{q}\frac{d[Q]}{dt}. \tag{3.2}$$

The reaction rate is generally a function of the initial abundances of the reactants and is related to them through a proportionality coefficient, the so-called rate constant k (which in turn is a function of temperature T):

$$v_R = -\frac{1}{m}\frac{d[M]}{dt} = k[M]^\mu[N]^\nu, \tag{3.3}$$

where $\mu + \nu$ is the order of the process, with μ and ν being not necessarily integers.

To the eye of a chemist, the notation of Eq. (3.1) is still inherently incomplete, as it barely accounts for what is going on in the real world when the species M and N react with each other. In everyday life, in fact, chemical reactions happen in a given context, involving the presence of surrounding gas or solvent, different physical conditions (such as temperature or pressure), and involve a huge number of molecules that can exist in different states of matter and that take part in processes in the gas phase, in solution or at the interface with solids. Still, the macroscopic phenomenological process taking place in nature or under the eyes of a chemist in the laboratory, is the result of the behavior of every single atom that is involved in the process, and must ultimately be modelable and rationalizable through the physics governing the involved nuclei and electrons.

The understanding and modeling of a chemical process on the basis of physical principles thus involves the scaling down from the macroscopic level (which relates to thermodynamics and kinetics treatments) to the microscopic one (which relates to dynamics treatments) [18]. In other words, reaction (3.1) results from a huge number of individual reactive encounters between the M and N molecules (each of which, from the physics perspective, can be viewed as a two-body collision – albeit a particular one, due to the fact that it leads to a reactive event involving bond breaking/forming), and one has to be able to simulate the evolution in time of these reactive encounters, and then to perform the correct averaging over a huge number of these processes in order to reproduce the macroscopic behavior of the reaction. The chemistry – or at least a large part of it – of reaction (3.1), is thus reduced to the description (and proper averaging) of multiple two-body reactive collisional events.

Part II of this book is entirely devoted to this topic, and aims at showing how an accurate treatment of the nuclear motion (microscopic level) can lead to the estimation of the rate constant $k(T)$ for a chemical reaction at a given temperature T (macroscopic level). For the sake of clarity, we shall confine the treatment to rarefied (very low pressure) gases in which single-collision processes occur with no exchange of energy (isolated systems) and no exchange of mass (closed systems) to the exterior and we shall focus on the simplest bimolecular reaction with a chemical bond being broken and a new one being formed:

$$A + BC \rightarrow AB + C, \tag{3.4}$$

where A, B, and C are now atomic species.

From a physical point of view, Eq. (3.4) is the problem of the reactive encounter of an atom A with a diatomic molecule BC. As we learnt in Chapter 2, by virtue of the Born–Oppenheimer approximation, this can be assimilated to the problem of the motion, or rearrangement in space, of the nuclei A, B and C subject to a potential energy represented by the electronic potential-energy surfaces (Eq. (2.30)). As will be discussed in Chapter 7, the nuclear motion can be treated either at a quantum level by solving directly Eq. (2.30), or classically through Newton's equation by reverting nuclei back to classical particles subject to an external potential.

Now, real-life reactions are typically much more complicated than reaction (3.4). Even in simplified physical conditions such as the already-mentioned low-pressure/low-density, chemical reactions may involve several different elementary processes of the type of Eq. (3.4), where different species evolve in a complex reaction mechanism (see Chapter 9). However, the reactions that we will focus on in Part II of this book are definitely not a mere abstract model totally at odds with reality. In fact, simple reactions such as those of Eq. (3.4) represent an important class of processes that occur in the extreme conditions of the interstellar medium in exactly the same simplified conditions that we shall assume. As we shall discuss in Chapters 5 and 10, despite their simplicity these reactions are highly involved in such a big question as the origin of life on Earth.

3.3 Chemical bonding

In the previous section, we examined how chemical reactions can be analyzed from a microscopic point of view in terms of the rearrangement in space of the involved atoms. Within the Born–Oppenheimer approximation, this further reduces to the dynamics of the involved nuclei subject to an external potential arising from the separate electronic problem. The resulting potential-energy surface, to which the entire Chapter 6 is devoted, ultimately drives the course of a chemical reaction and the related breaking/forming of chemical bonds. Within such a view, the electrons are, in a sense, 'collapsed' in the concept of the potential-energy surface and any trace of their motion is lost.

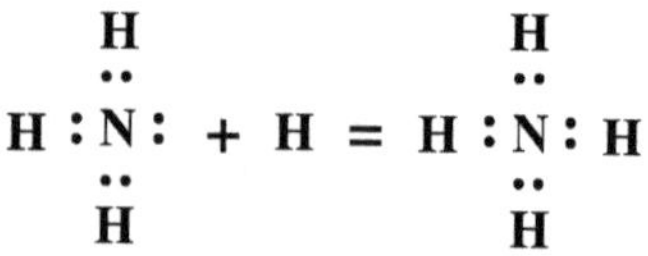

$$\begin{array}{ccc} \mathbf{H} & & \mathbf{H} \\ \cdot\cdot & & \cdot\cdot \\ \mathbf{H:N:} + \mathbf{H} & = & \mathbf{H:N:H} \\ \cdot\cdot & & \cdot\cdot \\ \mathbf{H} & & \mathbf{H} \end{array}$$

FIGURE 3.1 Lewis' original representation of the electron-pair sharing between ammonia (NH_3) and a proton (H^+) to form the ammonium cation (NH_4^+), satisfying the octet rule. Reprinted with permission from Ref. [19]. © 1916 American Chemical Society.

On the other hand, modern theory of chemical bonding starts with Lewis' intuition, in his seminal paper of 1916 [19], of chemical bonding resulting from the sharing of electron pairs between the bonding partners, and since that date chemical bonding is universally associated with the 'movement' of the electrons. Let us start by commenting on one of the original representations appearing in Lewis's paper, reproduced here in Fig. 3.1. In this figure, Lewis explicitly represented as dots the valence electrons of the atoms. Ammonia has three pairs of electrons involved in the three N–H bonds, and a so-called lone pair available for further chemical bonding with a proton H^+, which, in turn, has no electrons of its own. Adopting more modern conventions, nowadays we could write the same process as:

$$H_3N: + H^+ \rightarrow [H_3N - H]^+ , \qquad (3.5)$$

where the two dots in ammonia indicate a lone pair, which is transformed into a dash when it becomes a bonding pair. As is known, Lewis' representations and the octet rule are extremely popular among chemists due to their usefulness in rationalizing and predicting the chemistry of the elements.

Now, as already mentioned, Lewis' intuition predates the formulation of quantum mechanics and his representations are implicitly rooted in a classical view of physics, whereby electrons can indeed be drawn as black dots 'localized' – however schematically – in a precise point of the space. Still, as we have discussed in Chapter 2, classical physics is utterly inappropriate for the description of particles as light as the electron. In seeking a physical counterpart of representations such as that of Fig. 3.1, therefore, the problem of the motion, or rearrangement in space, of the electrons has to be completely cast in quantum terms. As we have already discussed, this also means giving up on knowing the precise knowledge of the position of the electrons in space, and thinking instead in terms of the associated probability density.

Part III of this book is devoted to the quantum treatment of the electrons, and Chapters 14–16 specifically to the analysis of their redistribution upon chemical bonding. A special role in this context is played by a three-dimensional function, the electron density $\rho(r)$, which can be easily worked out from the electronic wavefunction by integrating its square modulus over all electronic coordinates but one and multiplying by the number N of electrons:

$$\rho(r) = N \int |\psi(r, r_2, \ldots, r_N)|^2 \, dr_2 \ldots dr_N . \qquad (3.6)$$

While, as already discussed in Chapter 2, the electronic wavefunction is a multi-dimensional function of all the spatial coordinates of the electrons whose square modulus corresponds to the probability of finding the electrons in a spatial configuration $\{r_i\}$, the electron density is a three-dimensional function returning the fraction of electron charge contained in an infinitesimal volume of space surrounding a given point r.

In contrast to the wavefunction, the electron density is an experimental observable and thus represents a key quantity bridging the mathematical apparatus of quantum mechanics with reality through the concepts of electron delocalization and the electron cloud. As we shall see later, the quantum counterpart of the electron rearrangement underlying Lewis' representation may indeed be sought in the deformation of the electron density $\rho(r)$ around the atoms upon formation of a chemical bond.

Chapter 4

A brief historical account

While theoretical reflection has accompanied chemistry since its origins, the birth of modern theoretical and computational chemistry can be traced back to the early days of quantum mechanics, a little less than a century ago. On the one hand, this suggests that theoretical and computational chemistry is a relatively young discipline. On the other hand, as already mentioned, the origins of chemistry itself may be dated to around the end of the 18th century with the formulation of the law of conservation of mass. Accordingly, the life span of chemistry on the whole is only roughly double that of theoretical and computational chemistry.

In 1925, Erwin Schrödinger postulated his homonymous equation, which was published in 1926 [5]. A year later, Born and Oppenheimer developed a perturbation theory to understand how a molecule can exist in a stable state leading to the already-discussed separation of the nuclear and electronic motions and to the Born–Oppenheimer approximation of fixed nuclei [6]. The first attempts to solve the Schrödinger equation using hand-cranked mechanical calculators may be traced back to these years [20].

The Born–Oppenheimer approximation was so successful that a large part of current chemistry is investigated and rationalized within its terms. As already mentioned, the same approximation led to a crossroads, featuring the nuclear problem on one side and the electronic problem on the other side. A large number of researchers focused on this latter problem, giving rise to the dominant area in theoretical and computational chemistry of electronic-structure theory. While, in fact, on the one hand the Born–Oppenheimer approximation considerably simplified the molecular Schrödinger equation, on the other hand solving the resulting Schrödinger equation for the electronic problem turned out to be not a trivial task and it was soon clear that further approximations, which still today form part of the basic methods of theoretical and computational chemistry, had to be introduced. As we shall see in more detail in Chapter 12, the first of these was the orbital model, introduced by Hartree in 1928 [21], and further refined in 1929–30 by Fock and Slater [22,23] in order to properly account for the antisymmetry principle.

Even with the introduction of the orbital model, the mathematical formalism associated with the electronic problem was too complex and onerous to be effectively exploited, and the development in this field had to wait for the Second World War and the introduction of the first electronic computers, which were made available to scientists in the decade after. The physics of molecular systems, so far, had been mainly the preserve of physicists. However, in those years physicists gradually became interested in nuclear structure and from the mid-1950s onward the discipline was pursued primarily by chemists [20]. It is in those years that one of the cornerstone methods for an effective solution of the orbital problem was devised and summarized in important papers by Roothaan [24] and Hall [25]. As we shall see in further detail in Chapter 13, this involved the expansion of the sought orbitals in terms of known basis functions and the subsequent transformation of the problem from the world of mathematical functions to the world of vectors and matrices.

Within this algebraic method, the choice of the analytic form of the basis function is evidently a key aspect, and a central role was played in this regard by the British chemist Samuel Francis Boys [26], who introduced the Gaussian-type orbitals for the basis-set expansion. These functions turned out to be particularly suited for the treatment of chemical systems due to their property of being localized in space around the involved atoms (thus resembling the spatial distributions of the electrons in the isolated atoms) and to the ease of treatment that they introduced in the involved mathematical operations.

The following decades were very fertile, and astonishing progress was made, building on the Hartree–Fock method, to overcome the limitations of the orbital model, which inherently neglected electron correlation. Such progress, on the one hand, strongly relied on the enormous improvement and the greater affordability of computers. On the other hand, major developments were accomplished on the theoretical side, including the definition of small but accurate basis sets, the development of accurate approximate methods, and the calculation of analytic derivatives of the electronic energy [20]. During the 1970s, the first computer programs for electronic-structure calculations were developed, which were bound to find widespread usage among the scientific community (not necessarily the theoretical and computational one) in the following decades. At the same time, the foundations were laid of density-functional theory, which paved the way to a low-cost, still accurate, treatment of the electron correlation, thus extending the applicability of electronic-structure calculations to increasingly larger systems. Such extraordinary achievements led in 1998 to the awarding of the Nobel prize for Chemistry to Walter Kohn ('for his development of the density-functional theory') and John A. Pople ('for his development of computational methods in quantum chemistry'), which is probably the most concrete sign of the full acknowledgment of theoretical and computational chemistry as an important facet of chemistry.

By looking back at this brief survey of the evolution of theoretical and computational chemistry with a focus on electronic-structure theory and methods, if one were to pick the names of the mentioned scientists and place them on a timeline not necessarily to scale, but rather emphasizing the three main turning points highlighted above (1929–30: Born–Oppenheimer approximation; 1950–51: introduction of the algebraic method; 1998: Nobel prize to Kohn and Pople), one would produce the scheme depicted in Fig. 4.1, which seems to show

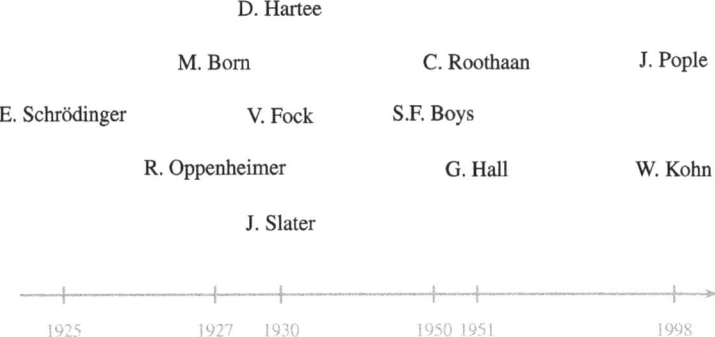

FIGURE 4.1 Timeline for the development of electronic-structure theory and methods reporting the name associated with the main contributions as a function of time. The timeline is not to scale but rather emphasizes the three turning points of 1927–30, 1950–51, and 1998 discussed in the text.

a formidable football team in a robust, defensive, 5–3–2 formation. While this scheme is certainly not exhaustive, it nonetheless provides a clear and compact picture of the main stages of the development of theoretical and computational chemistry with regard to electronic structure. It is worth noting here that most of the scientists reported in Fig. 4.1 are either physicists or mathematicians (or both), the only chemists being Boys and Pople, whose names are both tightly connected with the use of Gaussian-type orbitals – probably the most 'chemical' idea in the development of electronic-structure theory.

In the meantime, while, as already mentioned, a large number of researchers focused their efforts on the solution of the many-body electronic Schrödinger equation, another group of scientists turned themselves to the other side of the Born–Oppenheimer approximation, that dealing with nuclear dynamics. We shall briefly review the stages of the development of this area mainly in relation to the study of chemical-reaction dynamics, which is the subject of Part II of the book.

The initial steps in the area of chemical-reaction dynamics were made also in this case in the late 1920s, when the Born–Oppenheimer approximation permitted in principle to calculate the interaction energy between atoms as a function of the interatomic distances, i.e., the potential-energy surface [27]. As already mentioned, the potential-energy surface – typically the electronic ground-state one – represents the forces driving the motion of the atoms during a chemical reaction. In 1929, London formulated a semiempirical potential-energy surface governing the nuclear motions for the simplest atom–diatom exchange reaction, $H + H_2 \rightarrow H_2 + H$, and his approach was extended in the following few years by Eyring and Polanyi, who calculated numerically the potential-energy surface for several other systems. The first classical trajectory study of the dynamics of a chemical reaction followed in 1936, and was performed by Hirschfelder, Eyring and Topley for the $H + H_2$ reactive collisions using London's potential-energy surface [28].

However, the development of efficient quantum schemes for the treatment of the nuclear motion in chemical reactions also had to go through a long hiatus, partly for the same reasons already discussed when addressing the development of electronic-structure methods, i.e., the lack of an effective computing technology. Besides the introduction of electronic computers in the 1950s, a breakthrough in this area was represented by the development in the 1960s of crossed–molecular beam techniques [27], which allowed for the reproduction of single collisions between molecular species and a quantitative characterization of the reaction-dynamics properties such as the total and differential reactive cross section. The 1970s saw the growth of the first computer programs for quantum reactive scattering calculations and in 1976 Schatz and Kuppermann published the first quantum reactive scattering study of the H + H$_2$ reaction [29,30]. In the successive years, much progress was made in the development of rigorous time-dependent and time-independent methods for the solution of the nuclear Schrödinger equation in reactive scattering problems and many studies, though mainly limited for computational reasons to three- or four-atom systems, were published. Excellent reviews of the reactions studied from the early days up to recent times can be found in the works of Bowman and Schatz [31] and Althorpe and Clary [32].

For a long time, this area of research has certainly been a niche one, due to the fact that, as already mentioned, the treatment was limited to systems of only three or four atoms, which are evidently a minimal part of the possible reactions of chemical interest. However, the 1990s saw rapid progress in the development of a new approximate scheme, the multiconfiguration–time dependent Hartree (MCTDH) [33], that took inspiration from multiconfigurational–electronic structure methods. Active research is currently ongoing in this field and is likely to extend the range of applicability of the quantum modeling of chemical reactions to systems with many more degrees of freedom.

It should be stressed here that the survey made so far on the development of two important areas of theoretical and computational chemistry is far from being exhaustive and representative of the whole discipline. As already mentioned, in fact, it only focuses on the aspects of theoretical and computational chemistry that are covered in this book. It is thus worth mentioning here that other important threads of the discipline are those dealing with the treatment of liquids, solids, biological macromolecules and with the adoption of less accurate or detailed methods such as molecular mechanics, molecular dynamics, statistical mechanics, some of which are briefly reviewed in Chapter 11.

Before leaving this chapter and continuing our journey with a closer exploration of the treatment of nuclei and electrons, a final comment of historical character may be fruitful in framing the above-discussed developments in chemistry in the overall general cultural context of the last century. Undoubtedly, the advent of quantum mechanics revolutionized chemistry: the entire concepts of atoms, molecules, reactivity had to be redefined within the terms of the newly established theoretical paradigm. On the other hand, as briefly discussed in

Chapter 1, the acceptance of quantum mechanics involved such a radical change in the relation of man with reality that this could not but have a deep impact on other, even apparently remote, fields of knowledge. In 1925, the same year as the formulation of the Schrödinger equation, the poetry collection *Ossi di seppia* by the Nobel Prize in Literature for 1975 Eugenio Montale was published. One of the most famous poems of that collection reads[1]:

Maybe one morning, walking in dry, glassy air,
I'll turn, and see the miracle occur:
nothing at my back, the void
behind me, with a drunkard's terror.

Then, as if on a screen, trees houses hills
will suddenly collect for the usual illusion.
But it will be too late; and I'll walk on silent
among the men who don't look back, with my secret.

The author of this book has always been struck by the coincidence of the date of formulation of the founding equation of quantum mechanics and that of publication of these lines of poetry, and by the suggestive similarities that the therein referenced 'void', that precedes the experience of reality, shares with the quantum-mechanical wavefunction.

---

[1] The English translation is by Jonathan Galassi.

Part II

Nuclear dynamics and chemical reactions

Chapter 5

Reactive collisions

As discussed in Chapter 3, chemical reactions at the macroscopic scale involve a huge number of reactive events where atomic or molecular species recombine themselves according to a chemical equation in certain given external conditions. However, the observable properties of a chemical reaction are the result of the microscopic behavior of each of the involved atoms, and the event of their recombination in different chemical species must be traceable back to their microscopic dynamics.

Within the Born–Oppenheimer approximation, the dynamics of the atoms involved in a chemical reactions can be reduced to the problem of the motion of the related nuclei under the action of forces arising from the electronic potential-energy surfaces and it is often sufficient to take into account only the potential-energy surface of the electronic ground state. Starting from this chapter up to the end of Part II, we will explore the topic of chemical-reaction dynamics, i.e., how to work out observable properties of a chemical reaction from a proper modeling of the underlying nuclear dynamics over a single potential-energy surface. Thus from the mentioned crossroads to which the Born–Oppenheimer approximation led us at the end of Chapter 2, we shall take the road of the nuclear problem, leaving aside for the moment the electronic problem, which will be recovered in Part III of the book.

For this purpose, as already mentioned, we will focus on the simplest class of chemical reactions where a chemical bond is broken and a new chemical bond is formed, i.e., atom-exchange reactions between an atomic and a diatomic reacting partners. In order to convey the basic ideas behind the theoretical and computational modeling of chemical-reaction dynamics, we will assume simplified conditions of rarefied (very low pressure) gases in which single-collision processes occur with no exchange of energy (isolated systems) and no exchange of mass (closed systems) to the exterior.

Despite their simplicity, reactions of this type are important for a number of reasons. First, $A + BC$ reactions made the history of the development of the field of chemical-reaction dynamics. As already mentioned in Chapter 4, in fact, the first reaction dynamics study based on classical trajectories was performed by Hirschfelder, Eyring and Topley for the $H + H_2$ reactive collisions in 1936 [28], and the first quantum study was performed for the same reaction in 1976 by Schatz and Kuppermann [29,30]. Since then, several studies on reactions of the same class involving different elements were performed (see Refs. [31] and [32] for a review of the reactions studied from the early days up to recent times).

Chemistry at the Frontier with Physics and Computer Science
https://doi.org/10.1016/B978-0-32-390865-8.00014-3

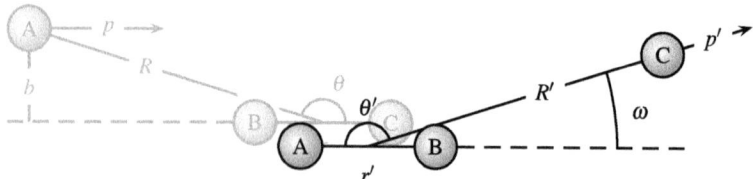

FIGURE 5.1 Initial stage (background light-color image) and final stage (foreground full-color image) of a reactive encounter between an atom A and a diatomic molecule BC. Some of the collision parameters are also notated, together with the reactant (R, r, θ) and product (R', r', θ') Jacobi coordinates discussed in Chapter 7.

Secondly, individual A + BC reactive collisions can be characterized experimentally through a sophisticated experimental technique that will be illustrated later (Section 5.2), so as to allow for a direct comparison of the calculated *versus* experimental observable properties. Finally, A + BC reactions represent one of the main forms of the chemistry that occurs in the extreme conditions of the interstellar medium. As we shall see in Section 5.3, such chemistry – commonly referred to as 'astrochemistry' – despite happening in a quite remote region of space, is tightly connected with the origins of the chemistry of life, and its understanding and proper modeling may contribute significantly to answer the challenging question of how life originated on our planet.

5.1 Atom–diatom collisions

We will thus be concerned from now on with gas-phase reactions of the type:

$$A + BC \rightarrow AB + C, \tag{5.1}$$

where A, B, and C notate atomic species. As discussed in Chapter 3, for the purpose of modeling reaction (5.1) from a theoretical and computational perspective, we have to scale down from the macroscopic level to the microscopic one and formulate the problem in terms of the physics governing the behavior of the three atoms.

In order to set up the chemical problem of reaction (5.1) from a physical point of view, let us forget for the moment about the quantum nature of matter and let us imagine that we can schematize reaction (5.1) by drawing the atoms as sphere-like particles as in Fig. 5.1. In the figure, the background light-color image represents the initial stage of the reaction (with the atoms in the reactant arrangement A + BC) and the foreground full-color image represents the final stage (with the atoms in the product arrangement AB + C). As is apparent from Fig. 5.1, the physical problem associated with reaction (5.1) is that of the reactive encounter between an atom A and a diatomic molecule BC leading to the rearrangement of the atoms in different chemical species through the breaking of the B–C bond and the forming of the A–B bond. In the first instance, it is thus the problem of the collision – albeit a particular one, in that its consequence is a

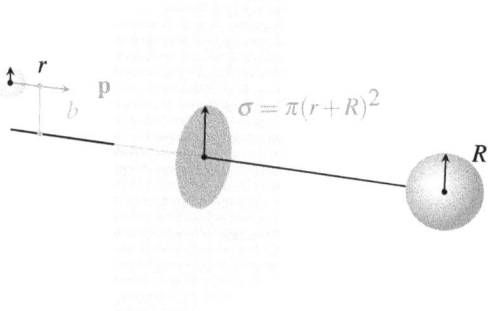

FIGURE 5.2 Scheme of the collision between two hard spheres of different radii r and R. The impact parameter b, initial momentum p, and cross section σ are also depicted.

chemical reaction – between the two bodies A and BC. It is thus in the territory of collision theory that we should move the first steps towards the modeling of reaction (5.1).

Now, in order to proceed, it will be useful to make a short digression briefly reviewing some basic aspects of collision theory. Let us do that for a much simpler case than reaction (5.1), a case with which the readers will certainly be familiar through their own direct experience: the case of the collision between two rigid spheres. Consider, for instance, the collision between two billiard balls of different size – a smaller one with radius r and a larger one with radius R – which is sketched in Fig. 5.2. For simplicity, let us assume that in an initial stage the large ball is still, and that the small ball is launched towards the large one. In other words, the small sphere acts as the projectile while the large sphere acts as the target. How can we quantitatively characterize the launch of the projectile toward the target? Two important collision parameters in this respect are the initial velocity, quantified by the momentum p in Fig. 5.2, and the 'aim' of the projectile, quantified by the impact parameter b in the same figure. This latter quantity is defined as the perpendicular distance to the closest approach to the target if the projectile were undeflected, or in other words the perpendicular offset of the trajectory of the incoming particle. Accordingly, the smaller the impact parameter b the better the aim, with the projectile perfectly hitting the center of the target for null impact parameter.

So far the launch, what about the collision? How can we characterize it? A third important quantity in collision theory is the so-called collisional cross section, which is commonly notated with the Greek letter σ. For two interacting particles in classical physics, their mutual cross section is the area transverse to their relative motion within which they must meet in order to scatter from each other. If the particles are hard spheres interacting upon contact, their scattering cross section is related to their geometric features. In the case of the two billiard balls of Fig. 5.2, simple geometrical considerations lead to the estimation of the cross section as the area of a circle (also depicted in Fig. 5.2) of radius equal to the sum of the radii of the two spheres, i.e., $\sigma = \pi(r + R)^2$. For two colliding hard spheres, the collisional cross section is thus a collision quantity related to the properties of the interacting partners and connected with the likelihood that the two partners will collide with each other.

Let us return now to atoms and molecules and, in analogy to the considered hard-sphere collisional system, let us imagine that A is the projectile and that BC is the target. What can we bring back from the simple example of the collision between two billiard balls to the problem of our reaction, involving atomic and molecular species? A quick visual comparison between Figs. 5.1 and 5.2 easily shows that the initial momentum of the projectile and the impact parameter are useful quantities that are straightforwardly recovered and adaptable to the molecular collision.

But what about the cross section? In the case of the collision between two hard spheres, it is easy to infer the shape and extension of the cross section as it is relatively easy to predict the trajectory of a launch with a given impact parameter b. Even prior to such an inference is the fact that we do not strive to describe what collision between the two hard spheres means: the two balls come into contact when their distance is equal to the sum of their radii, and as a result their motion is deflected. On the other hand, in the case of the encounter between an atom and a molecule, it is now rather difficult to define what a collision – and, as a consequence, the associated cross section – does mean. In fact, even though we depicted the atoms like spheres in Fig. 5.1, they are not properly rigid spherical bodies and certainly do not behave as such. However, assuming that we can keep the classical representation of nuclei as sphere-like localized particles with little harm (as is indeed often the case, as we shall see later), it turns out that we can recover the concept of cross section also for the collisional event of reaction (5.1), subject to a *caveat* involving two distinctions with respect to the hard-sphere model.

The first distinction is that, as already mentioned, atoms and molecules behave differently from rigid spheres and this difference may be succinctly stated as follows: while, as we know from our everyday experience, rigid spheres do not 'feel' each other until they 'touch' each other, the atoms feel one another even at large distances. More formally, within a Born–Oppenheimer view in which nuclei are reverted back to classical particles, the nuclei of the atoms involved in a chemical reaction are subject to an interaction potential[1] that we know from Chapter 2 derives from the presence of the electrons and is associated with the solution of the electronic problem. As is well known, quite generally the interatomic potential is made of a long-range attraction term and a short-term repulsion term, and as a result features a potential well at a given interatomic 'equilibrium' distance.

---

[1] Note that, strictly speaking, an interaction potential is of course also present in the case of the two hard spheres, though it has a very simple form in that it is null everywhere except for distances equal to or less than the sum of the two radii, where the potential suddenly becomes infinite.

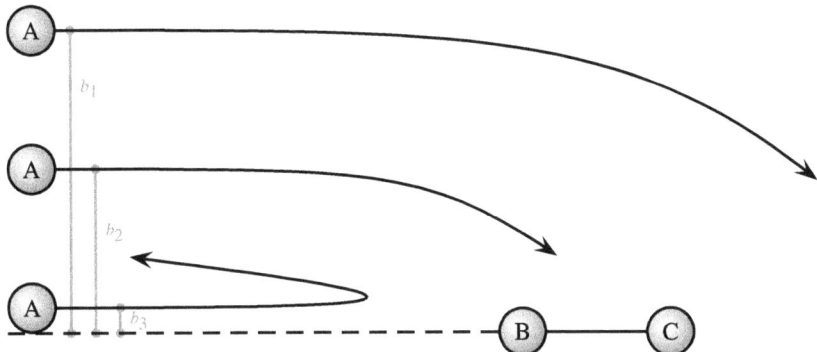

FIGURE 5.3 Schematic illustration of the different fates of three trajectories of an atom A approaching a BC molecule with different impact parameters.

Let us delve a little into the consequences of the presence of a nontrivial interaction potential between the atoms involved in a chemical reaction. On the basis of the above generic description of the interatomic potential, we should be able to predict, at least to a schematic extent, the possible fates of a projectile atom launched towards a target molecule. Let us consider the case of the three encounters between atom A and diatom BC occurring with different impact parameters b_1, b_2, and b_3 that are shown in Fig. 5.3. The launch of A with a large impact parameters (b_1) will likely see the atom A initially traveling in straight motion along its original direction. However, as soon as the atom A feels in a nonnegligible way the long-range attraction exerted by the BC molecule, the trajectory of its motion will be deflected, until the atom reaches a far enough region where the attractive forces due to atoms B and C are negligible. An opposite behavior might have the trajectory with small impact parameter b_3, where the A atom reaches the vicinity of the B atom and is reflected back due to the short-range repulsion forces. A third and final case is that of the launch with impact parameter b_2, which brings the atom A into the vicinity of the B and C atoms. In such conditions a chemical reaction between the colliding partners might now take place, and here we come to the second distinction.[2]

The second distinction between the collision of two hard spheres and the atom–diatom one is that while for the two billiard balls collision means that the two bodies come into contact and, as a result, their motion undergoes a deflection, in the case of an atom–diatom reactive process the relevant collisions are those that lead to reaction, i.e., those upon which the B–C bond breaks, a new A–B bond is formed, and the newly formed species AB and C are eventually scattered in some direction. Accordingly, an associated 'reactive' cross section can now be defined as the area of a plane transverse to the not yet deflected

[2] Note that the fate of the trajectories depicted in Fig. 5.3 is purely schematic. There is in principle no straightforward association between, for instance, short impact parameters and backward reflection, and a chemical reaction can indeed occur upon encounters with null or short impact parameter.

projectile A that the projectile has to cross in order to reach the target in such a way that the reaction takes place.

The reactive cross section is a central quantity in chemical-reaction dynamics as it bridges the microscopical collisional event with phenomenological quantities such as the rate constant on the one hand, and – as we shall see in Section 5.2 – theory and experiment on the other hand. The reactive cross section is, in fact, determined by the microscopic features of the reactive collision but at the same time is linked with macroscopic aspects such as the probability of the reactive event whence observable properties of the reaction can be extracted. In contrast to the hard-sphere model, the presence of a nontrivial interaction potential makes it now difficult to predict the exact trajectory of the collision and, as a consequence, the shape and the extension of the reactive cross section. Therefore, the reactive cross section has to be extracted by accurate dynamics calculations. As we shall see in Chapter 7, this can be done both within classical and quantum mechanics. In the classical approach (which is the approach that underlies the above discussion together with Fig. 5.3), it is easy to see that the cross section can be worked out by calculating a batch of trajectories for different initial conditions and monitoring which of them leads to reaction.

In the following chapters, we will discuss in more detail the features of the interaction potential in atom–diatom reactive systems (Chapter 6) and the classical and quantum treatment of the nuclear motion on a potential-energy surface (Chapter 7). Before leaving this chapter, however, we will briefly discuss two contexts where simple reactions of the type (5.1) do materialize.

5.2 The experimental perspective: crossed molecular beams

As already mentioned in Chapter 4, a breakthrough in the development of the field of chemical-reaction dynamics was the introduction in the 1960s of crossed–molecular beam experiments. Rather astonishingly, after decades of theoretical speculation on nuclear dynamics, for the first time it was possible to delve into the microscopic, detailed mechanisms of chemical reactions from an experimental point of view. The development of this technique was such a revolution in this respect that the Nobel Prize in Chemistry for 1986 was awarded jointly to its pioneers Dudley R. Herschbach, Yuan T. Lee and John C. Polanyi 'for their contributions concerning the dynamics of chemical elementary processes.'

The idea was to build an experimental apparatus capable of analyzing a chemical reaction beyond the macroscopic phenomenological treatment traditionally pertaining to thermodynamics, but rather from the microscopic perspective of the individual reactive collisions. Now, we certainly do not have available a magic pair of pliers capable of picking one single atom and launching it against a single molecule in a sufficiently undisturbed context. Yet, individual molecular collisions can indeed be reproduced and analyzed with the aid of an ingenious device that we shall briefly describe in the following, while referring the interested reader to Ref. [27] for further details.

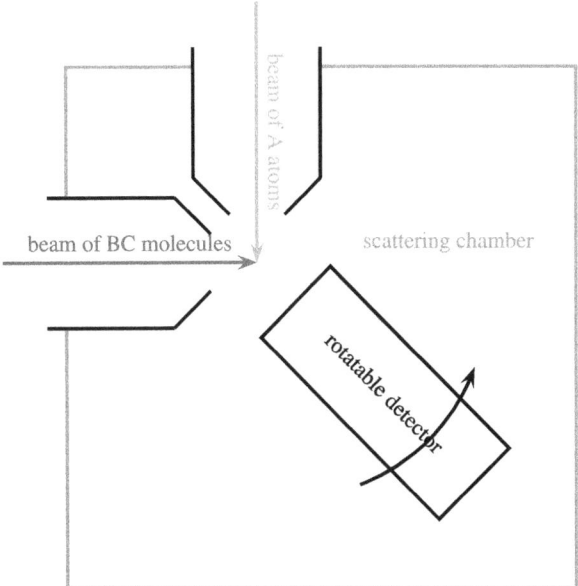

FIGURE 5.4 Scheme of a crossed–molecular beam machine. Atomic or molecular beams are generated by the beam sources and made to intersect in the scattering chamber where a rotating detector analyzes the collision products.

In crossed–molecular beam machines a 'beam' of A atoms is generated (see Fig. 5.4). These atoms, which travel in the same direction with a velocity as uniform as possible, are introduced into the so-called scattering chamber in which a vacuum has been previously made with the aid of a vacuum-pump system. The situation is not so different from that of a relatively busy highway: the cars travel in the same direction with an almost uniform speed. There are many cars, but (luckily!) each travels on its own, and the relative speed between the cars is minimal and null on the average. At the same time, a second beam of BC molecules is generated from a source that is oriented at a right angle with the source of the A atoms. The two beams are formed by expansion from pulsed valves in source chambers that are separated from the scattering chamber by bulkheads, with the only aperture at the tip of the skimmer. The resulting collimated molecular beams cross at right angles in the center of the scattering chamber (hence the name 'crossed molecular beams'). Then, what happens at the intersection between the beams is a most dangerous 'red light roulette', where the A atoms collide with the BC molecules but, due to the special conditions imposed by the generation of the beams, each collision has its own life and does not interfere with the others. It is thus possible to study a large number of individual collisions characterized by different initial conditions.

The products of the reactive collisions are then analyzed by a rotatable detector, the most common detection scheme being electron-impact (or photo) ionization mass spectroscopy, probing the velocity of the detected species by time-of-flight measurements. In this way, not only the kinetic energy and internal quantum-state distributions of the atomic or molecular products can be measured, but also, most importantly, their dependence on the scattering angle, i.e., the direction in which the molecules are scattered (the reaction cross section itself can be obtained experimentally with such angular resolution – the so-called 'differential cross section'). In fact, crossed–molecular beam machines with rotatable detectors coupled to time-of-flight spectroscopy have permitted so far the determination of state-averaged, state-specific or in some cases even state-to-state differential cross sections [27].

One of the key requirements in studying a chemical reaction in crossed–molecular beam experiments is to generate sufficiently intense beams of the reactant species. While the generation of supersonic beams of stable species is straightforward, the production of sufficiently intense beams of unstable species such as atoms and radicals is instead troublesome. Over the past decades, progress in the beam-generation techniques has led to a rapid increase in the availability of atomic and molecular beams, including virtually any atomic species and several important diatomic and polyatomic radicals, and thus allowing for the investigation of reactions of relevance in combustion chemistry, atmospheric chemistry, and astrochemistry.

5.3   The chemistry of the interstellar medium

Crossed–molecular beam experiments are not the only context where reactions of the type (5.1) can actually occur. Indeed, individual collisions between simple atomic or molecular species in conditions similar to those recreated in crossed–molecular beam experiments also occur in nature, though in the rather far and singular context of the interstellar medium.

In remote regions in our and other galaxies in the universe, in fact, accumulation of gas-phase sparse matter, plasma and dust, form the so-called diffuse interstellar clouds. These are characterized by extreme physical conditions featuring very low temperatures, pressures, and density of matter. As a consequence, the chemistry of an interstellar cloud is rather different from the chemistry occurring on our planet [34]. In particular, it mainly consists of two main classes of processes: gas-phase barrierless reactions involving ions or radicals on the one hand, and heterogeneous or multiphase processes involving dust grains and icy mantles on the other hand. Among the first class of processes, those involving the reactive collision of an atom A with a diatomic molecule BC represent an important type of reaction.

Now, one of the grand challenges of astrochemistry is that of developing a complete chemical model of an interstellar cloud capable of quantitatively accounting for the chemical evolution in time of its constituting atomic and molecular species. Such a model would, for instance, allow researchers to estimate the age of a molecular cloud based on measurements of its chemical

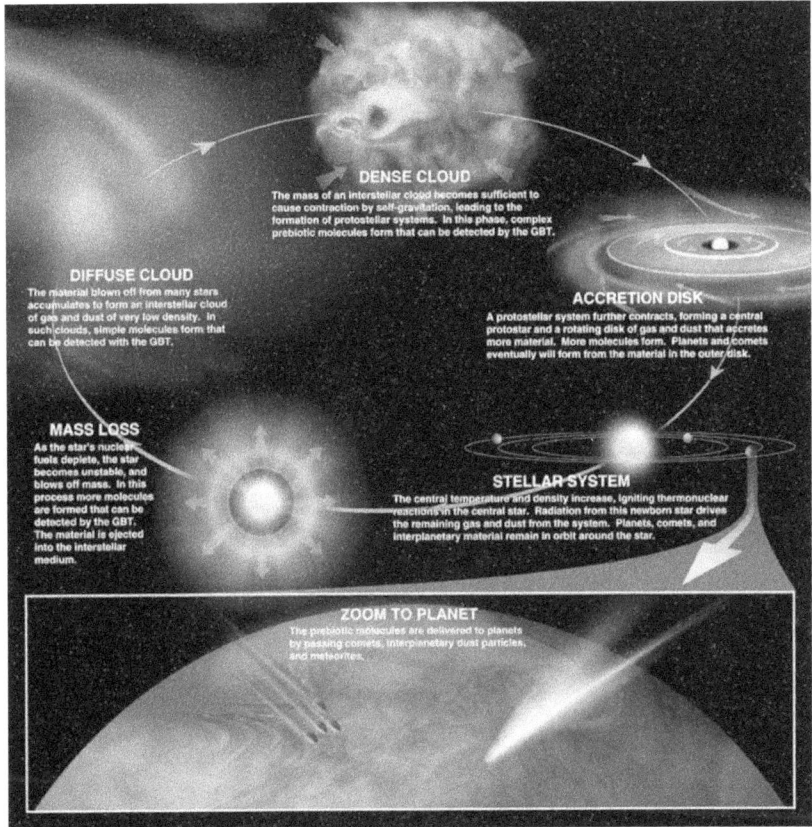

FIGURE 5.5 Illustration of the life cycle of a star highlighting the possible delivery on Earth of prebiotic molecules deriving from interstellar clouds. Credit: Bill Saxton, NRAO/AUI/NSF.

composition. Much more than that, an accurate modeling of the chemistry of the interstellar clouds might also shed light on such a big question as that of the origins of life on Earth.

Interstellar clouds are, in fact, part of the life cycle of a star, which can be summarized through the illustration in Fig. 5.5. As already mentioned, accumulation of gas, plasma and dust in the interstellar space forms diffuse interstellar clouds. These gradually evolve into dense clouds, and when the mass of the interstellar cloud becomes sufficiently large, it causes contraction by self-gravitation and leads to formation of protostellar systems. A protostellar system contracts further, forming a central protostar and a rotating disk of gas and dust, whereby planets and comets will eventually form. Then the central temperature and density increase, igniting thermonuclear reactions in the central star and radiation from this newborn star drives the remaining gas and dust from the system, while planets, comets, and interplanetary material remain in orbit around the star. As the star's nuclear fuels deplete, the star becomes unstable, and blows off mass. The material is ejected into the interstellar medium and is ready to enter the cycle again forming diffuse interstellar clouds by accumulation.

Now, according to one of the prevalent theories on the origins of life on Earth – that of the exogenous delivery as opposed to endogenous production (see, for instance, Ref. [35]) – the molecules that are at the basis of the chemistry of life (so-called prebiotic molecules) could have been delivered to our planet by comets, interplanetary dust particles, and meteorites forming during the life cycle of a star. These molecules start forming in dense interstellar clouds and their formation has its roots in the very simple chemistry of the diffuse interstellar clouds where, as already mentioned, reactions of the type $A + BC \rightarrow AB + C$ such as those that we will be focusing on in the following chapters play a key role.

Chapter 6

The potential-energy surface

Within the Born–Oppenheimer approximation, the modeling of chemical-reaction dynamics requires solving the equations of motion of nuclei on one or a manifold of potential-energy surfaces associated with the electronic states of a molecular system. We shall see in the next chapter that this can be done either at the classical level or at the quantum level. In both cases, the dynamical results are analyzed in order to extract observable properties of the reaction at hand to be compared with the outcome of crossed–molecular beam experiments.

As pointed out in Chapter 2, the potential-energy surface results from the solution of the electronic problem at fixed positions of the nuclei (the solution of this problem will be the subject of Part III of this book). Ideally, accurate values of the electronic energies defining a potential-energy surface should be available at all the system geometries (or 'nuclear configurations') needed during the dynamical treatment. While examples have been reported of so-called 'direct dynamics', where the electronic energies are calculated on the fly as required by the dynamics calculations [36–38], the most common and efficient approach is to cast the potential-energy surface in a suitable analytic-functional form. This can either be a semiempirical expression (such as the popular London–Eyring–Polanyi–Sato, LEPS [39–42]) or, more commonly, the result of the fitting or interpolation of a set of electronic energies computed at a finite number of nuclear configurations by high level–electronic structure methods.

In this chapter we will illustrate some analytic representations of the potential-energy surface for a three-atom reactive system (Section 6.1) and address the issue of the sampling of the nuclear configuration space in order to generate an informative set of electronic energies useful for the fitting/interpolation procedure (Section 6.2). Finally, focusing on a potential-energy surface for the $H + H_2$ reaction generated according to the previously outlined methods, we will cover the topic of the visualization and analysis of the potential-energy surface for the purpose of gaining predictive insight into the features of the reactive process (Section 6.3)

6.1 Analytical formulations of the potential-energy surface

The topic of analytic representations of potential-energy surfaces for few-atom reactions, especially with regard to the most suitable functional forms, has been reviewed over the years by Sathyamurthy [43], Truhlar [44], Schatz [45,46] and several others [47–49]. As already mentioned, quite generally the scheme for

Chemistry at the Frontier with Physics and Computer Science
https://doi.org/10.1016/B978-0-32-390865-8.00015-5

generating a potential-energy surface for use in dynamics calculations of few-atom reactions is to calculate a finite set of electronic energies at many nuclear configurations and use this information for generating an analytic functional formulation returning the value of the electronic energy at any desired point of the nuclear configuration space. This second step can be tackled through two main classes of methods. Analytic representation of the potential-energy surface can in fact be either local (see for instance Collins' modified Shepard [50–54] and Schatz's interpolant moving least squares (IMLS) [55,56] schemes) or global. In local schemes, the fitted/interpolated value of the potential-energy surface in a given point of configuration space depends on the values of the neighboring energies, while in global schemes the fitted/interpolated value of the potential-energy surface in a given point of configuration space depends on the values of the whole set of available energies.

While a complete account of local and global fitting/interpolation schemes can be found in the above-cited reviews, we will convey the basic ideas behind the two classes of schemes by focusing on the modified-Shepard local interpolation scheme [50] and on the global fitting scheme proposed by Aguado and Paniagua [57,58].

In the modified-Shepard local interpolation scheme the potential-energy surface of a polyatomic system of N atoms is formulated as a weighed sum of second-order Taylor expansions around a set of N_{Sh} computed electronic energies:

$$V(\boldsymbol{q}) = \sum_{i=1}^{N_{Sh}} w_i(\boldsymbol{q}) T_i(\boldsymbol{q}) , \tag{6.1}$$

where $\boldsymbol{q}$ is the vector of the $N_q = 3N - 6$ internal coordinates ($N_q = 1$ for diatomic systems, $N_q = 3$ for triatomic systems), $w_i(\boldsymbol{q})$ are weight functions and $T_i(\boldsymbol{q})$ are second-order Taylor expansions around the ith electronic energy. These are given by

$$T_i(\boldsymbol{q}) = V_i + \sum_{\alpha=1}^{N_q} \Delta q_\alpha \frac{\partial V}{\partial q_\alpha}\bigg|_i + \frac{1}{2!} \sum_{\alpha=1}^{N_q} \sum_{\beta=1}^{N_q} \Delta q_\alpha \Delta q_\beta \frac{\partial^2 V}{\partial q_\alpha \partial q_\beta}\bigg|_i , \tag{6.2}$$

where Δq_α is the difference between coordinate α of the desired $\boldsymbol{q}$ space point and coordinate α of the geometry of the ith electronic energy. The following inverse-distance weighting can be adopted

$$w_i = \frac{1}{d_i} \bigg/ \sum_{j=1}^{N_{Sh}} \frac{1}{d_j} , \tag{6.3}$$

where

$$d_i = \sqrt{\sum_{\alpha=1}^{N_q} (\Delta q_\alpha)^2} \qquad (6.4)$$

is the distance between space point $\mathbf{q}$ and the space point associated with the ith computed electronic energy. This ensures that a larger weight in the summation of Eq. (6.1) is attributed to electronic energies computed at points that are close to the $\mathbf{q}$ space point (thus yielding a better Taylor expansion).

As is evident, the modified-Shepard interpolation is a local scheme in that the resulting potential energy in a given space point will depend on the values of the neighboring available energies, while the effect of the presence of distant energies will be nullified by the inverse-distance weighting of Eq. (6.3). On an opposite front, in a global scheme such as that proposed by Aguado and Paniagua, all available energies concur in determining the value of the analytic potential-energy surface in a given point space. The Aguado–Paniagua scheme is described in detail in Refs. [57] and [59] and computer programs for three- and four-body potential fitting have been made available by the authors [58,60, 61]. Its main features as to two- and three-body potential-energy surface fitting are as follows. A diatomic (two-body) potential is given the functional form

$$V^{(2)}(r) = c_0 \frac{e^{-\alpha r}}{r} + \sum_{i=1}^{I} c_i (r e^{-\gamma^{(2)} r})^i , \qquad (6.5)$$

where α, c_0 and $\gamma^{(2)} > 0$ and the linear $c_0, c_1, \ldots, c_I$ and nonlinear α and $\gamma^{(2)}$ parameters are determined by fitting a set of values of the diatomic potential energy. Following a many-body expansion [62] and omitting for clarity the constant one-body terms, a three-body potential is written as a sum of three two-body and one three-body terms:

$$V(r_1, r_2, r_3) = V_1^{(2)}(r_1) + V_2^{(2)}(r_2) + V_3^{(2)}(r_3) + V^{(3)}(r_1, r_2, r_3) , \qquad (6.6)$$

where subscripts 1, 2 and 3 label the three possible diatoms and the three-body term is given by

$$V^{(3)}(r_1, r_2, r_3) = \sum_{ijk}^{M} d_{ijk} (r_1 e^{-\gamma_1^{(3)} r_1})^i (r_2 e^{-\gamma_2^{(3)} r_2})^j (r_3 e^{-\gamma_3^{(3)} r_3})^k , \qquad (6.7)$$

with $i + j + k \neq i \neq j \neq k$ and $i + j + k \leq M$. Linear d_{ijk} and nonlinear $\gamma_1^{(3)}$, $\gamma_2^{(3)}$ and $\gamma_3^{(3)}$ parameters in the equation above are determined by fitting the values of the electronic energy of the triatomic system minus the diatomic potentials at the corresponding internuclear distances.

The advantage of global methods is that an appropriate behavior is in principle ensured by the choice of a functional form that is tailored to the type of system that it is meant to model. For a small set of sampled electronic energies, this may give better results over local methods, whose reliability in certain regions of space strongly depends on the presence of neighboring computed electronic energies. On the other hand, for increasing sets of sampled energies, the mentioned advantage of global functional forms may indeed turn into a bias as the employed analytic expression may impose a rigidity due to the fact that its flexibility is ultimately limited by the finite number of fitting parameters, whereas the accuracy of a local method such as the modified-Shepard interpolation is free to increase with no limitations for an increasing number of available energies.

No matter which fitting or interpolation method one chooses, the starting point in the analytic formulation of a potential-energy surface is inescapably the collection of a set of electronic energies. This raises a couple of questions that will be addressed in the following section: how many energies should one compute, and at which molecular geometries?

6.2 Configuration-space sampling

While the topic of analytic formulation of potential-energy surfaces has been widely addressed in the literature, less attention has been paid to the issue of properly sampling the configuration space when generating the electronic energies to be fitted or interpolated – in other words, the choice of the geometries at which one has to calculate the electronic energy prior to the fitting or interpolation of the potential-energy surface. At this stage, choosing an as small as possible yet informative set of points that optimally samples the nuclear configuration space is a crucial point. Accurate electronic-structure calculations are in fact usually computationally expensive. Moreover, some of the points could not converge to the desired electronic state and the analysis of a wisely reduced set of points could avoid unwanted difficulties [63,64]. This is indeed one good reason for opting for analytic representations of the potential-energy surface rather than direct methods.

Information on this subject in the cited reviews do not go beyond hints and, in fact, besides iterative sampling methods based on classical trajectories [51, 65] or successive-degree fitting [66], to the best of the author's knowledge no general and established ways of selecting suitable sets of nuclear configurations are available.

The author of this book has recently proposed a general and flexible scheme for an effective sampling of the nuclear configuration space in potential-energy surface fitting or interpolation that we shall briefly review in the following. The scheme, known as space reduced–bond order (SRBO) [67], is based on a space-reduced formulation of the so-called bond-order variables allowing for a balanced representation of the attractive and repulsive regions of a diatom configuration space.

As already mentioned, the first step when crafting a new potential-energy surface is the calculation of a set of electronic energies that will be either fitted or interpolated by a suitable functional form. Prior to this, however, one faces the problem of choosing at which nuclear geometries the calculations should be performed, or in other words how the nuclear configuration space should be properly sampled in order to produce an accurate potential-energy surface. Focusing for the moment on the one-dimensional case of a diatom, a straight-forward and rather conventional way is to adopt as a coordinate the internuclear distance r, fix two boundary values r_{min} and r_{max}, and compute the electronic energies at the points of an evenly spaced grid on the segment of r between r_{min} and r_{max}. For higher-dimensionality problems, multidimensional grids are in order. For the specific three-dimensional[1] case of an atom–diatom exchange reaction considered here, a common choice for the set of the three internal coordinates is the two interatomic distances of the reactant and product diatom (r_{BC} and r_{AB}, respectively) plus the angle ($\Phi = \widehat{ABC}$) formed by the corresponding bonds. The resulting three-dimensional grid is set up using as building blocks two uniform grids (as already defined for the diatom case) for r_{BC} and r_{AB} and an angular grid for Φ.

In the SRBO approach, a more effective sampling specifically tailored to the physics of atoms and molecules is ensured by replacing the interatomic distances with carefully reformulated so-called bond-order variables. Based on the 'bond-order' concept dating back to the time of Pauling [68], bond-order variables are defined as the exponential of the weighed diatomic displacement:

$$n = e^{-\beta(r-r_e)} , \qquad (6.8)$$

where β is a constant related to one or more diatomic force constants, r_e is the equilibrium diatomic distance and r is the already-mentioned internuclear distance of the diatom. As concisely pointed out by Laganà for the first time in a letter of 1991 [69], the appeal of bond-order variables for use in dynamics studies lies in the property of the bond-order space of having the origin at infinite internuclear distances and of being confined within zero (for $r \to \infty$) and the limiting value $e^{\beta r_e}$ (for $r \to 0$).

A further property of the bond-order variables directly following from their definition (Eq. (6.8)) is that of confining the attractive and repulsive regions of a diatom configuration space in the $(0, 1)$ and $(1, e^{\beta r_e})$ range, respectively, with unity in bond-order space representing the equilibrium point. In fact, bond-order variables have been used in formulating polynomial representations of potential-energy surfaces [70,71] and rotating diatomic-like ones (like the rotating bond-order, ROBO, [69] and the largest-angle generalization of the rotating bond order, LAGROBO [72]). In addition, bond-order variables have been used for producing relaxed representations of potential-energy surfaces evidencing process channels [73] and developing the related quantum reactive scattering formalism [74,75].

---

[1] It is worth recalling here that the relevant coordinates in this context are the internal coordinates, which are in general $3N - 6$ for an N-atom system, thus 1 and 3 for a two- and a three-atom system, respectively.

As already mentioned, in configuration-space sampling for potential-energy surface fitting one usually excludes the strongly repulsive (dynamically inaccessible) regions and the near-asymptotic ones by introducing two boundary distances r_{min} and r_{max}. These boundary values map onto their analogs in bond-order space n_{max} and n_{min}, respectively. Now, a balanced representation of the sampled attractive and repulsive region of the diatom configuration space can then be obtained in bond-order space by relaxing the condition that β is linked to the diatom spectroscopic properties and allow it to vary until a desired attractive over repulsive space ratio f is reached [75]

$$f = \frac{1 - n_{min}}{n_{max} - 1} = \frac{1 - e^{-\beta(r_{max} - r_e)}}{e^{-\beta(r_{min} - r_e)} - 1} \,. \tag{6.9}$$

The resulting bond-order variables are refereed to as space reduced–bond order (SRBO) variables, in that once f has been fixed the same space relation between the attractive and repulsive regions of configuration space will hold for any considered diatom.[2]

Reasonable values of r_{min} and r_{max} (and thus n_{max} and n_{min}) tailored to a specific diatom can be obtained by modeling the diatom potential with a Morse function [76]

$$V(r) = D_e \left[1 - \exp\left(-\sqrt{\frac{k_e}{2D_e}} (r - r_e) \right) \right]^2, \tag{6.10}$$

whose parameters (the equilibrium distance r_e, the bond-dissociation energy D_e and the force constant at equilibrium $k_e = d^2 V / dr^2 \big|_{r_e}$) can be obtained by simple geometry optimization through standard electronic-structure calculations, and by tuning the factors $V_{fact} = V(r_{min})/D_e$ and $V_{thrs} = [D_e - V(r_{max})]/D_e$ associated with the mentioned boundary values. Fig. 6.1, for example, shows the Morse potential for the H_2 diatom as a function of the bond-length variable r (top panel) and of two SRBO variables yielding either $f = 1$ (center panel, $\beta = 0.813 \, a_0^{-1}$) or $f = 2$ (bottom panel, $\beta = 0.464 \, a_0^{-1}$) where boundaries r_{min} and r_{max} (n_{max} and n_{min}) were chosen by setting $V_{fact} = 2.0$ and $V_{thrs} = 0.001$. In this way, boundaries are obtained to the sampled configuration space that are linked to a diatom-specific physical model rather than being arbitrary, i.e., the white area in Fig. 6.1 is limited at short distances where the Morse potential is twice the dissociation energy and at long distances where it has reached the dissociation energy within 1‰.

Once an optimal attractive over repulsive space-ratio parameter f has been chosen and boundaries r_{min} and r_{max} have been set, the nuclear configuration space of the diatomic molecule can be conveniently sampled by an evenly

---

[2] It is worth adding here that SRBO variables including the whole accessible physical space may be defined by taking $r_{max} = \infty$ ($n_{min} = 0$) and $r_{min} = 0$ ($n_{max} = e^{\beta r_e}$).

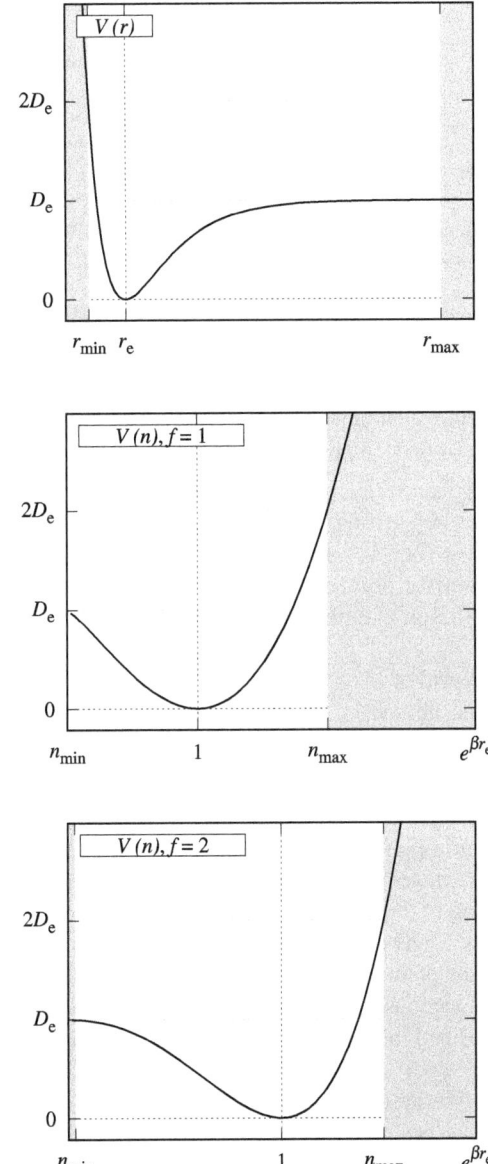

FIGURE 6.1 Morse potential for the H_2 diatom as a function of the interatomic distance r (top panel) and of two SRBO variables $n = e^{-\beta(r-r_e)}$ yielding $f = 1$ (center panel) and $f = 2$ (bottom panel). Reprinted with permission from Ref. [67]. © 2015 American Chemical Society.

spaced grid on the segment of the resulting SRBO variable comprised between n_{min} and n_{max}.

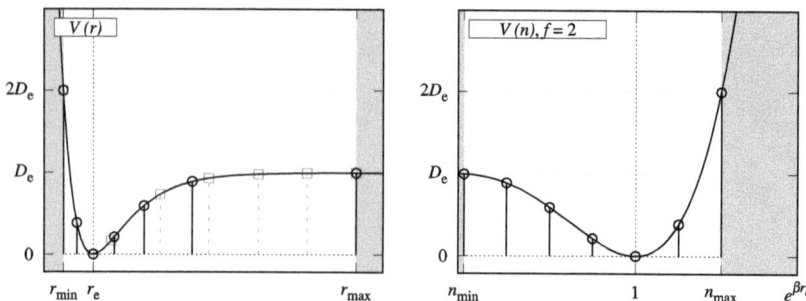

FIGURE 6.2 Points of a 7-point (2, 4) SRBO grid in physical space (left panel) and in bond-order space (right panel). Reprinted with permission from Ref. [67]. © 2015 American Chemical Society.

Thus summarizing, a suitable SRBO uniform grid can be generated according to the following four-step procedure:

1. fix the boundary values r_{min} and r_{max} in physical space (n_{min} and n_{max} in the bond-order space) by setting V_{fact} and V_{thrs};
2. choose the couple (N_r, N_a) of grid points in the repulsive (r) and attractive (a) region, respectively, of the diatom configuration space;
3. compute the SRBO mapping parameter β yielding $f = N_a/N_r$ from Eq. (6.9);
4. set up a uniform grid of $N_r + N_a + 1$ (the added 1 accounts for the equilibrium geometry that is, obviously, neither attractive nor repulsive) points on the segment of n ranging from n_{min} to n_{max}.

This type of SRBO grid may be quickly referenced as (N_r, N_a), like the 7-point (2, 4) shown in the right-hand side panel of Fig. 6.2. In the same figure, the corresponding grid points in physical space are also shown in the left-hand-side panel. The advantages of an SRBO sampling over the traditional one based on uniform sampling of the interatomic distance now appear clear from a visual inspection of Fig. 6.2. The points of an SRBO grid are concentrated in the high-gradient region of the potential and, as also discussed in Ref. [75], better sample the most important regions of configurations than the points of a uniform grid of the same size defined in physical space (which are also shown in Fig. 6.2 as gray squares).

For the three-atom system of our A + BC → AB + C reaction, a three-dimensional grid can be conveniently set up by using as building blocks two grids defined over the SRBO reactant and product bond-order variables, n_{BC} and n_{AB}, respectively, plus an angular grid in the already-mentioned angle $\Phi = \widehat{ABC}$. A two-dimensional representation of the potential-energy surface at fixed angle $\Phi = 180°$ for the H + H$_2$ reaction is shown in Fig. 6.3 as a function of the two interatomic distances r_{BC} and r_{AB}. The blue (dark gray in print version) points represent the computed electronic energies (see Ref. [67] for details on their calculation) over the two-dimensional SRBO grid, while the gray mesh shows the analytic representation of the potential-energy surface obtained by

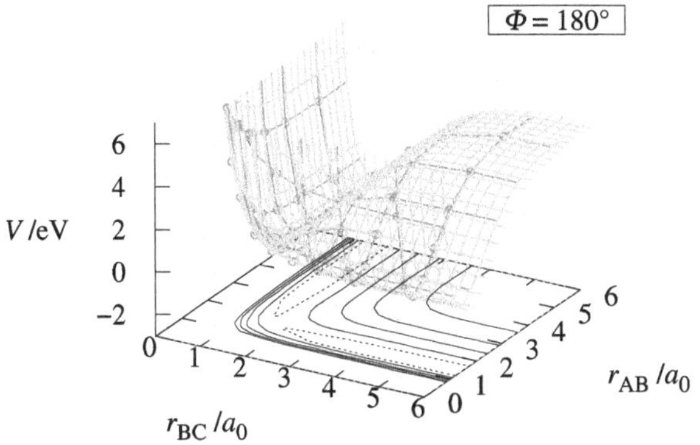

FIGURE 6.3 Potential-energy surface for collinear H + H$_2$ sampled on a two-dimensional SRBO grid built with two 10-point (3, 6) SRBO grid (blue (dark gray in print version) circles) and fitted to an Aguado–Paniagua analytic formulation (gray mesh) (see Ref. [67] for details on the fitting). A contour map is also shown at the base with contours at 1, 2, 3, and 4 eV (solid lines) plus a value (close to the barrier height) of 0.4 eV (dashed line).

the Aguado–Paniagua global fitting scheme. As is apparent from the figure, the distribution of the computed electronic energies indicates a fair coverage of the molecular geometry space by the SRBO grid used for the calculations, with a denser concentration in the high-gradient regions of the potential. The figure also shows the smoothness of the fitted potential-energy surface (free of spurious features).

As outlined above, SRBO variables are easily constructed from the values of the diatom equilibrium distance r_e, bond-dissociation energy D_e and force constant at equilibrium k_e according to a desired value of the parameter f, i.e., the ratio between the attractive and repulsive range of the diatom configuration space. As discussed in Ref. [67], benchmark calculations indicate that optimal SRBO performances are reached when β is set such as to yield an f value equal to 2, whereby the number of points sampling the attractive region is twice the number of points sampling the repulsive one. In the same work, the SRBO sampling is shown to generally outperform the traditional interatomic-distance sampling in converging the fitted potential-energy surface to the computed electronic energies with an increasing number of grid points, independently of the adopted fitting/interpolation scheme.

The conceptual advantage of the illustrated SRBO configuration-space sampling lies in the fact that the resulting choice of geometries has built-in a force-based metric rather than a purely geometric one as in the traditional interatomic-distance sampling. As we shall see in Chapter 8, this advantage can also be conveniently exploited in solving nuclear-dynamics problem such as the vibrational eigenvalue problem of a diatomic molecule.

6.3   Visualization and analysis: the H + H₂ reaction

Once a proper set of electronic energies has been computed and an analytic potential-energy surface has been generated either by global fitting or local interpolation methods, all that remains in order to simulate the reactive event is to implement a method of solution for the equations – either classical or quantum – governing the nuclear motion. This will be the subject of Chapters 7 and 8. However, an additional stage is often most helpful both for the setup of the dynamics calculations and for the rationalization of their outcome. Such an additional stage is the visualization and analysis of the topological features of the potential-energy surface with the aim of gaining, in a preliminary way, some useful insights into the dynamical features of the process. The outcome of dynamics calculations is in fact a direct consequence of the 'shape' of the generated potential-energy surface, as this is the only 'parameter', along with the nuclear masses, that differentiates one system from another, and often – at least to a qualitative extent – some dynamical features of the reaction can be inferred by a visual inspection of the topology of the potential-energy surface.

Now a potential-energy surface is generally a complicated mathematical object. It is indeed a $(3N - 6)$-dimensional hypersurface depending on all the internal coordinates of an N-atom system. Even for a simple system such as that of our A + BC reaction featuring a number of atoms N as small as 3, and thus only 3 internal coordinates, the resulting three-dimensional potential-energy surface cannot be entirely easily visualized as, even exploiting three-dimensional graphics, a further dimension would be required for plotting the values of the function itself.

The most convenient strategy for a fruitful visualization of a potential-energy surface is thus generally that of opportunely reducing the dimensionality of the surface to two coordinates, typically associated with the degrees of freedom that are mostly involved in the physical process of interest. These two coordinates are used as the 'latitude' and the 'longitude' and the associated value of the potential-energy surface as the 'elevation'. The potential-energy surface can then be studied through surface or contour plots analogous to those used for representing reliefs in geographic maps. A first and simple example of dimensionality reduction is indeed exactly the one that has been adopted for generating Fig. 6.3. In that case, one of the three internal coordinates r_{BC}, r_{AB}, and Φ has been frozen to a fixed value ($\Phi = 180°$) and a three-dimensional plot has been built by plotting the value of the potential energy as a function of the remaining r_{BC} and r_{AB} coordinates.

The resulting plot is indeed highly informative. The collinear ($\Phi = 180°$) H + H₂ atom-exchange reaction features a potential-energy surface that resembles a two-armed canyon with the two arms arranged at a right angle. One of these arms (high r_{AB} and $r_{BC} \simeq r_{eq}$) corresponds to the reactant (or entrance) channel, the other arm (high r_{BC} and $r_{AB} \simeq r_{eq}$) corresponds to the product (or exit) channel. Far into the reactant (or into the product) region, cuts of the potential-energy surface transverse to the canyon valley path coincide with the

H_2 dissociation curve. In order for the reaction to occur, the system has to travel from one channel to the other. This involves reaching a region of strong interaction (low values of both r_{BC} and r_{AB}) and overcoming a saddle point slightly higher than 0.4 eV (this can be deduced by the isocontour lines that are also reported in Fig. 6.3). Such a saddle point is also commonly called a reaction barrier.

While, as already stressed, such analysis is certainly informative, it is severely limited to one possible atom–diatom approach – the collinear one – while unfortunately leaving out all the remaining possible noncollinear reaction mechanisms. Information on these alternative reaction paths may be recovered by building additional plots similar to that of Fig. 6.3 for different values of Φ. Alternatively, if one is interested in the energetically favored reaction path, a three-dimensional plot similar to that of Fig. 6.3 could be built by 'relaxing' – rather than freezing – the angle Φ. In other words, an informative three-dimensional plot could be obtained by minimizing the potential energy with respect to Φ and plotting it as a function of r_{BC} and r_{AB}. An even more drastic simplification is that of producing a one-dimensional plot of the potential-energy profile as a function of the minimum-energy path (the path adhering to the bottom of the canyon valley path in Fig. 6.3), i.e., the potential energy at each value of a suitable reaction coordinate minimized with respect to all the remaining coordinates (see Fig. 10.3 in Chapter 10 for an example of the minimum-energy profile of a more complex reaction than $H + H_2$).

While the above-mentioned plots are widely adopted and provide a useful description of fixed-angle or minimum-energy reaction paths, information on possible competing reaction mechanisms involving different atom–diatom approaching or scattering angles is inescapably lost and the complexity of the reaction mechanism can indeed remain hidden. A more complete picture is instead given by three-dimensional plots where the two coordinates used for the plot are carefully crafted so as to be meaningful with respect to the process under study and the potential is minimized with respect to all the remaining coordinates. Accordingly, more useful to the end of determining and rationalizing the outcomes of the dynamical calculations performed on a given reactive channel are the so-called rectangular relaxed plots [73,77], where the potential-energy surface for a given atom–diatom exchange reaction is represented as a function of a 'reaction-progress', η, and a 'reaction-mechanism', Φ, coordinate based on the following coordinate change:

$$\eta = \arctan(r_{BC}/r_{AB}) \tag{6.11}$$

$$\Phi = \widehat{ABC} \tag{6.12}$$

$$\rho = \sqrt{r_{BC}^2 + r_{AB}^2} \, . \tag{6.13}$$

The first of these coordinates, η, is a suitable reaction-progress coordinate in that it quantifies the ratio between the breaking over of the forming-bond distance.

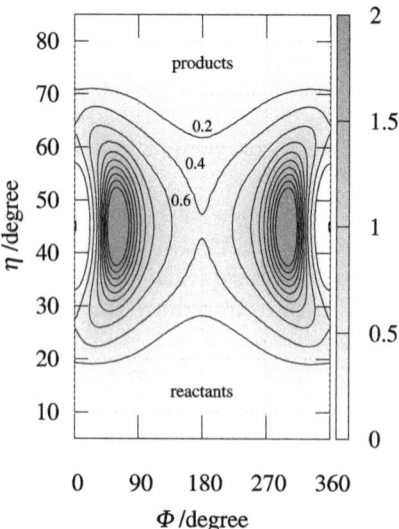

FIGURE 6.4 Rectangular relaxed plot of the potential-energy surface for the H + H$_2$ reaction. The η coordinate quantifies the progress of the reaction, while the Φ coordinate, which determines the shape of the triangle formed by the three atoms, provides information on the mechanism.

The second coordinate, Φ, accounts for the shape of the associated triangle providing an indication on the detailed mechanism by which the reaction occurs, as it relates to the approaching angle of A towards BC and the scattering angle of C from the newly formed AB diatom. The third coordinate, ρ, is an 'overall-size' coordinate with less informative content, providing information on the tightness of the collision partners while the system moves from one asymptote to the other.

A rectangular relaxed plot (see Fig. 6.4) is obtained by plotting for each couple (η, Φ) the value of the potential-energy minimized with respect to ρ, i.e.:

$$\min_{\rho} V(\eta, \Phi) \,. \tag{6.14}$$

The name rectangular derives from the shape of the domain of points usually adopted in such a representation, where reaction progress is emphasized by using the long side of the rectangle for coordinate η and the short side for coordinate Φ. Moving along the horizontal axis of the rectangle in Fig. 6.4, the Φ angle changes, providing information about the detailed mechanism of the reaction, while moving along the vertical axis the η angle changes, quantifying the progress of the reaction. This representation has the advantage of enveloping together all the fixed Φ minimum-energy paths providing a qualitative and quantitative overall view of the possible flux to products.

The rectangular relaxed plot for the H + H$_2$ potential-energy surface is shown in Fig. 6.4. The bottom area of the rectangle corresponds to the reac-

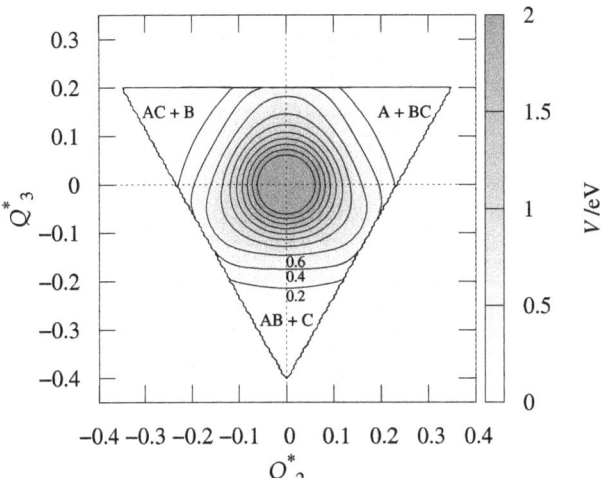

FIGURE 6.5 Triangular relaxed plot of the potential-energy surface for the H + H$_2$ reaction. The three edge regions correspond to the three asymptotic arrangements, while the inner sides describe reaction intermediates. The plot scans all possible reaction paths, from collinear (sides of the triangle) to perpendicular (passing through the center of the triangle) atom–diatom approaches.

tant region, while the top area corresponds to the product region. The progress of the reactive process is monitored by the values of the reaction coordinate η, while the angle Φ accounts for the detailed mechanism. In particular, the plot shows that the overall minimum reaction is the collinear one, while noncollinear approaches involve an increasingly high reaction barrier. The additional white areas for $\eta \simeq 45°$ and Φ approaching either 0° or 360° correspond to the third possible arrangement AC + B (which in the case of the H + H$_2$ reaction is equivalent to A + BC and AB + C from an energetic point of view).

Rectangular relaxed plots focus on a given reactive process from a nuclear arrangement to another and, as we shall see in an actual case in Chapter 10, are highly useful when studying a specific reaction channel. For generic A + BC reactions involving different A, B and C atoms and AB + C and AC + B product sets, it might be useful to opt for a relaxed representation that does not favor one process but rather accounts for the connections between all possible three arrangements on an equal footing. This can be elegantly achieved through the so-called triangular relaxed plots proposed by Varandas [78] in which use is made of the reduced-symmetry coordinates Q_2^* and Q_3^* defined as

$$Q_2^* = \frac{s_2/s_3}{\sqrt{2}} \qquad Q_3^* = \frac{2s_1 - s_2 - s_3}{\sqrt{6}} \, , \qquad (6.15)$$

where $s_i = r_i / \sum_{j=1}^{3} r_j$ (with $r_1 = r_{AB}$, $r_2 = r_{BC}$, $r_3 = r_{AC}$) and the relaxed variable is the perimeter of the molecular triangle, i.e., the sum of the three internuclear distances. The resulting triangular relaxed plot for the H + H$_2$ reactive system is shown Fig. 6.5. The three edge regions correspond to the three asymptotic arrangements, while the inner sides describe reaction intermediates. The plot scans all possible reaction paths, from collinear (sides of the triangle) to perpendicular (passing through the center) atom–diatom approaches. This kind of plot is of course most useful when the A, B, and C atoms are different as it provides useful information on the energetics of the three nuclear arrangements and of the paths interconnecting them. For the H + H_2 reaction, the plot in Fig. 6.5 confirms the picture offered by that of Fig. 6.4, i.e., the energetically favored approach is the collinear one (sides of the triangle), and complements it by making it clearer that the highest reaction barrier is associated with the perpendicular approach (energy peak at the center of the triangle).

Chapter 7

Theoretical treatments

Once the potential-energy surface of a chemical reaction has been generated, one has to solve the equations of motion for the involved nuclei. This can be done either within classical mechanics, by considering nuclei as classical particles subject to forces deriving from the potential-energy surface and solving Newton's equations of motion, or within quantum mechanics by retaining the quantum nature of nuclei and solving the Schrödinger equation for the nuclear problem. This chapter covers both topics with reference to A + BC reactions. In particular, in Section 7.1 the calculation and analysis of classical trajectories is discussed. In Section 7.2, the quantum approach to reactive scattering is illustrated in its time-independent and time-dependent formulations. In Section 7.3, focus is placed on time-dependent methods based on the propagation and analysis of a wavepacket.

7.1 Classical trajectories

As already mentioned, the classical approach to the modeling of chemical-reaction dynamics is based on the solution of Newton's equation (Eq. (2.1)). The forces acting on the atoms are calculated by computing the derivatives of the potential-energy surface with respect to the atoms' positions. If the potential-energy surface has been cast in an analytic form, usually the calculation of its derivatives is a cheap and straightforward computational task. Thus once the forces acting on the atoms are known, the problem reduces to solving a set of second-order differential equations for some given initial conditions. To illustrate in detail the practical steps involved in the solution of Eq. (2.1) for the modeling of chemical reactions we will focus on a popular finite-difference scheme known as the Verlet algorithm [79], more precisely on its variant known as velocity Verlet.

For this purpose, let us first consider the case of the motion of a particle subject to a force in a one-dimensional space, and then generalize to the many particle–three dimensional case. In the one-dimensional case, the equation of motion is a second-order differential equation for the position of the particle as a function of time:

$$ f = m \frac{d^2 x}{dt^2} = m \ddot{x} \,. \tag{7.1} $$

In a finite-difference scheme, time is discretized using a constant time step $\Delta t = \tau$. Let us denote with subscript k the quantities at the kth time step: the time

https://doi.org/10.1016/B978-0-32-390865-8.00016-7

$t_k = k\tau$, the position of the particle $x_k = x(t_k)$, its velocity $v_k = v(t_k)$, and the force acting on it $f_k = f(x_k)$. In the central finite-difference approximation, the second derivative of the position after k time steps is

$$\ddot{x}_k = \frac{x_{k+1} + x_{k-1} - 2x_k}{\tau^2} . \tag{7.2}$$

Inserting Eq. (7.2) into the right-hand side of Eq. (7.1) leads to the basic equation of the Verlet algorithm:

$$x_{k+1} = 2x_k - x_{k-1} + \tau^2 \frac{f_k}{m} . \tag{7.3}$$

Since the differential equation is a second-order one, two points are required in order to get the recursion started. It also turns out that the algorithm is 'stable', i.e., the error made at any iteration tends to decay rather than magnify during the propagation of the algorithm.

A variant of the Verlet algorithm, which is identical to the original algorithm from a mathematical point of view in the sense that it generates the same trajectory (except for round-off errors due to the finite precision of computers), is the so-called velocity Verlet algorithm. Considering that the velocity of our particle is the derivative of its position with respect to time, the second-order differential Eq. (7.1) can be reduced to two first-order equations:

$$\dot{x}_k = v_k \tag{7.4}$$

$$\dot{v}_k = \frac{f_k}{m} . \tag{7.5}$$

Now, expanding x_{k+1} and v_{k+1} through a Taylor series about x_k and v_k, respectively, and using Eqs. (7.4) and (7.5) in the resulting expressions, one obtains the following finite-difference scheme:

$$x_{k+1} = x_k + \tau v_k + \frac{\tau^2}{2} \frac{f_k}{m} + O(\tau^3) \tag{7.6}$$

$$v_{k+1} = v_k + \frac{\tau}{2m}(f_k + f_{k+1}) + O(\tau^3) , \tag{7.7}$$

forming the basis of the velocity Verlet algorithm and offering a recursion scheme for the coordinate-velocity pair in terms of the forces at two successive stages.

Such a scheme can be easily generalized to the many atom–higher dimensional case. For a system of N particles in three-dimensional space, the potential is a function of the positions of the particles $\{r_a\}$. The force acting on the ath atom will then be:

$$f^{(a)} = -\nabla^{(a)} V(\{r_a\}) , \tag{7.8}$$

where superscript a on the gradient operator indicates that the derivatives should be taken with respect to the coordinates of the ath atom. For each separate Cartesian component, the above-discussed Verlet recursive scheme will apply, with the x, y, and z components of the forces being, respectively,

$$f^{(a,x)} = -\frac{d}{dx^{(a)}} V(\{r_a\}) \tag{7.9}$$

$$f^{(a,y)} = -\frac{d}{dy^{(a)}} V(\{r_a\}) \tag{7.10}$$

$$f^{(a,z)} = -\frac{d}{dz^{(a)}} V(\{r_a\}) . \tag{7.11}$$

Thus a distinct recursion relation rooted in the velocity Verlet algorithm can be used for each coordinate–velocity pair but these are all coupled through the force, which depends on the coordinates of all the atoms.

Focusing on the x component, for instance, the final algorithm, with given initial x_k and v_k and a known expression for $f(x)$, is:

1. Calculate $x_{k+1}^{(a)}$:

$$x_{k+1}^{(a)} = x_k^{(a)} + \tau v_k^{(a,x)} + \tau^2 \frac{f_k^{(a,x)}}{2m_a} . \tag{7.12}$$

2. Evaluate $f_{k+1}^{(a,x)}$
3. Calculate $v_{k+1}^{(a,x)}$:

$$v_{k+1}^{(a,x)} = v_k^{(a,x)} + \frac{\tau}{2m_a} \left(f_k^{(a,x)} + f_{k+1}^{(a,x)} \right) . \tag{7.13}$$

4. Assign the value of $x_{k+1}^{(a)}$ to $x_k^{(a)}$ and go back go back to step 1.

Analogous schemes can be written for the y and z components.

An easy and fruitful exercise for the eager student is to implement the velocity Verlet algorithm for a system of N atoms interacting through pairwise additive Lennard-Jones potentials:

$$V(\{r_a\}) = \sum_a^N \sum_{b<a}^N V_{\text{LJ}}(r_{ab}) , \tag{7.14}$$

where r_{ab} is the distance between atoms a and b

$$r_{ab} = \sqrt{x_{ab}^2 + y_{ab}^2 + z_{ab}^2} , \tag{7.15}$$

with $x_{ab} = x_a - x_b$, and the Lennard-Jones potential is given by:

$$V_{\text{LJ}}(r) = 4\epsilon \left[\left(\frac{\sigma}{r}\right)^{12} - \left(\frac{\sigma}{r}\right)^6 \right] , \tag{7.16}$$

featuring the well-known short-range repulsion term and long-range attraction term. In this case, the derivative of the pair potential with respect to each Cartesian component of the position of an atom can be written in terms of V', i.e., the derivative of the Lennard-Jones potential with respect to the argument of the same function. For the x component, for example:

$$\frac{\partial V_{LJ}(r_{ab})}{\partial x_{ab}} = V'_{LJ}(r_{ab})\frac{\partial r_{ab}}{\partial x_{ab}} = \frac{x_{ab}}{r_{ab}}V'_{LJ}(r_{ab}), \tag{7.17}$$

where $V'_{LJ}(r)$ is easily evaluated as:

$$V'_{LJ}(r) = 4\epsilon\left[-12\frac{\sigma^{12}}{r^{13}} + 6\frac{\sigma^6}{r^7}\right]. \tag{7.18}$$

Accordingly, the x component of the force acting on the ath atom to be used in step 2 of the above-summarized algorithm is:

$$f_{k+1}^{(a,x)} = -\sum_{b\neq a}\frac{x_{ab}}{r_{ab}}V'_{LJ}(r_{ab}). \tag{7.19}$$

A more general and realistic formulation for the interaction potential of a many-atom system popular in dynamics simulation of systems of hundreds or thousands of atoms will be discussed in Chapter 11. For our A + BC reactive system, at any rate, the potential-energy surface will be a higher-level one based on accurate electronic-structure calculations and cast in one of the analytic formulations already discussed in Chapter 6. Simulating the reactive event will thus essentially involve reiterating the above-outlined algorithm for the x, y, and z components of each of the three atoms using the forces derived from the triatomic potential-energy surface.

However, the outcome of a single trajectory is insufficient to offer a description of the features of the reactive event. To this end, a statistically significant sample of trajectories has to be integrated and the related outcomes averaged. This is at the heart of the so-called quasiclassical trajectory (QCT) method [80]. The term 'quasiclassical' denotes that, despite the nuclei being treated as classical particles, some aspects pertaining to quantum dynamics are approximately recovered for a more meaningful comparison with experiment – that inherently has to do with quantum phenomena. In particular, the reactant molecules are selected before the collisions at discrete internal energy states corresponding to quantum states, and after the collision a 'quantization' of the internal energy is also enforced on the product molecule.

The quasiclassical trajectory method assumes that the nuclei involved in a chemical reaction move according to the laws of classical mechanics on the potential-energy surface of the system. By integrating the classical equations of motions, for instance through the above-discussed velocity Verlet algorithm, the coordinates and velocity of all nuclei are calculated at any time step during the collision. The integration is carried out until the fragments produced by the collision are sufficiently separated. Then the species produced are identified and the channel (nonreactive, reactive or dissociative) is assigned. The result of the integration is a set of final values for coordinates and velocities of all the nuclei. By using this information, the final properties of the system (like, for instance, the internal energies and rotational angular momenta of the molecules, the relative translational energy, the scattering angle) are evaluated.

As already mentioned, in quasiclassical trajectory calculations one usually runs batches of trajectories. Each trajectory is characterized by a properly chosen pair of initial values of the relative atom–diatom collision (or translational, 'tr') energy E_{tr} and impact parameter b. The trajectory is also characterized by a set of angles determining the orientation of the diatom and by two quantum-like internal states (the vibrational v and rotational j ones) of the reactant diatom. The reactive properties can all be traced back to the fraction of trajectories associated with reactive events (i.e., trajectories ending in the product channel) $N_{v,j}^{R}(E_{tr}, b)$ out of the integrated total $N_{v,j}^{tot}(E_{tr}, b)$ trajectories, for collisions starting from a given set of E_{tr}, b, v, and j values with the remaining variables being selected randomly. This fraction gives the state-specific opacity function

$$P_{v,j}(E_{tr}, b) = \frac{N_{v,j}^{R}(E_{tr}, b)}{N_{v,j}^{tot}(E_{tr}, b)}, \tag{7.20}$$

while the following fraction, obtained after summing over b, gives the state-specific reactive probability

$$P_{v,j}(E_{tr}) = \frac{N_{v,j}^{R}(E_{tr})}{N_{v,j}^{tot}(E_{tr})}. \tag{7.21}$$

Dynamical information can be given an even more detailed formulation if the labels associated with products (the primed quantities E_{tr}', b', v', and j') are explicitly considered. When needed, product properties can be worked out of the classical variables at the end of the trajectory (to determine the quantum-like rovibrational states the energies of the final diatom are worked out by semiclassically discretizing the action integral of the vibrational motion at the asymptotic cuts of the potential-energy surface). As we shall see in Chapter 10, the resulting state-to-state quantities, as well as the above-mentioned state-specific ones, are most useful for rationalizing the reactive properties of the considered process and supporting the formulation of related reaction mechanisms.

Within the above-outlined framework, the reaction cross section can be easily estimated as

$$\sigma = \pi b_{max}^{2} \frac{N^{R}}{N^{tot}}, \tag{7.22}$$

where N^{tot} is the total number of integrated trajectories, N^R the number of re-active trajectories, b_{max} is the maximum value of the impact parameter leading to reaction, and the error in this estimation is proportional to $(N^R)^{-1/2}$.

7.2 The quantum approach

The above-discussed reaction probability is also a key quantity in bridging the classical view with the quantum view. The central quantity in the quantum theory of chemical-reaction dynamics is in fact the quantum-mechanical probability amplitude or scattering-matrix element $S_{f,i}(E)$ for a transition from an initial channel i of the reactants to a final channel f of the products as a function of the total energy E. The indices i and f are composite indices that relate to a certain nuclear arrangement (A + BC, AB + C, or AC + B) and to the rovibrational state of the related diatomic molecules. The quantum reaction probability $P_{f,i}(E)$ from a state i of the reactants to a state f of the products at total energy E is, in fact, given by the square modulus of the scattering-matrix element $S_{f,i}(E)$. All the observables properties of the reactive process can indeed be traced back, by means of well-established formulae, to the knowledge of the scattering matrix at a sufficiently large number of energies and for a sufficiently large number of reactant and product channels.

In order to illustrate the basic formalism of the quantum treatment of atom–diatom reactions, we will focus on the simple two-dimensional case of a collinear reaction. The treatment of the full three-dimensional case involves indeed quite cumbersome expressions and the interested reader is referred to the references cited in the following for a complete account of that topic. On the contrary, collinear reactions have been clearly and concisely reviewed by Manolopoulos [81], whose work we will closely follow here with minor adaptations in the formalism.

After separation of the motion of the center of mass of the system (which is irrelevant to the calculation of the reactive properties), the problem can be entirely cast using internal coordinates. A useful set of internal coordinates for reactive scattering problems are the so-called Jacobi coordinates, which are arrangement dependent. For the reactant arrangement A + BC, these are defined as the BC interatomic distance r_{BC} and the distance R_A between the atom A and the center of mass of the BC molecule. Analogously, for each product arrangement AB + C and AC + B, a set of Jacobi coordinates r_{AB}, R_C and r_{AC}, R_B, respectively, can be defined. Reactant and product Jacobi coordinates for the collinear reaction A + BC $\rightarrow$ AB + C are depicted in Fig. 7.1. The usefulness of the Jacobi coordinates for the treatment of atom–diatom reactions ultimately resides in the fact that the R coordinate relates to the atom–diatom translational motion and the r coordinate to the vibrational motion of the diatomic molecule.

For the reactive scattering treatment, it is actually convenient to work with a more suitable set of internal coordinates, the so-called mass-scaled Jacobi coordinates. The reactant mass-scaled Jacobi coordinates for our reaction, for

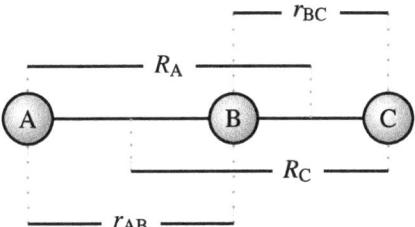

FIGURE 7.1 Reactant (r_{BC}, R_A) and product (r_{AB}, R_C) Jacobi coordinates for the collinear A + BC → AB + C reaction.

instance, are defined as

$$r_a = \lambda_a^{-1} r_{BC} \tag{7.23}$$

$$R_a = \lambda_a R_A , \tag{7.24}$$

where we used the lowercase letter (a) corresponding to the isolated atom (A) to label the reactant arrangement (A + BC) and where λ_a is an arrangement-dependent mass-scaling factor

$$\lambda_a = \left(\frac{m_A m_{BC}^2}{m_B m_C m_{ABC}} \right)^{\frac{1}{4}} . \tag{7.25}$$

Analogous sets of mass-scaled Jacobi coordinates can be written for the b and c product arrangements, corresponding to AC + B and AB + C, respectively. The reason for opting for mass-scaled Jacobi coordinates is that the Hamiltonian for the collinear atom–diatom reaction can in this case be written equivalently for all nuclear arrangements:

$$\hat{H} = -\frac{\hbar^2}{2\mu} \left(\frac{\partial^2}{\partial R_\alpha^2} + \frac{\partial^2}{\partial r_\alpha^2} \right) + V(R_\alpha, r_\alpha) , \tag{7.26}$$

with α being either a, b, or c, using the same reduced mass:

$$\mu = \left(\frac{m_A m_B m_C}{m_{ABC}} \right)^{\frac{1}{2}} . \tag{7.27}$$

Now, as implicitly already mentioned when commenting on Fig. 6.3 in the previous chapter, at large R_α, $\hat{H}$ separates into the sum of a translational–kinetic energy operator and a vibrational Hamiltonian $\hat{H}_\alpha$:

$$\hat{H} \overset{R_\alpha \to \infty}{\sim} -\frac{\hbar^2}{2\mu} \frac{\partial^2}{\partial R_\alpha^2} + \hat{H}_\alpha \tag{7.28}$$

$$\hat{H}_\alpha = -\frac{\hbar^2}{2\mu} \frac{\partial^2}{\partial r_\alpha^2} + V_\alpha(r_\alpha) , \tag{7.29}$$

where $V_\alpha(r_\alpha)$ is the diatomic potential (i.e., the dissociation curve) for the diatomic molecule of the arrangement α. Accordingly, asymptotic solutions of the Schrödinger equation in each arrangement must be linear combinations of incoming ($-$) and outgoing ($+$) wavefunctions of the form

$$\gamma^\pm_{E\alpha v}(R_\alpha, r_\alpha) = u_{\alpha v}^{-\frac{1}{2}} e^{\pm i k_{\alpha v} R_\alpha} \gamma_{\alpha v}(r_\alpha), \tag{7.30}$$

where $\gamma_{\alpha v}(r_\alpha)$ is an eigenfunction of the vibrational Hamiltonian

$$\hat{H}_\alpha \gamma_{\alpha v}(r_\alpha) = \varepsilon_{\alpha v} \gamma_{\alpha v}(r_\alpha), \tag{7.31}$$

and $k_{\alpha v} = (2\mu(E - \varepsilon_{\alpha v}))^{\frac{1}{2}}/\hbar$ and $u_{\alpha v} = \hbar k_{\alpha v}/\mu$ are the asymptotic wavenumber and velocity in channel αv. Now, the reactive scattering wavefunctions that define the scattering matrix are the solutions $\Gamma_{E\alpha v}$ of the Schrödinger equation:

$$\hat{H} \Gamma_{E\alpha v} = E \Gamma_{E\alpha v}, \tag{7.32}$$

satisfying the following asymptotic boundary conditions in each arrangement

$$\Gamma_{E\alpha v} \overset{R_{\alpha'} \to \infty}{\sim} \gamma^-_{E\alpha v}(R_\alpha, r_\alpha)\delta_{\alpha\alpha'} - \sum_{v'} S_{\alpha'v',\alpha v}(E)\gamma^+_{E\alpha'v'}(R'_\alpha, r'_\alpha). \tag{7.33}$$

The boundary condition of Eq. (7.33) is simply a linear combination of functions of the form given in Eq. (7.30), as required by the above-discussed separation, and corresponds to the physical situation of an incoming wave in channel αv and outgoing scattered waves in all energetically accessible channels $\alpha' v'$ at energy E. The coefficients of the scattered waves are exactly the scattering-matrix elements. These include reactive–scattering matrix elements of the form $S_{cv',\alpha v}(E)$, inelastic–scattering matrix elements of the form $S_{\alpha v',\alpha v}(E)$, and elastic–scattering matrix elements of the form $S_{\alpha v,\alpha v}(E)$.

Methods for calculating the scattering matrix based on the solution of the time-independent Schrödinger equation (Eq. (7.32)) are commonly referred to as time-independent methods. After a careful choice of suitable coordinates (ranging from the more intuitive natural collision coordinates [82,83] to the mathematically more convenient hyperspherical ones), typically time independent–reactive scattering calculations involve the solution of a set of coupled second-order linear differential equations resulting from the expansion of the wavefunction in terms of analytical basis functions, and the propagation of the wavefunction along a spatial continuity variable. A final analysis is then carried out when the wavefunction has been propagated up to an atom–diatom distance large enough to consider the potential asymptotic.

Alternatively, the scattering-matrix elements can be obtained by the following simple expression involving the Fourier transform of a correlation function $C_{cv',\alpha v}(t)$ between an initial reactant wavepacket $\chi_{\alpha v}$ and a final product wavepacket $\chi_{cv'}$ (both expressed in some sets of internal coordinates, the dependence on which has here been dropped for better clarity):

$$S_{cv',av}(E) = -\frac{1}{\langle \chi_{cv'}|\gamma^+_{Ecv'}\rangle\langle\gamma^-_{Eav}|\chi_{av}\rangle} \int_0^\infty e^{iEt/\hbar}C_{cv',av}(t)\,dt\,, \qquad (7.34)$$

where

$$C_{cv',av}(t) = \langle \chi_{cv'}|e^{-i\hat{H}t/\hbar}|\chi_{av}\rangle\,. \qquad (7.35)$$

The interested reader is referred to the works of Dai and Zang [84], Tannor and Weeks [85], and the already quoted Manolopoulos [81] for derivations of the expression of Eq. (7.34) for the scattering-matrix elements. By inspection of Eqs. (7.34) and (7.35), the reader will recognize that the principal task here is that of calculating the time evolution of a wavefunction (Eq. (2.18)), i.e., calculating $\chi_{av}(t) = e^{-i\hat{H}t/\hbar}\chi_{av}(0)$, descending from the time-dependent formulation of the Schrödinger equation (Eq. (2.15)). Techniques based on the time evolution of a nuclear wavepacket are called time-dependent techniques and have found, in recent years, widespread application to diverse problems such as inelastic and reactive scattering, gas–surface scattering, photodissociation, and transition-state spectroscopy.

Since Schatz and Kuppermann's first calculation on $H + H_2$ in 1976 [29,30], much progress has been made in the development of both time-independent [86–89] and time-dependent [90–93] methods and many studies, though mainly limited for computational reasons to three- or four-atom systems, have been published (see the already cited Refs. [31] and [32] for a review of the reactions studied from the early days up to recent times). As already outlined above, quite generally in both methods the nuclear wavefunction is propagated using either time or a spatial coordinate as the continuity variable, and then matched to proper asymptotic conditions to get the elements of the scattering matrix [94], corresponding to the quantum mechanical–reaction probability amplitude, out of which observable properties can be extracted to be compared with the outcome of crossed–molecular beam experiments.

Time-independent and time-dependent techniques, however, differ considerably. Time-dependent problems are, in fact, initial-value problems: the wavepacket, whose formulation is dictated by the scattering boundary conditions, is set up in some initial conditions. It is then evolved in time and the reaction properties are extracted either by projecting the final wavefunction onto individual internal states of the products (as in Eq. (7.35)) to obtain state-to-state quantities, or alternatively by computing the flux across a dividing line or a surface far away from the interaction region, if one is interested only in state-specific quantities. A single wavepacket calculation thus yields dynamical information over a wide range of energies, at the price that one only reactant state at a time is taken in consideration.

A time-independent calculation, on the contrary, includes all the product and reactant channels below a given total energy, but a single calculation only yields the scattering-matrix elements at a single energy. While being generally more robust and reliable, time-independent methods involve the solution of a system of close-coupled equations at each energy and this usually makes the calculations significantly heavy from a computational point of view. Time-dependent wavepacket methods, on the other hand, are extremely simple and efficient, however, their reliability becomes questionable at low translational energies because of the required relatively long propagation times and because, as we will discuss in the next section, they approximate the scattering boundary conditions with an empirical absorbing potential rather than applying them correctly at each energy.

7.3 Wavepacket methods

As reported by Balakrishnan, Kalyanaraman, and Sathyamurthy in their comprehensive review [90], the first work to appear in the literature about the use of time-dependent methods in order to solve the Schrödinger equation was that of Mazur and Robin [95], who in 1959 solved the case of a model atom–diatom collinear exchange reaction. Ten years later, McCullough and Wyatt [96] addressed the reactive problem of $H + H_2$ using an implicit propagation scheme. However, the true step forward for an efficient numerical integration of the time-dependent Schrödinger equation was made by Kosloff and collaborators [97,98]. Their work also pointed out the analogies between classical mechanics of reactive collisions and the time evolution of wavepackets in quantum mechanics, making more intuitive the meaning of the quantistic features of scattering processes. To provide the reader with an outlook on the main computational aspects involved in reactive scattering wavepacket calculations, we will illustrate in the following some of the methods described in the above-mentioned review [90] with specific reference to the collinear $A + BC \rightarrow AB + C$ problem.

As already outlined in the previous section, in time dependent–reactive scattering calculations an initial wavepacket is set up and discretized (or 'collocated') on the points of a grid at the beginning of the dynamical event. It is then evolved in time and analyzed throughout the propagation until the event can be considered ended, in order to extract relevant dynamical quantities. For the purpose of simplifying the formalism, we will indicate with (r, R) the reactant Jacobi coordinates (the nonmass-scaled ones), and with (r', R') the analogous coordinates for the product arrangement. The initial wavefunction for an atom–diatom collision is conveniently expressed in reactant Jacobi coordinates (r, R) and built as a product of a normalized Gaussian function (related to the translational motion of the isolated atom with respect to the center of mass of the molecule), the eigenfunction $\gamma_v(r)$ describing the vibrational eigenstate of the diatom, and a phase factor of the form $e^{-ik(R-R_0)}$, which gives the initial wavepacket a relative kinetic energy towards the strong interaction region:

$$\chi(R, r) = N_g e^{-\alpha(R-R_0)^2} e^{-ik(R-R_0)} \gamma_v(r) . \tag{7.36}$$

In Eq. (7.36), N_g is the Gaussian normalization constant, k determines the average relative momentum or kinetic energy of the collision partners, and α relates to the spatial localization of the wavepacket that is inversely proportional to the range of represented kinetic energies.

Once the initial wavepacket is built, the time-dependent Schrödinger equation (Eq. (2.15)) has to be solved. To perform the time propagation of the wavefunction Kosloff and Kosloff [97] adopted a second-order differencing scheme based on the following equation:

$$i\hbar \frac{\chi(t_{k+1}) - \chi(t_{k-1})}{2\Delta t} = \hat{H}\chi(t_k), \qquad (7.37)$$

which allows for the wavepacket at a given time step to be evaluated from the previous two:

$$\chi(t_{k+1}) = \chi(t_{k-1}) - \frac{2i\Delta t}{\hbar}\hat{H}\chi(t_k). \qquad (7.38)$$

Besides its conceptual simplicity, among the advantages of this method is the fact that it is easy to implement and that it can easily be used also with time-dependent Hamiltonians. However, the algorithm turns out to be numerically unstable unless a small time increment is used. Otherwise, in fact, contributions from higher order in Δt increase and errors accumulate in a way that the solution becomes unstable after a few steps of the propagation.

A more efficient approach to wavepacket propagation is that proposed by Tal-Ezer and Kosloff [98] that is based on the Chebyshev polynomial expansion of the time-evolution operator. This, in fact, takes the simple form of Eq. (2.17) when the Hamiltonian of the system is time independent and thus the wavepacket at time t can be obtained from the wavepacket at time $t = 0$ through:

$$\chi(R, r, t) = e^{\frac{-i\hat{H}t}{\hbar}} \chi(R, r, t = 0). \qquad (7.39)$$

Since the Chebyshev polynomial is bounded in the interval $[-1, 1]$, the Hamiltonian has to be preliminarily renormalized by shifting its eigenvalues to this range. When using uniform grids in the Jacobi coordinates this can be achieved by considering that the maximum and minimum energies that can be represented on the grid are

$$E_{max} = \frac{\hbar^2 \pi^2}{2\mu_r(\Delta r)^2} + \frac{\hbar^2 \pi^2}{2\mu_R(\Delta R)^2} + V_{max} \qquad (7.40)$$

$$E_{min} = V_{min}, \qquad (7.41)$$

where Δr and ΔR are the grid spacings along r and R, respectively, and μ_r and μ_R are the reduced masses associated with the (B, C) and (A, BC) pairs, respectively. The Hamiltonian can then be renormalized by shifting its eigenvalues to

[−1, 1] using the expression

$$\hat{H}_{\text{norm}} = \frac{\hat{H} - \hat{I}\bar{E}}{\Delta E} ,$$

(7.42)

where the energy mean value $\bar{E}$ and the energy range ΔE are calculated as follows:

$$\bar{E} = \frac{1}{2}(E_{\text{max}} + E_{\text{min}})$$

(7.43)

$$\Delta E = \frac{1}{2}(E_{\text{max}} - E_{\text{min}}) .$$

(7.44)

The time-evolution operator can then be rearranged as

$$e^{\frac{-i\hat{H}t}{\hbar}} = e^{\frac{-i\bar{E}t}{\hbar}} e^{\frac{-i\Delta Et\hat{H}_{\text{norm}}}{\hbar}} ,$$

(7.45)

and expanded in a complex Chebyshev series, as follows:

$$e^{\frac{-i\hat{H}t}{\hbar}} = e^{\frac{-i\bar{E}t}{\hbar}} \sum_{n=0}^{\infty} C_n J_n(\alpha_J) T_n(-i\hat{H}_{\text{norm}}) ,$$

(7.46)

where $C_n = 1$ for $n = 0$ and $C_n = 2$ for $n \geq 0$, and $\alpha_J = \frac{\Delta Et}{\hbar}$. In Eq. (7.46), the functions J_n are Bessel functions of the first kind of order n, and $T_n(-i\hat{H}_{\text{norm}})$ are complex Chebyshev polynomials satisfying the recurrence relation

$$\varphi_{n+1} = -2i\hat{H}_{\text{norm}}\varphi_n + \varphi_{n-1} ,$$

(7.47)

with $\varphi_n = T_n(-i\hat{H}_{\text{norm}})\chi(0)$ and the condition $\varphi_1 = \chi(0)$ and $\varphi_2 = -i\hat{H}_{\text{norm}}\chi(0)$. As the Bessel function $J_n(\alpha_J)$ falls off to zero exponentially for $n > \alpha_J$, it is sufficient to include only a few extra terms above $n = \alpha_J$ for convergence, so that the method turns out to be particularly efficient. A further interesting implication of the method is that there are no restrictions on the step size for t and, in fact, the entire evolution of the wavepacket could even be completed in a single time step. It is, however, often convenient to discretize the time propagation so as to collect and analyze useful intermediate information on the collisional dynamics.

As is evident from Eqs. (7.38) and (7.46), whether one uses a finite-difference scheme or the Chebyshev expansion of the time-evolution operator, the time evolution of a wavepacket involves the evaluation of $\hat{H}\chi$, where the Hamiltonian has the form:

$$\hat{H} = \hat{T} + V(R,r) = -\frac{\hbar^2}{2\mu_r}\frac{\partial^2}{\partial r^2} - \frac{\hbar^2}{2\mu_R}\frac{\partial^2}{\partial R^2} + V(R,r) .$$

(7.48)

This consists of two parts: $\hat{T}\chi$ and $V\chi$, where $\hat{T}$ is the kinetic-energy operator and $\hat{V}$ the potential. The latter is obtained by a simple multiplication of χ at each grid point by V since the potential-energy operator is diagonal in the co-ordinate representation, while the evaluation of $\hat{T}\chi$ is less straightforward since the kinetic-energy operator is nonlocal in the coordinate representation. Mc-Cullough and Wyatt [96,99,100] used a three point–finite difference scheme for computing the Laplacian of the wavepacket. In this method the second derivative of the wavepacket $\chi(x)$ with respect to the generic coordinate x at $x = x_i$, for example, is evaluated as follows:

$$\frac{d^2\chi(x_i)}{dx^2} = \frac{\chi(x_{i+1}) - 2\chi(x_i) + \chi(x_{i-1})}{(\Delta x)^2}. \tag{7.49}$$

For the same purpose, subsequent work [101–103] used a higher order–finite difference scheme. However, finite-difference methods in this context turned out to be affected by large error accumulation.

The introduction of the fast Fourier transform (FFT) method by Feit et al. [104–106] and Kosloff and Kosloff [97] for computing the action of the kinetic-energy part of the Hamiltonian on the wavepacket marked a dramatic development in this respect. The FFT method for computing the second derivative of the wavefunction involves Fourier transforming the coordinate space wavefunction to momentum space

$$\chi(k) = \frac{1}{\sqrt{2\pi}} \int_{-\infty}^{\infty} \chi(x) e^{-ikx} dx = \mathcal{F}[\chi(x)], \tag{7.50}$$

multiplying it by $-k^2$ (where k is the wavenumber), and transforming it back to the coordinate space by an inverse Fourier transform

$$\chi(x) = \frac{1}{\sqrt{2\pi}} \int_{-\infty}^{\infty} \chi(k) e^{ikx} dk = \mathcal{F}^{-1}[\chi(k)]. \tag{7.51}$$

Thanks to the use of Fourier transforms, the first and second derivatives of the wavefunction can be calculated as follows:

$$\frac{d\chi(x)}{dx} = \frac{1}{\sqrt{2\pi}} \int_{-\infty}^{\infty} \chi(k)(ik) e^{ikx} dk = \mathcal{F}^{-1}[(ik)\chi(k)] \tag{7.52}$$

$$\frac{d^2\chi(x)}{dx^2} = \frac{1}{\sqrt{2\pi}} \int_{-\infty}^{\infty} \chi(k)(ik)^2 e^{ikx} dk = \mathcal{F}^{-1}[-k^2\chi(k)]. \tag{7.53}$$

In this way, the action of the kinetic-energy operator on the wavepacket is evaluated accurately, just as in the case of the potential energy operator in the coordinate space.

The method is extremely efficient from a computational point view. In fact, its computational cost scales slowly (as $N\log N$) with the grid size N, and highly vectorized or parallelized FFT computational libraries are currently easily available. On the other hand, the method requires the wavepacket to satisfy periodic boundary conditions and is exact only for band-limited functions. If the function is not band limited or if boundary conditions are not satisfied, as usually happens at some point during the propagation, then it is necessary to carefully 'switch off' the wavepacket through an artificial 'absorption' potential in order to prevent it from reaching the edges of the grid and introducing errors due to the wavepacket wrapping around at the opposite side of the grid.

Throughout the propagation, at each time step or at constant intervals, the wavepacket is analyzed through Eq. (7.35). As already mentioned, the analysis involves the projection of the wavepacket onto product vibrational eigenstates and is thus conveniently performed along one-dimensional cuts of the wavepacket taken at a fixed analysis line $R' = R'_\infty$ far into the product region, where R' is the translational product Jacobi coordinates and R'_∞ is sufficiently large to consider the potential asymptotic. For this reason, after setting up the initial reactant wavepacket in reactant Jacobi coordinates, it is usually convenient to map it onto a grid defined in the product Jacobi coordinates and carry out the entire propagation on this grid. The resulting time-dependent (Eq. (7.35)) data are converted into the energy dependent–scattering matrix elements (Eq. (7.34)).

An illustration of three different stages of a wavepacket propagation for the collinear $H + H_2$ ($v = 0$) reaction is shown in Fig. 7.2 where the square modulus of the wavepacket (corresponding to the probability density for the position of the nuclei) is plotted as a function of the product Jacobi coordinates r' and R'. On the base of each plot, contours lines of the $H + H_2$ potential-energy surface are also reported using the same values as in Fig. 6.3. The reader will immediately recognize the already-discussed reactant and product arms of the reaction 'canyon', though these appear now to form an acute (rather than right) angle due to the fact that they are plotted on the plane of the product Jacobi coordinates rather than of the interatomic distances.

In the first stage (top panel), the wavepacket is set up as the product of a Gaussian function along the translational coordinate R and the vibrational ground-state eigenfunction of H_2. Accordingly the wavepacket has a bell-like shape and is located far into the reactant channel.[1] In a later stage of the propagation (middle panel), the wavepacket has reached the strong interaction region and lost its original shape. The system is now more of a triatom than a couple of reactive partners. At this stage, part of the wavepacket has already passed the saddle point (where the transition state of the reaction is located). In a further stage of the propagation (bottom) panel, the wavepacket has spread all along the two arms of the canyon. Part of the wavepacket has reached the product region (relating to the classical counterpart of reactive trajectories), part has been

---

[1] Note that the oscillations that are seen in the figure at this stage are due to the discretization of the wavepacket and to the adopted graphical representation and are not a feature of the wavepacket, which, as mentioned, has a bell-like shape.

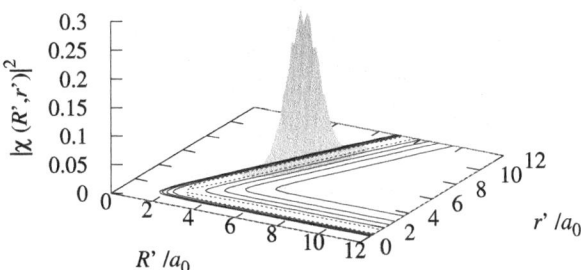

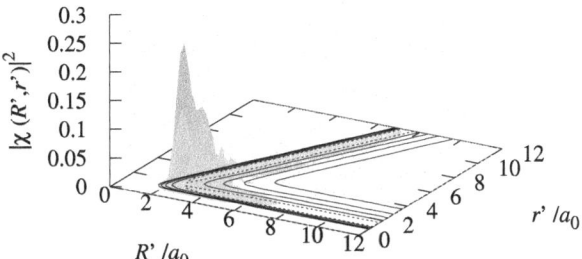

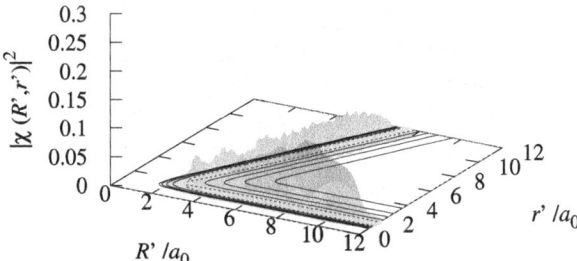

FIGURE 7.2 Propagation of a wavepacket for the collinear H + H$_2$ reaction on a grid defined in the product Jacobi coordinates. A wavepacket for the reactant H$_2$ vibrational ground state is prepared in the initial stage (top panel). The wavepacket reaches the strong interaction region in an intermediate stage (middle panel). The wavepacket spreads into the reactant and product channel in a successive stage (bottom panel). A contour map of the potential-energy surface is also shown at the base with contours at 1, 2, 3, and 4 eV (solid lines) plus a value (close to the barrier height) of 0.4 eV (dashed line).

reflected back into the reactant region (relating to the classical counterpart of nonreactive trajectories). As the wavepacket reaches the edges of the grid, it starts to be absorbed by the artificial absorbing potential, as is evident from the 'quenched' shape of the wavepacket near the two exit borders of the grid.

Chapter 8

From theory to computing: collinear reactive scattering with real wavepackets

This chapter illustrates with some detail the implementation of a method for the calculation of the scattering-matrix elements for a collinear reaction of the type A + BC (v) → AB (v') + C. The implementation is based on a method developed by Gray and Balint-Kurti [107] for propagating and analyzing a real-valued wavepacket, and is thus simpler than standard methodologies handling complex-valued wavepackets. While the use of complex algebra cannot be avoided in the final stage of the reactive analysis, in Gray and Balint-Kurti's method the entire propagation is, in fact, carried out using real numbers. We will briefly outline the method in Section 8.1, discuss some details of its implementation in Section 8.2, and devote Section 8.3 to the preparatory stage of solving the diatomic vibrational problem.

8.1 The real-wavepacket method

The starting point of wavepacket-based methods, assuming a time-independent Hamiltonian, is the equation for the time evolution at time $t + \tau$ of a wavepacket at time t:

$$\chi(t + \tau) = e^{\frac{-i\hat{H}\tau}{\hbar}} \chi(t), \tag{8.1}$$

where the dependence of χ on the spatial coordinates is implied and has been omitted to make the formalism simpler. The i in Eq. (8.1) forces the wavepacket to be in general a complex function. As pointed out by Gray [108], writing the backwards $(\tau \rightarrow -\tau)$ counterpart of this equation

$$\chi(t - \tau) = e^{\frac{+i\hat{H}\tau}{\hbar}} \chi(t), \tag{8.2}$$

and adding it to Eq. (8.1) one obtains:

$$\chi(t + \tau) = -\chi(t - \tau) + 2\cos\left(\frac{\hat{H}\tau}{\hbar}\right) \chi(t), \tag{8.3}$$

Chemistry at the Frontier with Physics and Computer Science
https://doi.org/10.1016/B978-0-32-390865-8.00017-9

which offers a three-term recursion scheme with discrete time steps τ allowing us to compute the wavepacket at a given time step from the knowledge of the wavepacket at the previous two. Unlike Eq. (8.1), this equation does not involve i, and thus allows for the real (q) and imaginary (w) part of the wavepacket $\chi = q + iw$ to be propagated independently. Accordingly, one obtains for the real part:

$$q(t+\tau) = -q(t-\tau) + 2\cos\left(\frac{\hat{H}\tau}{\hbar}\right) q(t) , \qquad (8.4)$$

and an analogous equation for the imaginary part w. Thus, after a first special iteration still referencing the imaginary part of the wavepacket:

$$q(\tau) = \cos\left(\frac{\hat{H}\tau}{\hbar}\right) q(0) + \sin\left(\frac{\hat{H}\tau}{\hbar}\right) w(0) , \qquad (8.5)$$

one can continue evolving the real part by repeatedly applying Eq. (8.4) with discrete time steps τ.

As detailed in Ref. [107], the iterative scheme of Eq. (8.4) can be made even simpler with no effect on the results of the extraction of the observable properties of the reactive process from the wavepacket evolution, by using a modified time-dependent Schrödinger equation

$$i\hbar\frac{\partial \chi(t)}{\partial t} = f(\hat{H})\chi(t) \qquad (8.6)$$

after adopting a functional mapping of the Hamiltonian

$$f(\hat{H}) = -\frac{\hbar}{\tau}\cos^{-1}(\hat{H}_s) , \qquad (8.7)$$

where $\hat{H}_s$ is the same scaled and shifted Hamiltonian with eigenvalues in the $[-1, 1]$ range that is used in the Chebyshev expansion (see Section 7.3) of the time-evolution operator:

$$\hat{H}_s = a_s\hat{H} + b_s, \qquad (8.8)$$

with $a_s = 2/\Delta E$, $b_s = -1 - a_s E_{min}$, and $\Delta E = E_{max} - E_{min}$. In so doing, in fact, the three-term recursion scheme of Eq. (8.4) can be rewritten as

$$q(t+\tau) = -q(t-\tau) + 2\hat{H}_s q(t) , \qquad (8.9)$$

involving the simple iterative application of the scaled Hamiltonian to the wavepacket, while the first iteration of Eq. (8.5) becomes

$$q(\tau) = \hat{H}_s q(0) - \sqrt{1 - \hat{H}_s^2}\, w(0) . \qquad (8.10)$$

While it is common in reactive-scattering calculations to employ a complex-valued incoming Gaussian wavepacket (Eq. (7.36)), Eq. (8.10) can be made simpler by employing a real-valued initial wavepacket of the form:

$$q(R, r, t = 0) = \pi^{-1/4} \alpha^{-1/2} e^{\frac{-(R-R_0)^2}{2\alpha^2}} \cos(k_0 R) \gamma_v(r), \qquad (8.11)$$

where Eq. (8.11) is an initial-state-selected (v) wavepacket with translational part centered at $R = R_0$ containing equal parts of a Gaussian wavepacket moving inward in R with wavevector $-k_0$ and outward in R with wavevector $+k_0$. With this choice for the initial wavepacket, the first iteration becomes simply:

$$q(\tau) = \hat{H}_s q(0). \qquad (8.12)$$

Once the real-valued initial wavepacket of Eq. (8.11) has been prepared, it can be evolved through Eqs. (8.12) and (8.9) featuring no complex algebra.

As already discussed in Section 7.2 of Chapter 7, dynamical information on the reactive properties of the studied process is elaborated from the analysis of the wavepacket throughout its propagation. We refer the interested reader to the original Ref. [107] for details on the derivation of the following expressions, and give here only the final results. Within the Gray and Balint-Kurti method, state-to-state reactive information is obtained by analyzing the wavepacket amplitude $q(r', t; R' = R'_\infty)$ along an analysis line R'_∞ far into the product channel. In particular, at each time step the following time-dependent coefficients are collected

$$C^q_{cv', av}(t) = \int_0^\infty \gamma_{v'}(r') q(r', t; R' = R'_\infty) \, dr', \qquad (8.13)$$

involving the projection of one-dimensional cuts of the wavepacket onto product vibrational eigenfunctions.

If the dynamics of the wavepacket has been generated by the functional mapping $f(\hat{H})$, then the energy-dependent scattering-matrix elements can be computed by $f(E)$-dependent coefficients obtained by half-Fourier transform of the time-dependent coefficients:

$$A^q_{cv', av}(f) = \frac{1}{2\pi} \int_0^\infty dt \, e^{\frac{ift}{\hbar}} C^q_{cv', av}(t), \qquad (8.14)$$

where $f(E) = -\frac{\hbar}{\tau} \cos^{-1}(E_s)$, and $E_s = a_s E + b_s$. In terms of these coefficients, the scattering-matrix elements read:

$$S_{cv', av}(E) = -\frac{\hbar^2 a_s}{\tau (1 - E_s^2)^{1/2}} \left(\frac{k_{cv'}, k_{av}}{\mu_c \mu_a} \right)^{1/2} e^{-ik_{cv'} R_\infty} \frac{2 A^q_{cv', av}(f)}{\bar{g}(-k_{av})}, \qquad (8.15)$$

where k_{av} and $k_{cv'}$ are wavevector components associated with the reactant and product channels and a total energy E, μ_a and μ_c are the reduced masses of the

(A, BC) and (AB, C) pairs, respectively, and $\bar{g}(k)$, due to the specific formulation of the initial wavepacket, takes the form:

$$\bar{g}(k) = \frac{1}{2}\sqrt{\frac{\alpha}{2\pi^{3/2}}} \left[e^{-i(k-k_0)R_0} e^{\frac{-\alpha^2(k-k_0)^2}{2}} + e^{-i(k+k_0)R_0} e^{\frac{-\alpha^2(k+k_0)^2}{2}} \right]. \quad (8.16)$$

8.2 Computational aspects

As can be easily assessed, the method outlined in the previous section involves two main computational tasks, i) the calculation of the reactant and product diatomic vibrational functions, γ_v and $\gamma_{v'}$, featured in Eqs. (8.11) and (8.13), and ii) the evaluation of the action of the (scaled) Hamiltonian on the wavepacket, required by the recursive scheme for the propagation. As already mentioned in Section 7.3 of Chapter 7, an efficient route to an accurate evaluation of the action of the Hamiltonian on the wavepacket is based on fast Fourier transform (FFT) techniques, which in turn require the wavepacket to be represented on an evenly spaced grid defined in the Jacobi coordinates either of the reactants or of the products. The adoption of a discrete representation of the wavepacket on uniform grids also allows for an easy and efficient solution of the vibrational-eigenvalue problem. We shall defer this last topic to the next section (Section 8.3), while focusing here on the evaluation of $\hat{H}_s \chi$ and on some other aspects related to the implementation of the method.

A note is due, at this point, on the so-called 'coordinate problem', which is crucial in all reactive-scattering calculations (see again the already cited Ref. [81] on this), i.e., choosing a coordinate representation that better suits the evolution of the reactive process. While the initial wavepacket (Eq. (8.11)) is in fact better expressed using the reactant Jacobi coordinates, the evaluation of the state-to-state, time-dependent coefficients of Eq. (8.13) is better performed on the line R'_∞ of a product Jacobi grid. This means that if the initial wavepacket has been discretized over a grid of reactant Jacobi coordinates (r, R) and the evolution is started on this grid, then a transformation of the wavepacket from the grid of the reactant to that of the product Jacobi coordinates (r', R') before the wavepacket reaches the analysis region is in order. A convenient alternative to the coordinate switch, one that can be easily adopted in the simple collinear case, is to express directly the initial wavepacket on the (r', R') grid and carry out the entire propagation on the product Jacobi grid. For the simple collinear case, this means that for each point of the (r', R') grid:

$$r'_i = i\,\Delta r' \quad (i = 1, 2, \ldots, N_{r'}) \quad\quad (8.17)$$
$$R'_i = i\,\Delta R' \quad (i = 1, 2, \ldots, N_{R'}), \quad\quad (8.18)$$

one computes the corresponding values of r and R and then uses Eq. (8.11) to evaluate the amplitude of the wavepacket at the points of the (r', R') grid. Using the masses of the collinear $H + H_2$ system considered below, for example, one has $R = R'/2 + 3r'/4$ and $r = R' - r'/2$.

Before the propagation starts, one has to scale the Hamiltonian according to Eq. (8.8). As already mentioned in Section 7.3, the adoption of evenly spaced grids imposes finite boundaries to the spectrum of the Hamiltonian, so that E_{max} and E_{min} can be worked out by the discretization step used for the grid. In particular $E_{min} = V_{min}$, where V_{min} is the minimum value of the potential-energy surface contained in the grid (in the calculations reported below on H + H$_2$ the zero of the potential-energy surface is set at the bottom of the asymptotic H + H$_2$ valley), while $E_{max} = \dfrac{\hbar^2 \pi^2}{2\mu_{r'}(\Delta r')^2} + \dfrac{\hbar^2 \pi^2}{2\mu_{R'}(\Delta R')^2} + V_{max}$, where V_{max} is the maximum value of the potential-energy surface contained in the grid and $\mu_{r'}$ and $\mu_{R'}$ are the reduced masses associated with the (A, B) and (AB, C) pairs, respectively.

According to Eq. (8.8), the scaled Hamiltonian is

$$\hat{H}_s = a_s \hat{H} + b_s = a_s \left[\hat{T} + V(R', r') \right] + b_s , \tag{8.19}$$

where in product Jacobi coordinates the kinetic term reads:

$$\hat{T} = -\frac{\hbar^2}{2\mu_{r'}} \frac{\partial^2}{\partial r'^2} - \frac{\hbar^2}{2\mu_{R'}} \frac{\partial^2}{\partial R'^2}. \tag{8.20}$$

The action of the potential term is readily evaluated by a simple multiplication of the wavepacket by the value of the potential at each point of the grid. As discussed in Section 7.3 of the previous chapter, the action of the kinetic term can instead be efficiently evaluated through the FFT method as follows:

$$
\begin{aligned}
\hat{T} q(R', r') = &-\frac{\hbar^2}{2\mu_{R'}} \mathcal{F}_{R'}^{-1} \left[-k_{R'}^2 \mathcal{F}_{R'} \left[q(R', r') \right] \right] \\
&-\frac{\hbar^2}{2\mu_{r'}} \mathcal{F}_{r'}^{-1} \left[-k_{r'}^2 \mathcal{F}_{r'} \left[q(R', r') \right] \right].
\end{aligned}
\tag{8.21}
$$

One issue when using the FFT method is that, as already mentioned in Section 7.3 of the previous chapter, the wavepacket must be damped before it reaches the border of the grid, or it will enter the grid from the opposite side and errors will be introduced in the calculation. This can be done by introducing an absorption potential at the edges of the grid defined as follows:

$$\hat{A}(r', R') = A_{r'}(r') A_{R'}(R') , \tag{8.22}$$

where

$$A_{r'}(r') = \begin{cases} e^{-C_{abs}^{r'}(r' - r'_{abs})^2} & \text{for } r' > r'_{abs} \\ 1 & \text{for } r' \leq r'_{abs} \end{cases} \tag{8.23}$$

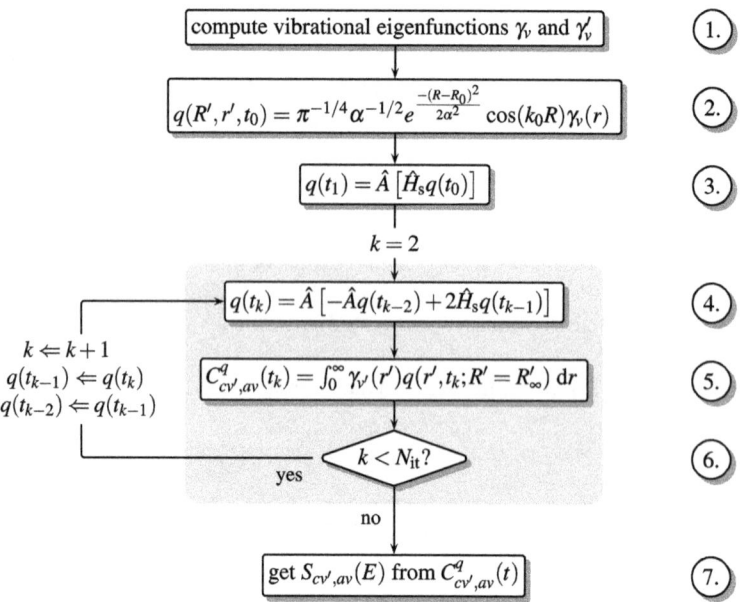

FIGURE 8.1 Flowchart for a collinear–reactive scattering calculation using the real-wavepacket method. The iterative section is highlighted with a blue (dark gray in print version) frame.

and

$$
A_{R'}(R') = \begin{cases} e^{-C_{\mathrm{abs}}^{R'}(R'-R'_{\mathrm{abs}})^2]} & \text{for } R' > R'_{\mathrm{abs}} \\ 1 & \text{for } R' \leq R'_{\mathrm{abs}}. \end{cases} \tag{8.24}
$$

As discussed in Ref. [108], the wavepacket absorption is introduced into the calculations by modifying the recursive scheme of Eq. (8.9) as follows:

$$
q(t+\tau) = \hat{A}\left(-\hat{A}q(t-\tau) + 2\hat{H}_s q(t)\right). \tag{8.25}
$$

The algorithm for the entire calculation may thus be summarized as in the flowchart of Fig. 8.1. In a preliminary stage that will be addressed in the following section, the vibrational eigenvalue problem is solved for the reactant and product molecules (step 1). Then the wavepacket is set up on the product Jacobi grid (step 2). After performing a first special iteration (step 3), the three-term recursive scheme is iteratively applied up to a given maximum number N_{it} of iterations, and at each time step the coefficients $C_{cv',av}^q(t)$ are evaluated (steps 4–6). Finally, once the iterative procedure is complete, the scattering-matrix elements are computed from the collected time-dependent data (step 7). The FFT of the wavepacket can be easily performed using freely available computational routines such as the Fastest Fourier Transform in the West FFTW

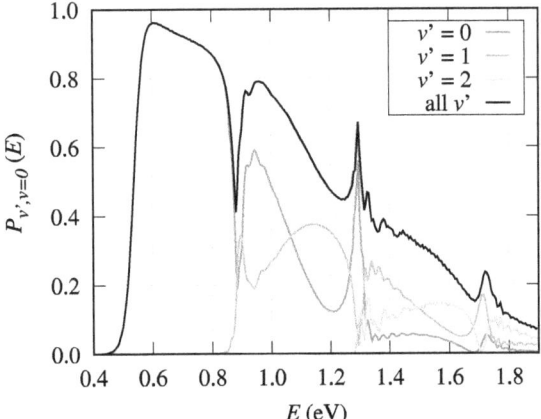

FIGURE 8.2 State-specific (black color) and state-to-state (blue (dark gray in print version) color) reaction probability as a function of total energy for the collinear H + H$_2$ ($v = 0$) $\rightarrow$ H$_2$ (v') + H reaction on the Liu–Siegbahn–Truhlar–Horowitz (LSTH) surface [109–112].

(https://www.fftw.org/). With this information at hand and with the details given in the following section on the solution of the vibrational problem, students with intermediate-level experience in scientific programming should be able to implement their own program for collinear–reactive scattering calculations.

For illustrative purposes, state-specific and state-to-state reaction probabilities $P_{v',v}(E)$ as a function of total energy E for the collinear H + H$_2$ ($v = 0$) $\rightarrow$ H$_2$ (v') + H reaction on the Liu–Siegbahn–Truhlar–Horowitz (LSTH) surface [109–112] are reported in Fig. 8.2. These were obtained by the author using a slightly modified version of the algorithm propagating the wavepacket on the nonorthogonal coordinate grid (r, r'), which leads to an additional term featuring mixed derivatives in the Hamiltonian. Results were obtained setting $\tau = 1$ in all equations, as suggested in Ref. [107], and using the parameters reported in Table 8.1. The figure shows the presence of an energy threshold to reaction close to the value of the height of the potential barrier and the opening of additional vibrational product channels with increasing total energies.

8.3 The vibrational eigenvalue problem

As shown in Fig. 8.1, the first stage of a collinear–reactive scattering calculation is the solution of a diatomic vibrational eigenvalue problem of the form:

$$\left[-\frac{\hbar^2}{2\mu} \frac{\mathrm{d}^2}{\mathrm{d}r^2} + V(r) \right] \gamma_v(r) = \varepsilon_v \gamma_v(r) , \qquad (8.26)$$

where μ is the reduced mass of the diatom. The problem has to be solved for the reactant molecule (one of the solutions will be used for the setup of the initial wavepacket) and for the product molecule (the solutions will be used for the

TABLE 8.1 Summary of the parameters used in the calculations on collinear H + H$_2$. Values are given in atomic units unless otherwise specified.

r_{max}	12.0	grid length
r'_{max}	12.0	grid length
N_r	128	number of grid points
$N_{r'}$	128	number of grid points
V_{cut}	0.22	potential cutoff (if $V > V_{cut}$, then $V \Leftarrow V_{cut}$)
R_0	6.5	center of the Gaussian wavepacket
R'_∞	6.5	analysis line
r_{abs}	8.5	starting line for absorption
r'_{abs}	8.5	starting line for absorption
C^r_{abs}	4.0	absorption coefficient
$C^{r'}_{abs}$	4.0	absorption coefficient
$|k_0|$	8.49	wavepacket initial wavevector
α	0.5	Gaussian width parameter
N_{it}	5000	maximum number of iterations

calculations of the time-dependent coefficients resulting from the overlap of the product vibrational states with the wavepacket along the analysis line).

When using an evenly spaced grid in r:

$$r_i = i\,\Delta r \quad (i = 1, 2, \ldots, N), \tag{8.27}$$

an efficient way of solving Eq. (8.26) strictly related to the FFT method for evaluating the action of the Hamiltonian on a wavefunction, is the Fourier grid Hamiltonian (FGH) method [113,114], which is a special case of discrete variable representation (DVR). In the FGH method a set of functions defined at the grid points

$$\gamma_j(r_i) = \delta(r_i - r_j) \quad (i, j = 1, 2, \ldots, N) \tag{8.28}$$

is introduced, and in this basis, both $\hat{T}$ and $\hat{V}$ are represented by $N \times N$ matrices. The vibrational eigenvalues ε_v and eigenfunctions γ_v can then be obtained through diagonalization of the $N \times N$ real symmetric matrix $\boldsymbol{H} = \boldsymbol{T} + \boldsymbol{V}$, where the potential matrix is diagonal, while the elements of the kinetic-energy operator can be written explicitly as a function of the number of grid points N (which is assumed here to be even) and the total length L of the grid. In atomic units, these elements read:

$$T_{ii} = \frac{\pi^2}{\mu L^2} \frac{N^2 + 2}{6} \tag{8.29}$$

$$T_{ij} = (-1)^{i-j} \frac{\pi^2}{\mu L^2} \frac{1}{\sin^2((i-j)\pi/N)}. \qquad (8.30)$$

The eigenvectors resulting from the diagonalization of H store directly the values of the eigenfunctions $\gamma_v(r_j)$ at the grid points. As is the case for FFT-based methods, efficient computational libraries for performing linear algebra operations are also easily available, such as the linear algebra package LAPACK (http://www.netlib.org/lapack/).

While the above discussion completes the picture giving the final details for implementing the methodology discussed in this chapter, a further comment is worth adding here. When discussing potential-energy surface fitting, we showed that a better configuration-space sampling was provided by space reduced–bond order (SRBO) evenly spaced grids, rather than analogous grids defined in the interatomic distance r. The same idea could indeed be applied to the solution of the vibrational eigenvalue problem using so-called mapped Fourier grid methods, as outlined in Ref. [115]. We shall briefly explore this topic in the remainder of this section.

In general, when mapping (see also Ref. [116]) the original coordinate r onto another generic coordinate c

$$r = f(c) \qquad (8.31)$$

$$dr = J(c)\, dc \qquad (8.32)$$

$$J(c) = \frac{d}{dc} f(c), \qquad (8.33)$$

where J is the Jacobian of the transformation,[1] the new kinetic-energy operator in atomic units becomes:

$$\hat{T} = -\frac{1}{2\mu} \left(\frac{1}{J} \frac{d}{dc} \right)^2 = -\frac{1}{2\mu J^2} \frac{d^2}{dc^2} + \frac{J'}{2\mu J^3} \frac{d}{dc}, \qquad (8.34)$$

with $J' = dJ/dc$. As discussed in Ref. [115], a more symmetric and convenient representation of the kinetic-energy operator can be obtained by defining a new wavefunction γ' [115] such that

$$\gamma(c) = J^{-\frac{1}{2}} \gamma'(c). \qquad (8.35)$$

Accordingly, the new Schrödinger equation for $\gamma'(c)$ in atomic units becomes:

$$\left[-\frac{1}{4\mu} \left(\frac{1}{J^2} \frac{d^2}{dc^2} + \frac{d^2}{dc^2} \frac{1}{J^2} \right) + \bar{V} \right] \gamma'(c) = \varepsilon \gamma'(c), \qquad (8.36)$$

---

[1] The dependence of J on c is implied from now on, where not explicitly declared.

where

$$\bar{V}(c) = V(c) + \frac{1}{2\mu} \left(\frac{7}{4} \frac{(J')^2}{J^4} - \frac{1}{2} \frac{J''}{J^3} \right) . \tag{8.37}$$

When an evenly spaced grid is adopted for the new coordinate, the kinetic operator matrix elements become:

$$T_{ii} = \frac{\pi^2}{\mu L^2} \frac{N^2 + 2}{6} \frac{1}{J_i^2} \tag{8.38}$$

$$T_{ij} = (-1)^{i-j} \frac{\pi^2}{2\mu L^2} \frac{1}{\sin^2((i-j)\pi/N)} \left(\frac{1}{J_i^2} + \frac{1}{J_j^2} \right) . \tag{8.39}$$

For the bond-order mapping (Eq. (6.8)) case, where $J = -1/(\beta n)$, one has[2]

$$\bar{V}_{ii} = V_{ii} + \frac{1}{2\mu} \frac{3}{4} \beta^2 \tag{8.40}$$

$$T_{ii} = \frac{\pi^2}{\mu L^2} \frac{N^2 + 2}{6} (\beta^2 n_i^2) \tag{8.41}$$

$$T_{ij} = (-1)^{i-j} \frac{\pi^2}{2\mu L^2} \frac{1}{\sin^2((i-j)\pi/N)} \left(\beta^2 n_i^2 + \beta^2 n_j^2 \right) . \tag{8.42}$$

For illustrative purposes, some results are reported in Figs. 8.3 and 8.4 obtained for the H$_2$ diatom using a Morse potential

$$V_M(r) = D_{eq}(1 - e^{-\beta_M(r - r_{eq})})^2 , \tag{8.43}$$

with $D_{eq} = 0.1744$ Hartree, $\beta_M = 1.02764 \; a_0^{-1}$ and $r_{eq} = 1.40201 \; a_0$ as dictated by spectroscopic considerations associated with the formulation of the bond-order variables [70]. In particular, in Fig. 8.3 the vibrational ($v = 9$) eigenfunction computed using the FGH method on a 64-point grid up to $r = 12 \; a_0$ defined in both the interatomic distance r (or bond-length, BL, gray squares) and in an SRBO variable with $\beta = 0.5 \; a_0^{-1}$ yielding an attractive/repulsive space ratio $f \simeq 2$ (blue (dark gray in print version) circles), is compared to the related analytic solution for the Morse potential. Note that, despite the grids being defined up to $r = 12 \; a_0$, both Figs. 8.3 and 8.4 focus only on the $[0, 6] \; a_0$ range.

The results reported in Fig. 8.3 show that the qualitative behavior of the $v = 9$ vibrational eigenfunction is reproduced by the curves obtained through the FGH method using both a BL and SRBO sampling, yet a quantitative agreement is reached only by the results obtained through the SRBO sampling. The advantage of adopting an SRBO sampling over the conventional BL one is, however, even more evident in Fig. 8.4, where the same vibrational eigenfunctions of

---

[2] Eigenvectors obtained through diagonalization of $H = T + V$ should be converted from $\gamma'(c)$ to $\gamma(c)$ according to Eq. (8.35).

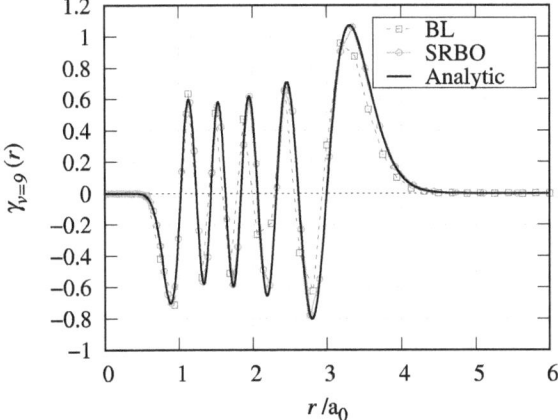

FIGURE 8.3 Vibrational ($v = 9$) eigenfunction for the H_2 molecule obtained through the FGH method using a Morse potential on a 64-point uniform grid defined both in the interatomic-distance (or bond length, BL) variable and in an SRBO (SRBO, $\beta = 0.5\, a_0^{-1}$) variable. The analytic solution is also reported for comparison.

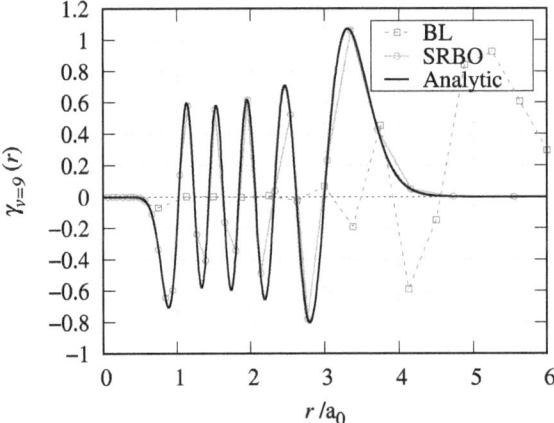

FIGURE 8.4 Vibrational ($v = 9$) eigenfunction for the H_2 molecule obtained through the FGH method using a Morse potential on a 32-point uniform grid defined both in the interatomic-distance (or bond length, BL) variable and in an SRBO (SRBO, $\beta = 0.5\, a_0^{-1}$) variable. The analytic solution is also reported for comparison.

Fig. 8.3 are now obtained using a 32-point grid, rather than 64, of equal length. The figure shows that, while due to the reduced number of points the sampling is too loose for the FGH method on a BL grid to provide even qualitatively accurate results, the curve obtained through the FGH method using an equivalent grid in the SRBO space still reproduces their analytical counterpart at the quantitative level.

As a final remark on the topic covered in this section, one may wonder whether the same advantage offered by the SRBO configuration-space sampling in relatively simple problems such as potential-energy surface fitting and the vibrational eigenvalue equation, may be exploited in the implementation of either time-dependent or time-independent reactive-scattering methods. While preliminary work has been reported with contributions from the present author [75] in this direction, an exploration of the suitability of SRBO grids for the solution of the dynamical problem has still to be fully addressed.

Chapter 9

From reaction dynamics to chemical kinetics

9.1 The reaction rate constant

Through Chapters 5 to 8, we discussed how a chemical reaction of the type $A + BC \rightarrow AB + C$ can be modeled in terms of the collisions between the reactant partners, either based on classical trajectories or on quantum wavefunctions. Both approaches lead to the estimation of a key quantity for bridging the microscopic world of individual collisions with the macroscopic, phenomenological one. Such a quantity is the reaction probability as a function of the collision energy, ranging from the highly detailed state-to-state one, relating to collisions starting from a given rovibrational state of the reactant molecule and ending in a given rovibrational state of the product molecule, to less detailed ones such as the state-specific (relating to a given reactant rovibrational state) or the cumulative (summing over all reactant and product states) reaction probabilities.

The reaction probability is ultimately related to the reactive cross section, and from the knowledge of these quantities (which, in the quantum framework, can be derived from the scattering-matrix elements) the calculation of the rate constant for a chemical reaction is straightforward. The thermal rate constant at a given temperature T is in fact given by the product of the reactive cross section and the average velocity of the colliding partners $\langle v \rangle$ at the temperature T, and in terms of quasiclassical trajectories, for instance, reads:

$$k(T) = \sigma \langle v \rangle = \left(\frac{8k_B T}{\pi \mu} \right)^{1/2} \pi b_{max}^2 \frac{N^R}{N^{tot}}, \tag{9.1}$$

where k_B and μ are the Boltzmann constant and the reduced mass of the reactants, respectively [117].

Now, in areas such as combustion chemistry, atmospheric chemistry, or astrochemistry, researchers are often interested in averaged quantities of a particular kind of a molecular ensemble, typically the canonical (as for $k(T)$) or the microcanonical one. While in principle these quantities should be derived from the outcome of dynamics calculations, it is often in practice convenient to adopt approximate theories, such as the transition-state theory in one of its flavors, that allow for the estimation of the rate constant simply from the energetics and internal structure of a limited set of molecular species located at important points (wells or saddle points) of the potential energy (hyper)surface. Dynamics calculations can indeed be computational expensive, not to mention that exact quantum-dynamical treatments are only feasible for systems with up to three or four atoms. Even when more affordable approximate quantum methods or quasiclassical trajectories could be used, in applications relevant to combustion chemistry, atmospheric chemistry, or astrochemistry most of the information generated by the dynamics would be discarded due to its excessively detailed nature. In these cases, one can usually rely on kinetic treatments, where the target is the evaluation of the rate constant of the chemical process at hand as a function of temperature or energy regardless of the detailed mechanism with which the reaction occurs.

In order to give an idea of how kinetic treatments are applied to actual cases, we will focus in the following on a class of astrochemical reactions involving more complex systems than the atom–diatom one considered insofar.

9.2 Kinetic treatment of astrochemical reactions

Consider the astrochemical reaction of the cyano radical (CN) and methanimine (CH_2NH) to form three possible isomers of cyanomethanimine:

$$CN + CH_2NH \rightarrow E\text{-}C\text{-}, Z\text{-}C\text{-}, \text{ and } N\text{-cyanomethanimine} . \tag{9.2}$$

The system is made up of $N = 7$ atoms and thus features a $3N - 6 = 15$-dimensional potential-energy surface. Typically, the first stage in a kinetic treatment of a reaction such as (9.2) is the calculation of the critical points of the potential-energy surface, which is usually achieved by combining chemical intuition of the possible reaction mechanisms with standard geometry-optimization methods available in electronic-structure computer programs. Such critical points include potential wells, which correspond to stable species known as reaction intermediates, and saddle points, which correspond to transition states and connect reaction intermediates with other intermediates or with products. Fig. 9.1 shows a scheme of a plausible reaction mechanism for the formation of the three isomers of cyanomethanimine from the cyano radical and methanimine, as hypothesized in Ref. [118]. The first step is a barrierless bimolecular association of the reactant partners to form a reaction intermediate (in the case of the reaction in Fig. 9.1 the bimolecular association can lead to two different possible intermediates). The formed intermediate can then evolve into other intermediates through unimolecular reactions, and eventually lead to reaction products through dissociation.

A common approach in calculating the kinetics of reactions such as Eq. (9.2) directly from the features of the potential-energy surface without undertaking dynamics calculations, is to model the barrierless association step with a capture model, whereby the translational energy of reactants must only surpass a long-range centrifugal barrier for reaction to occur, and the unimolecular steps with Rice–Ramsperger–Kassel–Marcus (RRKM) theory [120–122].

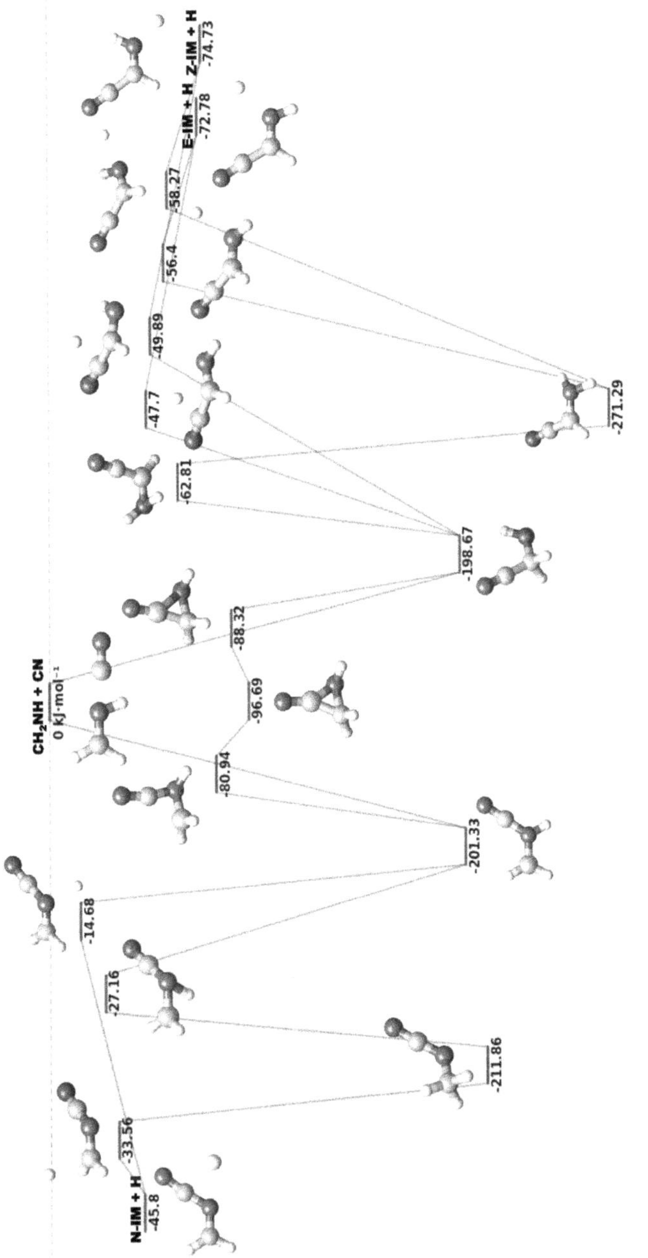

FIGURE 9.1 Scheme of a plausible reaction mechanism for the formation of cyanomethanimine isomers from the cyano radical and methanimine. E-IM stands for E-C-cyanomethanimine, Z-IM for Z-C-cyanomethanimine, and N-IM for N-cyanomethanimine. Relative energies corrected by the zero-point energy are also reported. Mechanism proposed in Ref. [118], with energies refined as in Ref. [119].

For neutral–neutral reacting partners, the capture microcanonical rate constant is given by the simple expression:

$$k(E) = 3\pi \frac{1}{\sqrt{\mu}} C_6^{\frac{1}{3}} \left(\frac{E}{2}\right)^{\frac{1}{6}}, \tag{9.3}$$

where C_6 is an effective van der Waals coefficient that can be computed by fitting the profile of the potential energy as a function of the distance between the reacting partners, E is the collision energy, and μ is the reduced mass associated with the two collision partners [123]. For the unimolecular steps, instead, use can be made of the RRKM theory, which is grounded on the following two assumptions: i) the available energy is assumed to be uniformly distributed throughout all degrees of freedom of the molecule and ii) the transition state is assumed to be a point of no-return for the reaction, i.e., if the system reaches the transition state, then it cannot go back to reactants and the reaction invariably occurs.

Within the RRKM scheme, the microcanonical rate constant for a unimolecular reaction is given by the equation [124]:

$$k(E) = \frac{N^{\ddagger}(E)}{h\rho(E)}, \tag{9.4}$$

where

$$N^{\ddagger}(E) = \int_0^E \rho^{\ddagger}(E')\, dE'. \tag{9.5}$$

In Eqs. (9.4) and (9.5), h is Planck's constants, $N^{\ddagger}(E)$ is the sum of states of the transition state (computed by excluding the normal mode with an imaginary frequency under the assumption that the motion along the reaction coordinates is separable from that of the other modes), and $\rho(E)$ and $\rho^{\ddagger}(E)$ are the density of states (i.e., the number of rovibrational states per energy interval) of the reactant molecule and transition state, respectively.

As is apparent, a central quantity in the RRKM framework is thus the molecular rovibrational density of states $\rho(E) = \Delta N(E)/\Delta E$, i.e., the number $\Delta N(E)$ of rovibrational states per energy interval ΔE within the interval $[E, E + \Delta E]$, of the involved molecular species.[1]

The rotational motion and the vibrational motion of the molecule are usually assumed to be approximately separable from one another (in other words, the energy of the molecule, E, is approximately the sum of the rotational energy,

---

[1] The partition function $Q(T)$ can also can be computed from the density of states $\rho(E)$ by a Laplace transform:

$$Q(T) = \int_0^{\infty} \rho(E) e^{-\frac{E}{k_B T}}\, dE, \tag{9.6}$$

with k_B being the Boltzmann constant and T being the absolute temperature.

E_r, and the vibrational energy, $E - E_r$). As a result, the rovibrational density of states can be written as a convolution of its rotational and vibrational counterparts [125]:

$$\rho(E) = \int_0^E \rho_r(E_r)\rho_v(E - E_r)\mathrm{d}E_r \ . \tag{9.7}$$

While a classical expression can be used for the rotational density of states

$$\rho(E) = \frac{1}{\sigma}\sqrt{\frac{4E}{ABC}} \ , \tag{9.8}$$

with σ being the rotation symmetry number and A, B, and C the three rotational constants, a quantum anharmonic treatment is recommended for the vibrational motions. Within second-order anharmonic perturbative approaches [126,127], the vibrational energy E (expressed in cm^{-1}) in terms of the n quantum numbers $\{v_i\} \equiv (v_1, \ldots, v_n)$ is given by

$$E = E_0 + \sum_{i=1}^n v_i \omega_i + \sum_{i,j=1}^n \chi_{ij}\left[v_i v_j + \frac{1}{2}\left(v_i + v_j\right)\right] . \tag{9.9}$$

In Eq. (9.9), E_0 is the zero-point vibrational energy, ω_i are the harmonic frequencies, and χ_{ij} are the elements of the anharmonic matrix (explicit function of cubic and semidiagonal quartic force constants, as well as Coriolis couplings).

While the density and sum of states of a molecule can be in principle computed by an exact state counting using the above expression, this becomes rapidly unaffordable from a computational point of view with increasing system size. Viable alternatives are the Stein–Rabinovitch [128] adaptation of the Beyer–Swinehart [129] counting algorithm to uncoupled anharmonic oscillators, or the Wang–Landau sampling method [130,131] as extended to the quantum anharmonic vibrational problem by Basire et al. [132] and further refined by Nguyen and Barker [133]. Within such scheme (the interested reader is referred to the original papers for further details) the density of states $\rho(E_j)$ is obtained iteratively for each jth bin of an equally spaced energy grid in a given energy range by performing a random walk in the space of quantum numbers $\{v_i\}$.

This level of treatment to obtain the microcanonical rate coefficients is generally sufficient for reactions that do not involve tunneling. Tunneling effects can be, however, included by adopting an improved version of Eq. (9.4), in which use is made of a modified version of the sum of states $N^{\ddagger}(E)$ of the transition state. A common and efficient way of including tunneling is in fact by means of the asymmetric Eckart barrier [134]. Within this model, the sum of state of the transition state is redefined by

$$N_{\text{tunn}}^{\ddagger}(E) = \int_{-V_0}^{E-V_0} \rho^{\ddagger}(E - E')P_{\text{tunn}}(E')\,\mathrm{d}E' \ , \tag{9.10}$$

where $N_{tunn}^{\ddagger}(E)$ is a tunneling-corrected version of the sum of state of the transition state and V_0 is the classical energy barrier for the forward reaction. The quantity $P_{tunn}(E')$ is the tunneling coefficient at the energy E', and is given by the expression

$$P_{tunn}(E') = \frac{\sinh(a)\sinh(b)}{\sinh^2((a+b)/2) + \cosh^2(c)}, \tag{9.11}$$

where, a, b and c are parameters defined by:

$$a = \frac{4\pi\sqrt{E'+V_0}}{h\nu_i(V_0^{-\frac{1}{2}} + V_1^{-\frac{1}{2}})}, \quad b = \frac{4\pi\sqrt{E'+V_1}}{h\nu_i(V_0^{-\frac{1}{2}} + V_1^{-\frac{1}{2}})}, \quad c = 2\pi\sqrt{\frac{V_0 V_1}{(h\nu_i)^2} - \frac{1}{16}}. \tag{9.12}$$

Here, V_1 is the classical energy barrier for the reverse reaction, and ν_i is the magnitude of the imaginary frequency of the saddle point.

When needed, the thermal rate constant can be easily computed from the microcanonical rate constant using the following equation:

$$k(T) = \frac{1}{Q(T)} \int_0^\infty k(E)\rho(E)e^{-E/k_B T} dE, \tag{9.13}$$

where $Q(T)$ is the partition function of the reactants.

Focusing again on the complex reaction mechanism of Fig. 9.1, once the rates of the involved elementary reaction steps have been determined, these have to be combined together according to the connections between the species in the reaction mechanism. The population of each species will in fact be determined by the rate of its formation from other species, and the rate of its evolution in other species, and all these rates are intertwined by the reaction mechanism. This task can be accomplished via so-called master-equation approaches.

9.3   Master-equation approaches

Within a master-equation approach (see, for instance, Ref. [135] or, closer to the approach followed here, Refs. [136,137]), to determine the time evolution of the relative abundance of the species involved in a chemical reaction, the microcanonical rate constants at a specified energy are opportunely combined in a matrix K, according to the mechanism of the reaction under study. In particular, the diagonal elements K_{ii} contain the loss rate of species i, while the off-diagonal elements K_{ij} contain the rate of formation of species i from species j. Then the rate of change in the concentration of each species is given by the vector differential equation:

$$\frac{dc}{dt} = Kc, \tag{9.14}$$

where c is the vector of the concentrations of the species at time t. This is a linear differential equation and can be solved by diagonalization of K. In terms of the eigenvector matrix Z and eigenvalue vector Λ, the solution of Eq. (9.14) reads:

$$c(t) = Ze^{\Lambda t} Z^{-1} c(0) , \tag{9.15}$$

where $c(0)$ is the concentration vector at $t = 0$. The projection of $c(t)$ at $t = \infty$ yields the so-called branching ratios, i.e., the relative proportions with which the reaction products are formed.

In this model, a fundamental hypothesis is that collisional relaxation occurs on time scales much shorter than those that characterize phenomenological kinetics [138]. It is worth adding here that a more general version of the master equation would involve diagonalizing a much larger matrix explicitly including collisional relaxation [135], for example through the simple so-called 'exponential down' model, as done in the MESMER computer program [139]. However, if collisional relaxation occurs on time scales much shorter than those of the phenomenological kinetics, the resulting eigenvalues would appear in two separated sets: one made up by so-called internal energy relaxation eigenvalues (IEREs) and one made up by so-called chemically significant eigenvalues (CSEs). These last eigenvalues, that relate to the phenomenological kinetics of interest in interstellar space and atmospheric studies, would be the same as those obtained by solving Eq. (9.14).

Chapter 10

Application: $C + CH^+ \rightarrow$ $C_2^+ + H$: an astrochemical reaction

In the previous chapters, based on prototype atom–diatom reactive systems, we discussed how chemical reactions can be modeled from a theoretical and computational perspective, from the evaluation of the interaction between the atoms to the modeling of the reaction kinetics. This chapter is meant to illustrate how the ideas and the approaches therein outlined – from the crafting of the potential-energy surface, to the dynamics simulation and the calculation of the rate constant – are put into action on a concrete case of study. For this purpose, we will focus on a simple astrochemical process, the atom–diatom reaction $C + CH^+ \rightarrow C_2^+ + H$, which has been the subject of recent work. Accordingly, the discussion will be based on the results reported in Refs. [140] and [141].

10.1   The $C + CH^+ \rightarrow C_2^+ + H$ reaction

As already mentioned in Section 5.3 of Chapter 5, the chemistry of the interstellar medium consists mainly of barrierless gas-phase collisions or processes occurring on the surface of ice or dust particles. One of the main challenges of astrochemistry is to predict the chemical evolution of an interstellar cloud over time, i.e., the evolution of the relative abundances of the different chemical species populating the cloud. For this purpose, astrochemists make use of kinetic models that combine together the rates of all the possible reactions (on the order of thousands) between the involved chemical species (on the order of hundreds). Kinetic parameters for these reactions are stored and made available for inclusion in the above-mentioned kinetic models in recently developed public online repositories such as the kinetic database for astrochemistry KIDA [142] (https://kida.astrochem-tools.org/, accessed 15 January 2022) or the UMIST (University of Manchester Institute of Science and Technology) database for astrochemistry UDfA [143] (http://udfa.ajmarkwick.net/, accessed 15 January 2022), where for bimolecular reactions such as the $C + CH^+$ one the kinetic data are encoded in the form of three parameters α, β and γ appearing in the popular Arrhenius–Kooij formula [144], also known to chemists as the modified Arrhenius equation [145]:

$$k(T) = \alpha (T/300)^\beta e^{-\gamma/T} . \tag{10.1}$$

Chemistry at the Frontier with Physics and Computer Science
https://doi.org/10.1016/B978-0-32-390865-8.00019-2

Many of the ion–molecule reactions in the above-mentioned databases are modeled by means of simple capture theories whereby, as discussed in Chapter 9, the translational energy of reactants must only surpass a long-range centrifugal barrier for reaction to occur. On the other hand, capture models are inherently approximate and, especially when experimental information is missing, it is desirable to assess their accuracy against the results of rigorous dynamical treatments. Among the many gas-phase barrierless processes occurring in the interstellar medium, an important class is that of the neutral–ion and neutral–neutral, atom–diatom reactions involving carbon and hydrogen atoms [34,146]. In order to provide accurate kinetic information on these processes, in the last twenty years several studies, either based on quasiclassical trajectories or quantum reactive scattering, have been published on the following three-atom systems: CH_2 (see [147–154] for reaction $C + H_2$), CH_2^+ (see [155] and [156] for reaction $C^+ + H_2$ and [157–159] for reaction $CH^+ + H$), and C_2H (see [160–163] for reaction $C + CH$).

We shall here focus on the positively charged variant of the latter system, C_2H^+, for which the lowest-energy atom–diatom reaction channel is the exoergic one

$$C \left({}^3P_0 \right) + CH^+ \left(X^1\Sigma^+ \right) \rightarrow C_2^+ \left(X^4\Sigma_g^- \right) + H \left({}^2S_{1/2} \right), \qquad (10.2)$$

where the symbols in parentheses label the electronic state of each species. Besides involving the methylidyne cation CH^+, which is ubiquitous throughout the interstellar space and was discovered in the diffuse interstellar medium as early as in 1941 [164], the importance of reaction (10.2) in astrochemical contexts lies in the fact that it represents a prototype reaction for formation of the carbon–carbon chemical bond, hence it is relevant in the chemistry of life. The dicarbon cation C_2^+, which was detected by the mass-spectroscopic sampling in comets Halley [165] and Giacobini–Zinner [166], is in fact incorporated in ion–molecule reactions for the production of hydrocarbons in interstellar clouds [167].

For reactions such as (10.2), capture-theory estimates of the rate constant are based on the Langevin capture model [168–171], which is the ion-neutral analog of the capture model discussed in Section 9.2 for neutral–neutral partners. In the Langevin capture model, the reaction rate constant is independent of temperature and is given by the expression

$$k = 2\pi e \sqrt{\frac{\alpha_D}{\mu}}, \qquad (10.3)$$

where e is the charge of the electron, α_D is the dipole polarizability (the polarizability volume in units of cm^3) of the neutral colliding partner and μ is the reduced mass of the reactants. Adopting a dipole-polarizability value for carbon of 1.64×10^{-24} cm^3 (which is very near to the most recent theoretical value of 1.67×10^{-24} cm^3 [172]) Prasad et al. [173] calculated a thermal rate constant of 1.2×10^{-9} cm^3 s^{-1} for the C + CH$^+$ reaction, thus leading to Arrhenius–Kooij parameters $\alpha = 1.2 \times 10^{-9}$ cm^3 s^{-1}, $\beta = 0$ and $\gamma = 0$ K. A second, more recent value of $\alpha = 1.14 \times 10^{-9}$ cm^3 s^{-1} has been proposed by Chabot et al. by taking into account the branching ratios of the C + CH$^+$ reaction to product sets C$_2^+$ + H (0.95) and C$_2$ + H$^+$ (0.05) obtained through a semiempirical model with reactants and products in their electronic ground states. The second reaction channel, however, if reactants and products are taken in their electronic ground state, is endoergic by 0.32 eV, and is thus expected not to be relevant at the temperatures of the cold and quiescent interstellar clouds.

Both values, at any rate, are independent of temperature and rooted in an inherently approximate model. We shall see in the following section that a more accurate modeling of the kinetics of reaction (10.2), alongside insightful information on its detailed mechanism, can be obtained by applying the methodology illustrated in the previous chapters.

10.2   The potential-energy surface

As extensively discussed in Chapter 5, the first step in modeling the dynamics of few-atom chemical reactions is the evaluation of the interaction potential between the involved atoms. This typically consists in a preliminary stage where an informative set of nuclear geometries (or configurations) of the system is selected, a second stage in which high-level electronic-structure calculations are run to evaluate the electronic energy of the system at the sampled nuclear configurations, and a third, final stage where the obtained energies are used to generate an analytic representation of the potential-energy surface.

Focusing on the first stage, in the above-quoted Refs. [140,141] the configuration-space sampling was performed according to the space reduced–bond order (SRBO) scheme [67] detailed in Section 6.2. In particular, for each of the two diatomic molecules CH$^+$ and C$_2^+$ a 10-point (3, 6) SRBO grid in coordinates r_{CH} and r_{CC}, respectively, was adopted after determining reasonable boundary values via a Morse modeling of the diatom potential through parameters V_{fact} and V_{thrs} set equal to 1.0 and 0.01, respectively. These two, one-dimensional, SRBO grids plus an additional evenly spaced grid in the angular coordinate $\widehat{C_A C_B H}$ ranging from 180° to 60° in steps of 30°, were then used as building blocks for setting up a three-dimensional grid of $10 \times 10 \times 5 = 500$ points sampling the configuration space of the process

$$C_A + C_B H^+ \rightarrow C_A C_B^+ + H, \tag{10.4}$$

where subscripts A and B have been used to distinguish the two carbon atoms. However, in order to obtain an accurate representation of the global potential-energy surface for the considered three-atom system, an additional set of nuclear configurations was taken into account in order to also model the alternative reaction channel:

$$C_A + HC_B^+ \rightarrow C_A H^+ + C_B. \tag{10.5}$$

For this purpose, use was made of analogous SRBO grids for r_{HC_B} and r_{C_AH} plus an angular grid for $\widehat{C_AHC_B}$ ranging from 180° to 60° in steps of 30°, which were used as building blocks for a three-dimensional grid with a smaller total number of points summing up to $\frac{10 \times (10+1)}{2} \times 5 = 275$ because of the specific symmetry properties of this channel.

In the second stage, high-level electronic-structure calculations were run to evaluate the ground-state electronic energies at the sampled geometries. In particular, ground-state electronic energies of CH^+, C_2^+ and triplet C_2H^+ were obtained by second-order multireference perturbation theory (MRPT) in the 'partially contracted' PC-NEVPT2 approach [174,175] through the computer program MOLPRO [176]. As detailed in Ref. [141], the obtained results are in agreement with previous electronic-structure calculations on diatomics CH^+ and C_2^+ [177–181] where the spectroscopic properties of the related ground states $^1\Sigma^+$ and $^4\Sigma_g^-$, respectively, were determined. Moreover, in agreement with previous studies on the triatomic molecule [182,183], the minimum-energy configuration of C_2H^+ is found to be collinear ($\widehat{CCH} = 180°$) with electronic ground state $^3\Pi$.

In the third and final stage, once the electronic energies are computed for the sampled nuclear configurations, the potential-energy surface of the system can be given an analytic form by fitting or interpolation schemes. In the cited works, an analytic representation of the potential-energy surface for the system including both the above-mentioned reactive channels, was obtained through an Aguado–Paniagua global fit of the three-body set of 775 points plus the two two-body sets of 10 points each. The fitting was performed using the GFIT3C program [58] adopting a sixth-degree polynomial fit for the two-body terms and a seventh-degree polynomial fit for the three-body term. Results of the electronic-structure calculations on the equilibrium properties of the considered diatoms and triatom are reported (labeled as 'computed') in Table 10.1 for a comparison with the values returned by the fitted potential-energy surface ('fitted'). Where available, the experimental data (reported in Refs. [184] and [185]) are also given in Table 10.1 ('exp.'). The data in the table show that the fitted potential-energy surface reproduces fairly well the data resulting from the electronic-structure calculations and experimental works, providing an exoergicity for the reaction of 1.64 eV.

As already stressed in Chapter 6, before launching the dynamics simulations, a preliminary analysis of the topological properties of the potential-energy surface may provide useful insights into the features of the reactive process, which can be helpful in setting up the dynamical calculations and in rationalizing their results. For this purpose, a two-dimensional representation of the fitted potential-energy surface as a function of the reactant, r_{C_BH}, and product, $r_{C_AC_B}$, interatomic distances at a fixed angle $\widehat{C_AC_BH} = 120°$ is shown in Fig. 10.1. In the figure, the blue (dark gray in print version) circles are the computed electronic energies, while the gray mesh represents the fitted potential-energy surface. At the base of the plot, isocontour lines are also shown for values

TABLE 10.1 Equilibrium distances r_e and dissociation energies D_e of CH$^+$, C$_2^+$ and C$_2$H$^+$ resulting from the electronic-structure calculations ('computed'), from the generated potential-energy surface ('fitted'), and from experimental works ('exp.')

	CH$^+$ ($X^1\Sigma^+$)	
	r_e /Å	D_e /eV
computed	1.13	4.08
fitted	1.13	4.08
exp. [184]	1.13	4.25

	C$_2^+$ ($X^4\Sigma_g^-$)	
	r_e /Å	D_e /eV
computed	1.41	5.73
fitted	1.42	5.72
exp. [185]	1.40	
exp. [184]	1.30	5.4

	C$_2$H$^+$ ($X^3\Pi$, $\widehat{C_A C_B H} = 180°$)		
	r_e^{CC} /Å	r_e^{CH} /Å	D_e /eV
computed	1.26	1.08	10.94
fitted	1.28	1.08	10.79

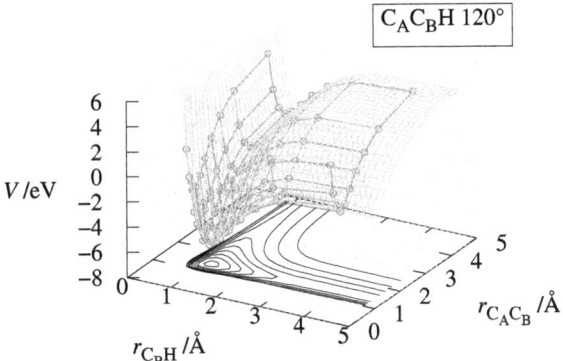

FIGURE 10.1 Two-dimensional representation of the potential-energy surface for the C + CH$^+$ → C$_2^+$ + H reactive channel at a fixed angle $\widehat{C_A C_B H} = 120°$. Blue circles: computed electronic energies. Gray mesh: fitted potential-energy surface. A contour map is also shown at the base with contours from −6 to 2 eV in steps of 1 eV. The energy zero is set to the bottom of the reactant valley. Reprinted with permission from Ref. [141]. © 2016 American Chemical Society.

ranging from -6 to 2 eV in steps of 1 eV. The entry channel for reaction (10.2) is located in the bottom-right area of the plot (low values of the reactant inter-atomic distance r_{C_BH}, high values of the distance between the isolated and the bound carbons $r_{C_AC_B}$). In contrast to the case of collinear H + H$_2$ discussed in Section 6.3, where the reactant and product channels are separated by a 'mountain pass' or reaction barrier, here the reaction proceeds barrierless towards a deep well (more than 6 eV, judging from the reported contour lines) in the strong interaction region at short distances in the bottom-left area of the plot. The ex-oergic product channel is reached in the top-left area of the plot. The depth and the shape of the potential well in the strong-interaction region can be better appreciated in the three-dimensional representation (circles and mesh), which also highlights the smoothness of the obtained potential-energy surface exhibiting no spurious structures.

Due to the peculiar representation adopted in Fig. 10.1, the analysis conducted so far is limited to the fixed angle $\widehat{C_AC_BH} = 120°$ reaction path. As discussed in Section 6.3, a more complete picture can be gained by combining the three internal coordinates of the system into two highly meaningful, process-oriented coordinates and plotting the potential as a function of these coordinates while relaxing it with respect to the remaining, less informative one. In so-called rectangular relaxed plots, the potential energy is plotted as a function of the η and Φ coordinates, which for reaction (10.2) are defined as

$$\eta = \arctan{(r_{C_BH}/r_{C_AC_B})} \tag{10.6}$$

$$\Phi = \widehat{C_AC_BH}, \tag{10.7}$$

after relaxation with respect to $\rho = \sqrt{r_{C_BH}^2 + r_{C_AC_B}^2}$. While the ρ coordinate measures the overall size of the system, the η acts as a reaction-progress co-ordinate of the $C_A + C_BH^+ \rightarrow C_AC_B^+ + H$ process ($\eta = 0°$ for the reactant asymptote and $\eta = 90°$ for the product asymptote). The Φ coordinate, instead, i.e., the angle formed by the reactant (breaking) and the product (newly formed) bond, informs on the reaction mechanism.

A rectangular relaxed plot for reaction (10.2) is given in Fig. 10.2. The reaction proceeds from the valley of the reactant channel (white area in the bottom region of the plot) to the valley of the product channel (light-blue (light gray in print version) area in the top region of the plot) which, as already mentioned, is lower in energy. The most favorable path to reaction is the collinear one, with C approaching CH$^+$ from the carbon side and leading to formation of the linear C$_2$H$^+$ intermediate (first well at η about 40°). Contours for bent approaches show a similar behavior though with a decreasing well depth reaching, for example, -5.6 eV at $\widehat{C_AC_BH} = 60°$. The collinear path remains the favorable one up to a value of η of about 50°, where the absolute minimum-energy path to products distorts from collinearity and branches symmetrically to a second well either at $\widehat{C_AC_BH} = 0°$ or 360°. This corresponds to the symmetric rotation of

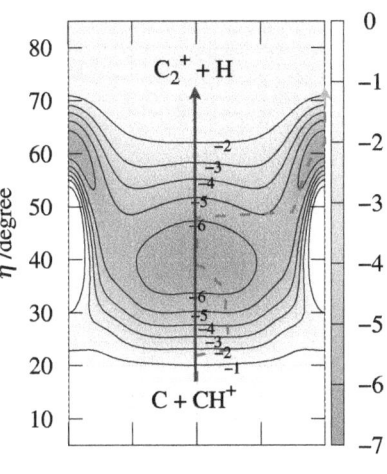

FIGURE 10.2   Rectangular relaxed plot for process C + CH$^+$ → C$_2^+$ + H. The absolute and collinear minimum-energy paths as a function of the reaction-progress coordinate η are also drawn as the blue (dark gray in print version) solid line and the red (gray in print version) dashed line, respectively. Energy zero is set to the bottom of the reactant valley. Reprinted with permission from Ref. [141]. © 2016 American Chemical Society.

the H atom about the carbon–carbon diatom leading, after overcoming a small barrier (corresponding to the 'mountain pass' between the two wells), to a structurally identical intermediate with the H atom binding the originally isolated C_A atom (second well at η about 60°) rather than C_B, as in the first well.

A more compact representation of the energy profile associated with these two alternative reaction paths is given in Fig. 10.3, which shows the minimum energy profile as a function of the reaction-progress coordinate η (i.e., the potential energy at each η value minimized with respect to all remaining coordinates). Both the absolute (blue (dark gray in print version) line) and the collinearly restrained (red (gray in print version) line) energy profiles are shown. The figure shows that the energy profile is almost identical for the two paths up to $\eta \sim 50°$. The depth of the energy well occurring at $\eta \sim 40°$ is 6.7 eV. However, from this point onward the reaction mechanism can branch, giving rise to the so-called 'microscopic branching', i.e., the competition between alternative reaction paths leading to the same product set. In particular, the system can either continue climbing along the collinear path and reach the reaction products, or proceed along an alternative path exploring a second well after overcoming a small energetic barrier.[1]

---

[1] Note that the barrier shown in Fig. 10.3 is not the same that the system would encounter by following the minimum-energy path from reactants to products. Fig. 10.3 shows in fact the minimum-energy profile as a function of η, which does not necessarily coincide with a reaction coordinate that

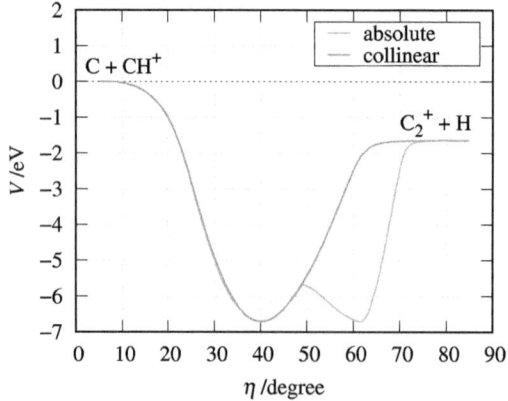

FIGURE 10.3 Absolute (blue (dark gray in print version) line) and collinear (red (gray in print version) line) minimum-energy profile for the reactive processes $C + CH^+ \rightarrow C_2^+ + H$ as a function of the reaction-progress coordinate η. The energy zero is set to the bottom of the reactant valley. Reprinted from Ref. [140]. ©2016 The Authors Published by Oxford University Press on behalf of the Royal Astronomical Society.

10.3 Dynamics and kinetics

Once an analytic formulation of the potential-energy surface has been generated, it is ready for use in either classical or quantum simulations of the reaction dynamics. A rigorous treatment of molecular reactive collisions should in principle include quantum effects, however, the large number of rovibrational states involved (mainly for the C_2^+ product molecule) and the presence of a deep well in the strong interaction region discourage the adoption of quantum-dynamics methods for reaction (10.2). Thus the quasiclassical trajectory method [80] becomes the most appropriate. In Refs. [140,141], an extensive quasiclassical trajectory computational campaign was performed using the computer program VENUS96 [186]. To provide the reader with an overview on practical aspects involved in the setup of this kind of calculation, some relevant computational details of those works will be reviewed in the following.

As discussed in Section 7.1, in order to compute the thermal rate constant $k(T)$ at a given temperature, a huge number of trajectories have to be integrated. The initial collision energies of the trajectories were selected according to the Boltzmann distribution at temperature T. The initial rovibrational states were selected also according to a Boltzmann distribution for temperature T over the manifold of states supported by the diatomic potential curve. In the cited works, thermal rate constants were calculated at 56 temperatures in the range from 5 K to 300 K, with a finer sampling for lower temperatures (steps of 1 K in the range

---

follows exactly the minimum-energy path. As can be evinced from Fig. 10.2, the rotational barrier along the true minimum-energy path, which is located at the 'mountain pass' linking the two wells, lies at an energy between −6 and −5 eV from the bottom of the reactant valley, and its height with respect to the bottom of the potential wells is thus seen to lie between 0.7 and 1.7 eV.

5–20 K, 2 K in the range 22–48 K and 10 K in the range 50–300 K), and a total of $N = 4 \times 10^6$ trajectories were integrated for each considered temperature, leading to an error on the rate constant below 1%. For the calculation of state-specific rate constants, the collision energies were also selected according to the Boltzmann distribution at temperature T but a single initial rovibrational state was set. State-specific rate constants were calculated starting from both the ground and the first excited vibrational state ($v = 0$ and $v = 1$) and from both the ground and the first excited rotational state ($j = 0$ and $j = 1$) using the same temperatures and number of trajectories used for the calculation of the thermal rates coefficients.

Particular care has to be taken in the setup of the parameters characterizing the start and the end of the trajectories in relation to the reactive event. As is well known, in fact, the interaction between an ion and a neutral particle extends to large distances. For this reason, initial and final distances to start and to end the trajectory integration were set to 25.0 Å, where the interaction between the fragments of the related channels results smaller than 3×10^{-7} eV. For analogous reasons, the maximum impact parameter was set to 20.0 Å in order to make sure that all reactive collisions are included in the integrated trajectories. A time step of 0.05 fs was used for the trajectory integration, guaranteeing a total energy conservation better than 2×10^{-7} eV even at the lowest considered temperature.

Dynamical results, also including the analysis of state-specific cross sections, are fully discussed in Refs. [140,141], where, for instance, the trend of the state-specific cross sections is analyzed, suggesting that the key driving factor to reaction is the gradient of the potential-energy surface in the entrance channel, while an increase of collision and internal energies reduces reactivity by acting as antagonist factors. We limit the discussion of the results here to the two relevant aspects of the microscopic branching of the reaction mechanism and the reaction kinetics.

As for the reaction mechanism, an instructive analysis is that of the reactive probability as a function of the impact parameter b, i.e., the opacity function (see Section 7.1, Eq. (7.20)), which in Refs. [140,141] is calculated using a histogrammic method with bins of 0.1 Å for b. Fig. 10.4 shows the state-specific opacity function $P_{v,j}(E_{tr}, b)$ for $v = 0$, $j = 0$, and translational energies 0.0013, 0.0390, 0.2600 eV, which correspond to the average translational energy for thermal distributions at ~ 10, 300, and 2000 K. The figure clearly suggests that, at the lowest considered collision energy (0.0013 eV) the reaction mechanism is of the capture type, whereby the colliding partners reach a relative distance close enough to allow the attractivity of the potential to dominate all antagonist actions and drive the system to react. This is clearly indicated by both the high value (13 Å) of b_{max} leading to reaction (the value of the sharp falloff of the opacity function) and the fairly high value (~ 0.8) of the height of the plateau (which would reach unity in a pure capture mechanism).

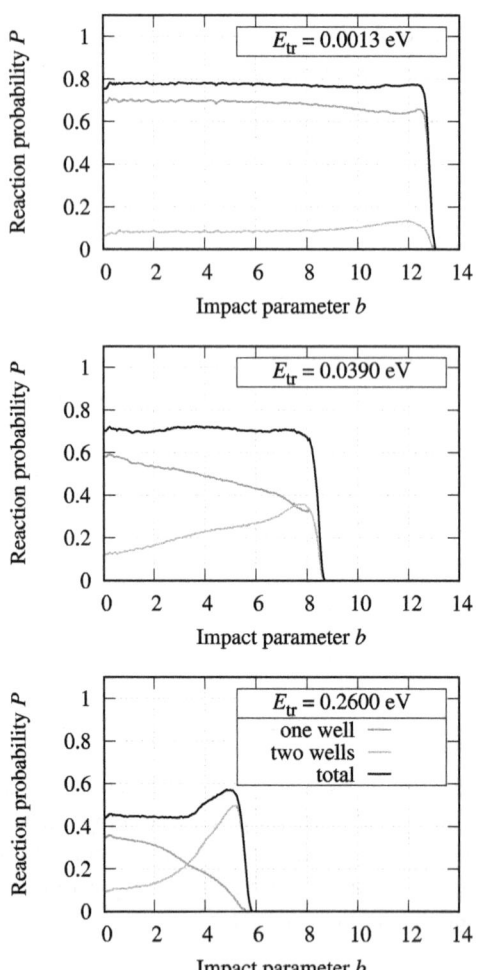

FIGURE 10.4 State-specific ($v = 0$, $j = 0$) opacity function computed at $E_{tr} = 0.0013$, 0.0390, 0.2600 eV and its components resulting from trajectories exploring either only the main well ('one well') or both the main well and the secondary well ('two wells'). Reprinted with permission from Ref. [141]. © 2016 American Chemical Society.

A similar behavior shows the opacity function calculated for $E_{tr} = 0.0390$ eV, though the somewhat less flat behavior alongside the lower values of both b_{max} and the plateau may suggest divergences from the mentioned capture mechanism. These divergences become apparent at the highest considered translational energy. In particular, at $E_{tr} = 0.2600$ eV both b_{max} and the plateau continue to reduce as the energy increases, and at the same time the opacity function features a peak at the right-hand corner, suggesting that an auxiliary mechanism comes into play in addition to the one driven by the potential-energy gradient.

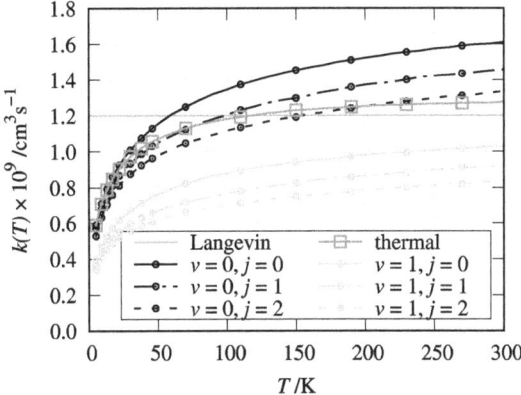

FIGURE 10.5 Thermal and state-specific rate constant for the reactive process C + CH⁺ (v, j) → C_2^+ + H as a function of the temperature T. The temperature independent–capture model value is also reported (red (gray in print version) line). Reprinted from Ref. [140]. ©2016 The Authors Published by Oxford University Press on behalf of the Royal Astronomical Society.

To better single out the nature of such auxiliary mechanism, the components of the opacity function associated with trajectories experiencing only the main well ('one well') and trajectories experiencing both the main and the secondary well ('two wells') are reported in the plots of Fig. 10.4 as blue and red (dark gray and gray in print version) curves, respectively. As can be clearly seen from the shape of these partial opacity functions, a capture-type mechanism with exploration of the sole main well is largely dominant at lower translational energies, while a different mechanism involving the exploration of both the potential wells becomes dominant at higher translational energies and impact parameters, clearly determining the peak in the right-hand corner of the opacity function at $E_{tr} = 0.2600$ eV.

As for the kinetics of the reaction, which is the most relevant information for astrochemistry, the computed thermal and state-specific rate constants are shown in Fig. 10.5 as a function of the temperature T. The temperature-invariant Langevin-model estimate of Prasad et al. [173] is also reported (horizontal red (gray in print version) line). As the figure shows, in going from $T = 5$ K to $T = 300$ K the computed thermal rate constant increases by more than a factor of two (from 0.6×10^{-9} to 1.3×10^{-9} cm³ s⁻¹) with a pronounced increase in the first 50 K. At 10 K, for instance, the thermal rate constant is 0.7×10^{-9} cm³ s⁻¹, about one half of the capture-model estimate – which, if included in astrochemical kinetic models would improperly enhance the destruction route of CH⁺ at that temperature. The figure also shows that the largest contribution to the value of the thermal rate constant is that associated with the ground ($v = 0$, $j = 0$) rovibrational state, and that an increase of the excitation of either the vibration or the rotation of the reactants has the effect of lowering the value of the rate constant.

   The Arrhenius–Kooij parameters for inclusion in astrochemical kinetic models were obtained from the calculated thermal rate constants via a nonlinear fit of the Arrhenius–Kooij expression (Eq. (10.1)), leading to best-fitting parameters $\alpha = 1.32 \times 10^{-9}$ cm^{3}s^{-1}, $\beta = 0.10$, and $\gamma = 2.19$ K. A slightly better fit (in terms of higher χ^2 and correlation coefficient) was obtained using a more flexible parametrized formulation of the temperature dependence of the rate constant, the so-called 'deformed Arrhenius' [187,188]:

$$k(T) = A\left[1 - d\frac{\epsilon}{RT}\right]^{\frac{1}{d}}, \qquad (10.8)$$

where R is the gas constant and the obtained best fitting parameters for the C + CH$^+$ reaction are $A = 1.34 \times 10^{-9}$ /cm^{3}s^{-1}, $d = -2.73$, and $\epsilon/R = 15.61$ K.

Chapter 11

Towards complexity

In the previous chapter, the astrochemical reaction of $C + CH^+$ served as a good example for a comprehensive illustration of the methodologies discussed in Chapters 5–9 for tackling the various stages of the modeling of the dynamics and kinetics of a chemical reaction. Even for a simple system such as $C + CH^+$, however, due to specific features of the related potential-energy surface that introduced a certain degree of complexity, the adoption of quantum techniques turned out to be challenging and it was necessary to resort to quasiclassical trajectory calculations. In this chapter, the final one of Part II of the book, we will discuss some of the main methodologies for coping with systems of increasing complexity. In particular, in Section 11.1 approximate quantum methods allowing for the treatment of systems featuring a few tens of degrees of freedom, or for an efficient evaluation of the thermal rate constant of a chemical reaction are addressed. In Section 11.2, popular approaches based on classical and statistical mechanics for modeling the dynamical properties of systems with hundreds or thousands of atoms are discussed. In Section 11.3, methodologies for the treatment of nuclear motion on more than one potential-energy surface in those cases when the Born–Oppenheimer approximation fails are illustrated.

11.1 Approximate quantum methods

Within the Born–Oppenheimer approximation, the time-independent and time-dependent techniques discussed in Chapter 7 offer in principle an exact treatment (besides errors introduced with their practical implementation on a computer) of the quantum dynamics of a system subject to a given potential-energy surface. These techniques are, however, computationally demanding and, in fact, their application is limited to three- or four-atom systems. When the number of degrees of freedom exceeds five or six, exact quantum-dynamics treatments are of little use and approximate schemes have to be adopted, either freezing and leaving out some degrees of freedom not directly involved in the reactive process but rather acting as spectator modes (see Ref. [189] for a recent work and Ref. [190] for an introduction to reduced-dimensionality treatments), or seeking approximations to the exact solution of the nuclear Schrödinger equation.

An efficient approximate quantum-mechanical treatment that has been recently developed is the so-called multiconfiguration–time dependent Hartree (MCTDH) [33,191,192], which falls in the family of time-dependent quantum techniques and has found widespread application in spectroscopy, photodynamics, and scattering problems. In the MCTDH scheme, based on ideas originally developed in electronic-structure theory that we will partly discuss in the forthcoming chapters of Part III of the book, the nuclear wavefunction of a molecular system is expressed in the form [193]:

$$\Gamma = \sum_I c_I(t)\Phi_I , \qquad (11.1)$$

with

$$\Phi_I = \prod_{i=1}^{N} \varphi_{Ii}(q_i,t) , \qquad (11.2)$$

i.e., as a linear combination of products of so-called single-particle functions or orbitals $\varphi_{Ii}(q_i,t)$, where N is the number of degrees of freedom of the system and q_i is the coordinate of the ith degree of freedom. The fact that both the expansion coefficients and the orbitals depend on time allows for a significant reduction in the basis set with respect to exact treatments. As a consequence, the computational requirements in terms of memory and computing time are much lighter. For the case, for instance, of a system with three degrees of freedom, in a standard wavepacket calculations using a 64-point grid for each of the three degrees of freedom, the total amount of numbers to be stored would amount to $64 \times 64 \times 64 = 262144$ for a single snapshot in time of the wavepacket, whereas in an MCTDH calculation, using 10 time-dependent orbitals and 64 points for each degree of freedom, one would have to store 1000 numbers for the coefficients plus 1920 numbers for all the orbitals, for a total amount of 2920 numbers. Accordingly, the MCTDH method is capable of computing the quantum dynamics of systems with a few tens of degrees of freedom. In a further recent development of the method, the so-called multilayer MCTDH (ML-MCTDH) [194,195], basis functions themselves are expressed in terms of lower-dimensionality functions, and this allows one to treat from a few hundreds to thousands of degrees of freedom.

One of the main disadvantages of the MCTDH method is that it requires the potential-energy surface of the system to be written as a sum of products of factors, one factor for each degree of freedom. This may pose serious problems as some fine details of the topology of the potential-energy surface may be lost as a consequence of the decomposition. This is circumvented in a class of methods known as Gaussian MCTDH (G-MCTDH), which are rooted in the seminal works of Heller [196] and use moving Gaussian functions as basis sets to describe nuclear wavefunctions. These schemes offer, in fact, the possibility of performing the so-called direct dynamics, whereby the potential energy and gradients are calculated on the fly (simultaneously with the nuclear dynamics) solving the electronic Schrödinger equation only at the nuclear geometries required by the propagation scheme. Moreover, the localized nature of Gaussian wavefunctions offers a direct connection with the classical picture of the dynamics.

On another front, moving closer again to few-atom systems such as those covered in the previous chapters, if one is interested only in the thermal reaction rate constant rather than in state-resolved quantities, an interesting alternative to full quantum calculations is the so-called ring polymer–molecular dynamics (RPMD). The RPMD method, which has been proposed by Craig and Manolopoulos in 2005 [197], is based on the classical isomorphism [198] between a quantum system and its classical ring-polymeric replica (harmonically coupled classical copies of the original system in the form of a necklace). In RPMD, the real-time quantum dynamics of a molecular system is approximated by purely classical molecular dynamics of the ring-polymer beads. While being purely classical molecular dynamics but in extended phase space, RPMD treats accurately and conserves in its real-time dynamics the quantum Boltzmann distribution, and the results are advantageously independent of the dividing surface used to separate reactants from products, which is particularly challenging to define when the reaction proceeds through a deep well such as the $C + CH^+$ process discussed in the previous chapter (see Ref. [199] for an application of the RPMD method to that reaction). Several gas-phase atom–diatom and polyatomic chemical reactions have been in recent years studied with the RPMD method (see Ref. [200] for a recent review). For prototype atom–diatom insertion chemical reactions, RPMD calculations provide thermal rate constants whose deviation from rigorous quantum dynamics results is within $\sim 15\%$ [201–203].

11.2 Molecular dynamics and stochastic approaches

Even though the MCTDH method has considerably extended the range of applicability of quantum-dynamical approaches, a vast class of chemical systems, especially those of relevance in biological chemistry, involve such a large number of atoms that quantum calculations are simply not feasible due to the featured overwhelming number of degrees of freedom. In these cases, one has no other choice than resorting to the integration of classical trajectories based on Newton's equation. In classical mechanics, the formalism of the dynamical treatment is relatively simple and, as discussed in Chapter 7, the focus shifts rather to the formulation of the potential-energy surface, the negative gradient of which gives the force acting on the particles, ultimately driving their trajectory.

In these contexts, the potential-energy surface is cast as the sum of n-body bonded and pairwise particle–particle interactions that are able to reproduce structural and conformational changes. The collection of the employed potential-energy functions (depending on some sets of internal coordinates of the system) and the associated parameters are frequently referred to as a force field (FF). Force fields are usually optimized for a given class of molecules (e.g., proteins, carbohydrates, etc.), and their parametrization is aimed at a good reproduction of experimental data (such as structural data or energetic data). In the force-field parametrization, experimental data are often supplemented with the results of high quality–electronic structure calculations.

In a similar spirit to that of the many-body expansion of the potential of Eq. (6.6) in Section 6.1, a force field is conventionally portioned in terms of 'bond' functions and 'nonbond' functions, as follows:

$$V = [V_{bonds} + V_{angles} + V_{dihedrals} + V_{impropers}]_{bond} + [V_{VdW} + V_{Coulomb}]_{nonbond}.$$
(11.3)

For example, the CHARMM force field [204] is formulated as follows:

$$V = \sum_{bonds} k_b (b - b_0)^2 + \sum_{angles} k_\theta (\theta - \theta_0)^2 + \sum_{dihedrals} k_\varphi [1 + \cos(n\varphi - \delta)]$$

$$+ \sum_{impropers} k_\omega (\omega - \omega_0)^2 + \sum_{Urey-Bradley} k_u (u - u_0)^2$$

$$+ \sum_{nonbonded} \epsilon \left[\left(\frac{r_{min_{ab}}}{r_{ab}} \right)^{12} - \left(\frac{r_{min_{ab}}}{r_{ab}} \right)^6 \right] + \frac{q_a q_b}{\epsilon r_{ab}}.$$
(11.4)

The first term in the potential-energy function relates to the bond stretches, where k_b is the bond-force constant and $b - b_0$ is the displacement from equilibrium. The second term accounts for the bond angles, where k_θ is the angle force constant and $\theta - \theta_0$ is the displacement from the equilibrium angle between three bonded atoms. The third term relates to the dihedrals (also known as torsion angles) where k_φ is the dihedral force constant, n is the multiplicity of the function, φ is the dihedral angle and δ is the phase shift. The fourth term accounts for the impropers, i.e., out-of-plane bending, where k_ω is the force constant and $\omega - \omega_0$ is the out-of-plane angle. The fifth term accounts for an additional Urey–Bradley component (crossterm accounting for angle bending using 1,3 nonbonded interactions), where k_U is the respective force constant and U is the distance between the 1,3 atoms in the harmonic potential. Non-bonded interactions between pairs of atoms (a, b) are represented by the last two terms. By definition, the nonbonded forces are only applied to atom pairs separated by at least three bonds. The van Der Waals (VdW) energy is calculated with a standard 12-6 Lennard-Jones potential and the electrostatic energy with a Coulombic potential (note that in the Lennard-Jones potential in Eq. (11.4), r_{min} does not correspond to the minimum of the potential, but rather to the point where the Lennard-Jones potential crosses the r-axis, i.e., the distance at which the Lennard-Jones potential is zero). Some of the functional forms usually employed for the most common terms in a force field are schematized in Fig. 11.1.

Calculations based on force fields – so-called 'empirical force-field calculations' – fall either into the class of molecular mechanics (MM) or molecular dynamics (MD). The MM method is generally employed to compute the relative energies of different geometries (conformations) of the same molecule that arise from rotations about chemical bonds as well as relative energies of intermolecular complexes, or, coupled to optimization techniques, energy minima of a given molecular system. Molecular dynamics aims instead at computing

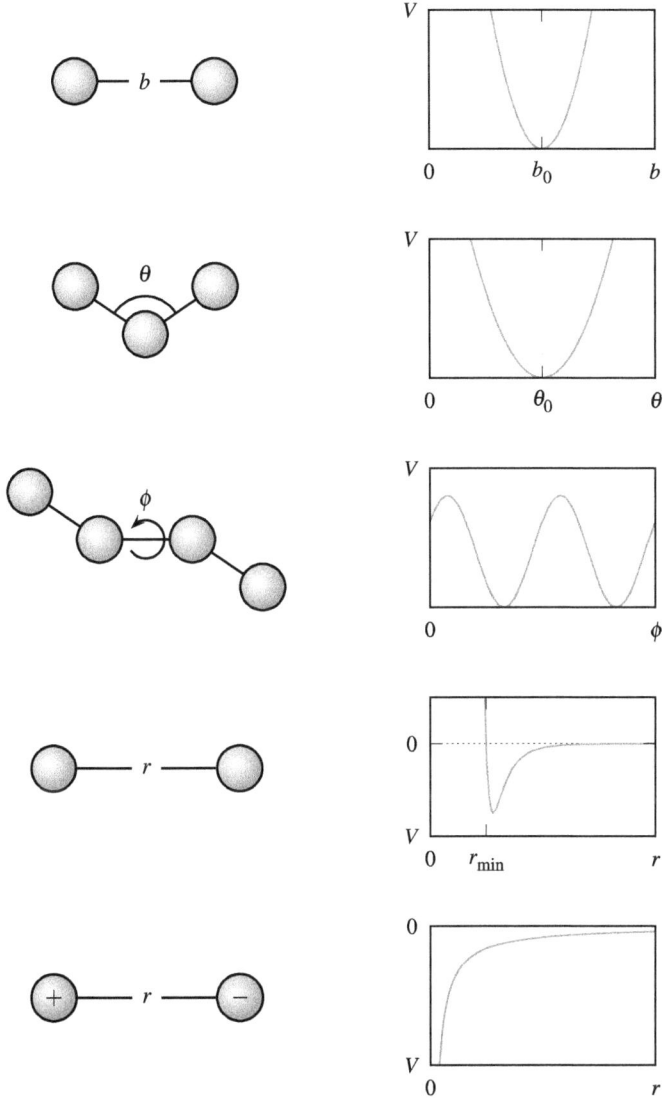

FIGURE 11.1 Schematic representation of the most common terms included in a force field.

properties that depend on the dynamical behavior of systems whereby, from the ergodic hypothesis from statistical mechanics, the statistical ensemble averages (or expectation values) are taken to be equal to time averages resulting from the integration of classical trajectories. These methods are indeed quite accurate for the estimation of certain molecular properties (those for which classical mechanics is appropriate), including conformational energies and binding affinity of small molecules to macromolecular receptors, and are useful in refining the structures derived from protein X-ray crystallography and protein–nuclear magnetic resonance spectroscopy.

When accurate thermodynamic information on an ensemble is not of interest, but rather the focus is on the dynamics of a subsystem immersed in a larger system (e.g., a solute in a solvent), the problem can be simplified by representing the larger system as a continuum interacting with the smaller system. In Langevin dynamics (see Ref. [205] for a more detailed discussion on the stochastic methods touched on in the remainder of this section), for example, the equation of motion of each particle is:

$$a(t) = -\zeta \, p(t) + \frac{1}{m}[f_{\text{intra}}(t) + f_{\text{continuum}}(t)], \qquad (11.5)$$

where the continuum is characterized by a microscopic friction coefficient, ζ, and a force, f, having one or more components (e.g., electrostatic and random collisional), while the intramolecular forces derive from a force field, and the position and momentum are propagated as in molecular-dynamics calculations.

A further simplification may be introduced on the basis of the consideration that the shape of a molecular system is roughly conserved over long time scales. Accordingly, in Brownian dynamics, one eliminates the momentum degrees of freedom by approximating as zero the momentum of each particle relative to the rotating center-of-mass reference frame. Introducing this approximation in Eq. (11.5) and integrating, one obtains the Brownian equation of motion:

$$r(t) = r(t_0) + \frac{1}{\zeta} \int_0^t [f_{\text{intra}}(t') + f_{\text{continuum}}(t')] \, dt', \qquad (11.6)$$

where only the position vector is propagated.

Within these schemes, where an increasing degree of stochastic behavior is introduced, due to the fact that a potentially very large surrounding system is represented by a continuum, much longer time scales can be accessed with respect to standard, fully deterministic molecular-dynamics simulations. Stochastic approaches are brought to their logical extreme by Monte Carlo methods, where the introduction of stochastic behavior is complete and one has no equation of motion to integrate, but rather has to carry out a suitable random selection of phase-space points and use statistical mechanics to predict average values of a property.

11.3 Beyond the Born–Oppenheimer approximation

Since the beginning of Part II, the discussion has so far moved within the limits of the Born–Oppenheimer approximation. In many chemical problems, it is indeed the case that the electrons respond nearly instantaneously to the motion of the nuclei, and this allows for a decoupling of the electronic and the nuclear

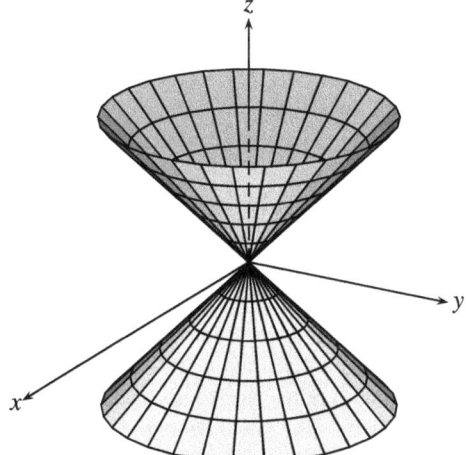

FIGURE 11.2   Schematic representation of an ideal conical intersection.

problems. In other important cases, however, especially when excited electronic states are involved, in the regions of nuclear configuration space where the separation between two adiabatic electronic states vanishes, i.e., in the vicinity of the so-called conical intersections, so-called nonadiabatic transitions between electronic states almost invariably occur.

A conical intersection is the set of points of nuclear-configuration space where potential-energy surfaces of two electronic states are degenerate (intersect) and the nonadiabatic couplings between these states are nonvanishing. For two-dimensional systems, there is only one degeneracy point and, as schematized in Fig. 11.2, the two potential-energy surfaces form a double cone in the region of the degeneracy. In polyatomic molecules of N atoms, where, as already mentioned, the internal coordinates are $(3N - 6)$, two potential-energy surfaces are allowed to cross along a $(3N - 8)$-dimensional subspace (intersection space) and two special coordinates can be found defining the so-called branching plane (the xy-plane in Fig. 11.2) containing the degeneracy point in the vicinity of which the potential-energy surfaces form again a double cone such as that of Fig. 11.2. Accordingly, movement in the branching plane removes the degeneracy. Upon movement in the remaining $(3N - 8)$ directions, instead, the energies of the ground and excited state remain degenerate.

In the vicinity of the conical intersections, the Born–Oppenheimer approximation breaks down, i.e., the nonadiabatic couplings are nonnegligible, and a simultaneous treatment of the nuclear motion on two or more potential-energy surfaces is required. As mentioned, these situations are the norm for problems that involve excited electronic states, from photochemistry to electron transfer, which are relevant to diverse application fields such as renewable energy, chemical synthesis, and bioimaging.

We saw in Section 2.3 of Chapter 2 that the molecular wavefunction can be expanded in an orthonormal electronic basis with nuclear-dependent coefficients (2.27), where the electronic functions are conveniently chosen as the eigenfunctions of the time-independent electronic Schrödinger equation for a given nuclear configuration $\{r_a\}$ (so-called adiabatic representation). As we discussed therein, insertion of such expansion into the Schrödinger equation (2.20) leads to a set of coupled equations of motion for the nuclear wave functions, Eq. (2.28) (insertion of Eq. (2.27) in a molecular time-dependent Schrödinger equation of the form of Eq. (2.15) leads to an analogous set of time-dependent equations). The first term of Eq. (2.28) (or of its time-dependent analog) describes the adiabatic motion of the nuclear component on the nth electronic state corresponding to the potential-energy surface $\varepsilon_n(\{r_a\})$. The second term accounts for the coupling with other electronic states.

While the expansion of Eq. (2.27) is formally exact only when including an infinite number of electronic states, in practice a few electronic states suffice to provide an excellent approximation, and, when the Born–Oppenheimer approximation is valid, the summation in Eq. (2.27) is limited to a single electronic term, usually taken as that of the ground electronic state. However, when two electronic states become nearly or exactly degenerate, then the Born–Oppenheimer approximation is no longer valid and neglecting the nonadiabatic couplings would lead to incorrect results. Thus, in these cases, one is forced to solve the equations for the nuclear motion on coupled potential-energy surfaces.

The methods for simulating nonadiabatic nuclear dynamics can be divided into two main families: full quantum methods, and methods employing propagation schemes based on classical trajectories. As for the quantum approaches, in principle numerically exact solutions for the nuclear motion on two or more potential-energy surfaces can be obtained through the quantum-dynamics methodologies already discussed, provided that both the electronic energies and nonadiabatic coupling matrix elements are known. The method of reference, in this context, is the MCTDH, that, as already mentioned, is applicable to molecules featuring up to a few tens of nuclear degrees of freedom (see for instance Ref. [206]). Alternative methods for nonadiabatic dynamics are those based on the already-mentioned expansion of the nuclear wave function in terms of a linear combination of traveling Gaussian basis functions (see Ref. [207] for a recent review).

In classical-trajectory based methods, instead, the nuclei are treated classically. A popular approach in this context is the so-called surface-hopping dynamics, originally developed by Tully and Preston [208] and further refined by Tully [209]. In surface-hopping approaches, the dynamics of a nuclear wavepacket is approximated by a swarm of classical trajectories that can hop from one electronic state to the other based on a stochastic algorithm (the probability that the trajectory will hop from one electronic state to the other responds to a set of electronic amplitudes that are integrated together with the classical trajectories). However, while the surface-hopping trajectory method retains all the advantages – especially in terms of computational requirements – of classical-based approaches, it cannot be fully derived from first principles and this makes uncertain the predictions of the situations in which the method will provide accurate results.

The topic covered in this final section of Part II of the book has led us back to the Born–Oppenheimer approximation and to the crossroads discussed at the end of Section 2.3, where the routes of the electronic and nuclear problem separate. Having so far explored several aspects related to the solution of the nuclear problem, it is now time to take the other route, that of the electronic problem, and look at the vast realm that lies behind what, in the context of nuclear dynamics, for a given molecular geometry ($\{\boldsymbol{r}_a\}$) reduces to a single number $\varepsilon_n(\{\boldsymbol{r}_a\})$.

Part III

Electronic structure and chemical bonding

Chapter 12

The wavefunction and the electron density

In Chapter 2 we saw how by virtue of the Born–Oppenheimer approximation the nuclear problem, exemplified by Eq. (2.30), and the electronic problem, exemplified by Eq. (2.25), can be treated separately. Part II was entirely devoted to the treatment of the nuclear problem, where the presence of the electrons is simply reduced to the collection of values $\varepsilon_n(\{r_a\})$, being the eigenvalues of an electronic problem for an electronic Hamiltonian $\hat{H}_e$ (Eq. (2.26)) at a fixed molecular geometry, and constituting the potential-energy surface governing the nuclear motion.

In Part III we will be concerned with the solution of the electronic problem. As we will be confined to molecular systems with fixed nuclear geometry, in order to make the formalism simpler we will i) adopt atomic units, ii) consider separately the last term of Eq. (2.26), the nucleus–nucleus repulsion term $\sum_{a>b} \frac{Z_a Z_b}{r_{ab}}$ (this is simply a constant term that can be added later to the calculated electronic energy) and iii) redefine the electronic Hamiltonian minus the nucleus–nucleus repulsion term simply as $\hat{H}$, and the associated electronic energy as E. Accordingly, we will deal with the time-independent Schrödinger equation for the electronic wavefunction of N electrons at a given nuclear configuration $\{r_a\}$:

$$\hat{H}\psi_n(\{r_i\}; \{r_a\}) = E_n \psi_n(\{r_i\}; \{r_a\}),\qquad(12.1)$$

where the Hamiltonian is:

$$\hat{H} = -\frac{1}{2}\sum_i \nabla_i^2 - \sum_{a,i}\frac{Z_a}{r_{ai}} + \sum_{i>j}\frac{1}{r_{ij}},\qquad(12.2)$$

and the electronic wavefunction is a function of the coordinates of all the N electrons $\{r_i\} \equiv (r_1, r_2, \ldots, r_N)$, depending only parametrically on the ensemble of the fixed nuclear coordinates $\{r_a\}$.

Even with the simplifications introduced by leaving out the nuclear degrees of freedom due to the Born–Oppenheimer approximation, Eq. (12.1) is extremely difficult to solve. Moreover, it is centered around a mathematical function that depends on $3N$ variables and is thus potentially quite complicated. While the physical meaning of the wavefunction is clear (in the sense that, as discussed in Chapter 2, its square modulus returns the probability density for

Chemistry at the Frontier with Physics and Computer Science
https://doi.org/10.1016/B978-0-32-390865-8.00022-2

the position of the electrons), it is actually difficult to practically handle, or even mentally conceive, such a complicated function.

As already sketched in the historical overview given in Chapter 4, attempts towards a simplification of the problem of Eq. (12.1) date back to the early days of quantum mechanics and are based on the same separation criterion that inspired the Born–Oppenheimer approximation. On the heels of the separation of the nuclei from the electrons of a given molecular system, one may in fact think to further separate the electrons from each other – with the rationale that, as is often the case, solving many small problems is easier than solving one large problem. This separation is the essence of the model known as 'Hartree–Fock', which was developed by Hartree, Fock and Slater in the late 1930s [21–23]. In spite of its limitations, descending from the fact that – as we shall see in the next section – it treats the electrons as uncorrelated particles, the Hartree–Fock model is still a fundamental paradigm in modern quantum chemistry, and many current methods recovering the electronic correlation and providing chemically accurate results are built on top of the Hartree–Fock model or are tightly related to it – as is the case of the Kohn–Sham implementation of density-functional theory.

The following sections are meant to give an account on the above-mentioned topics. In particular, the Hartree–Fock model will be illustrated in Section 12.1. A discussion on the electronic correlation and a brief overview of the so-called 'post-Hartree–Fock' methods follow in Section 12.2. Density-functional theory is introduced in Section 12.3 and the Kohn–Sham approach is discussed in relation to the previously illustrated Hartree–Fock model.

Before going ahead, however, it will be useful to add a few notes. First, a general comment on the separability of a Hamiltonian is in order. If a Hamiltonian $\hat{H}$ is separable in terms $\hat{H}_A$ and $\hat{H}_B$ depending each only on the coordinates of a subset A and B, respectively, of particles:

$$\hat{H} = \hat{H}_A + \hat{H}_B , \tag{12.3}$$

then the solution of the overall problem $\hat{H}\psi = E\psi$ can be built from the solutions of the separate two subproblems $\hat{H}_A\psi_A = E_A\psi_A$ and $\hat{H}_B\psi_B = E_B\psi_B$. In particular, following from the mathematical nature of the Schrödinger equation, the overall wavefunction is given by $\psi = \psi_A\psi_B$ and the total energy is $E = E_A + E_B$.

It will also be useful to recall two basic principles of quantum mechanics, which also follow from the mathematical nature of the involved equations: the superposition principle and the variational principle. The first one, the superposition principle, states that if ψ_1 and ψ_2 are two allowed wavefunctions for a system, then any linear combination of them $c_1\psi_1 + c_2\psi_2$, with c_1 and c_2 complex coefficients, is also an allowed wavefunction. The second one, the variational principle, states that an approximate wavefunction for a molecular system, when substituted into the Schrödinger equation, will always yield an energy higher than the true energy of the system. The more accurate the trial wavefunction is, the closer will the calculated energy be to the true energy. Computational methods that, as we will see, use this principle in order to obtain the best approximation to the correct wavefunction are called variational methods.

A final note is deserved on the antisymmetry principle. According to this principle, as further elaborated below, the wavefunction of the electrons must be antisymmetric, i.e., has to satisfy the requirement of changing sign upon the interchange of any two electrons. In general, all particles with half-integer spin (fermions) are described by antisymmetric wavefunctions, while all particles with zero or integer spin (bosons) are described by symmetric wavefunctions. Incidentally, we note that, besides taking into account the antisymmetry principle, during the development of the formalism in Section 12.1 the existence of spin will be ignored in a first stage, while its consequences will be briefly addressed at the end of the section.

12.1 The Hartree–Fock model

Due to the last term on the right-hand side of Eq. (12.2) (electron–electron repulsion), the electronic Hamiltonian cannot be separated in the sense of Eq. (12.3) in N terms depending each on the coordinates of a single electron. Let us nonetheless assume that the term $\sum_{i>j} \frac{1}{r_{ij}}$ could indeed be written or approximated as a sum of potentials $v(r_i)$ depending each on the coordinates of one electron. If we now group together the kinetic and nuclear attraction term of a single electron in the one-electron operator $\hat{h}(r)$

$$\hat{h}(r) = -\frac{1}{2}\nabla^2 - \sum_a \frac{Z_a}{|r - r_a|}, \qquad (12.4)$$

we could then separate the Hamiltonian into one-electron terms $\hat{F}_i$ depending each on the coordinates of only one electron:

$$\hat{H} = \sum_i \hat{F}_i(r_i), \qquad (12.5)$$

where $\hat{F}$ is the one-electron operator:

$$\hat{F}(r) = \hat{h}(r) + v(r). \qquad (12.6)$$

In this case, as noted above when considering the separability of a Hamiltonian, the solutions of the overall eigenvalue problem for $\hat{H}$, could be written as products of the solutions of one-electron subproblems for $\hat{F}$ of the form:

$$\hat{F}\varphi_i(r) = \varepsilon_i \varphi_i(r), \qquad (12.7)$$

where $\varphi_i(r) \equiv \varphi_i(x, y, z)$ are one-electron (or single-particle), three-dimensional functions called 'orbitals'. More precisely, solution of Eq. (12.7) would generate an infinite set of orthonormal orbitals ($\int \varphi_i^*(r)\varphi_j(r)\,dr = \delta_{ij}$) and the N-electron wavefunction could be constructed as any of the possible products $\{\Theta_\mu\}$ of N orbitals

$$\psi \simeq \Theta_\mu(r_1, r_2, \ldots, r_N) = \varphi_{\mu_1}(r_1)\varphi_{\mu_2}(r_2)\cdots\varphi_{\mu_N}(r_N),\qquad(12.8)$$

where μ identifies an N-tuple of distinct and ordered orbitals and is also referred to as the 'electronic configuration'. Such a model, besides drastically reducing the complexity of the equations to be solved, also offers the conceptual advantage that each electron is described by its own wavefunction and will thus have associated a particular probability density for the position.

Yet, as is easy to verify, a wavefunction of the form of Eq. (12.8) would violate the symmetry requirements imposed by the antisymmetry principle. As already mentioned, in fact, according to this principle electrons must be described by wavefunctions that are antisymmetric with respect to interchange of the coordinates of a pair of electrons. More formally, indicating as $\hat{P}_{ij}$ the operator that interchanges electron i with electron j, one must have:

$$\hat{P}_{ij}\psi(\{r_i\}) = -\psi(\{r_i\}),\qquad(12.9)$$

which is not verified if ψ is written as a product of orbitals. In this case, in fact, the interchange of two particles – for example, the swap of r_1 and r_2 in Eq. (12.8), i.e., assigning orbital μ_1 to electron 2 and orbital μ_2 to electron 1 – would leave the wavefunction unaltered.

An antisymmetric formulation of the N-electron wavefunction in terms of orbitals can instead be purposely prepared by taking appropriate linear combinations of products of orbitals. In particular, a suitable form is that of a normalized determinant Φ_μ:

$$\Phi_\mu(r_1, r_2, \ldots, r_N) \equiv |\varphi_{\mu_1}(r_1)\varphi_{\mu_2}(r_2)\cdots\varphi_{\mu_N}(r_N)|$$

$$= \frac{1}{\sqrt{N!}}\begin{vmatrix} \varphi_{\mu_1}(r_1) & \varphi_{\mu_2}(r_1) & \cdots & \varphi_{\mu_N}(r_1) \\ \varphi_{\mu_1}(r_2) & \varphi_{\mu_2}(r_2) & \cdots & \varphi_{\mu_N}(r_2) \\ \vdots & \vdots & \ddots & \vdots \\ \varphi_{\mu_1}(r_N) & \varphi_{\mu_2}(r_N) & \cdots & \varphi_{\mu_N}(r_N) \end{vmatrix}$$

$$= \frac{1}{\sqrt{N!}}\sum_P (-1)^p \hat{P}\Theta_\mu(r_1, r_2, \ldots, r_N),\qquad(12.10)$$

constructed from all possible permutations P of the orbitals in a given electronic configuration μ, with p being the parity of the permutation (number of interchanges). The function Φ_μ is called a Slater determinant and now satisfies Eq. (12.9).

Summing up, upon the assumption that the N-electron Hamiltonian could be separated into N one-electron terms, so far we arrived at a plausible form for the N-electron wavefunction in terms of N one-electron wavefunctions (orbitals) that are solutions of the eigenvalue problem for a one-electron operator $\hat{F}(r) = \hat{h}(r) + v(r)$. However, we do not know how to express, or reasonably approximate, the term $v(r)$ that should account for the electron–electron interactions. In this regard, it will be instructive to analyze the expression for the energy of a wavefunction approximated by a Slater determinant.

As discussed in Chapter 2, the average value for the energy of a wavefunction expressed as a Slater determinant can be obtained by inserting Eq. (12.10) and the Hamiltonian (12.2) into Eq. (2.9):

$$\langle E \rangle = \int \Phi_\mu^*(\{r_i\}) \left[-\frac{1}{2} \sum_i \nabla_i^2 - \sum_{a,i} \frac{Z_a}{r_{ai}} + \sum_{i>j} \frac{1}{r_{ij}} \right] \Phi_\mu(\{r_i\}) \, dr_1 \dots dr_N \, .$$

(12.11)

While this may seem rather intricate (an easy exercise is to consider the simple case of $N = 2$ electrons), elaboration of Eq. (12.11) leads to the following compact form for the average value of the energy of a Slater determinant:

$$E = \sum_i h_{ii} + \sum_{i>j} V_{ijij} - \sum_{i>j} V_{ijji} \, ,$$

(12.12)

where we implied that i and $j \in \mu$ and dropped any reference to the μth N-tuple. The elements of the two- and four-index matrices appearing in Eq. (12.12) have the general form

$$h_{ij} = \int \varphi_i^*(r) \left[-\frac{1}{2} \nabla^2 - \sum_a \frac{Z_a}{|r - r_a|} \right] \varphi_j(r) \, dr \, ,$$

(12.13)

and

$$V_{ijkl} = \int \varphi_i^*(r_1) \varphi_j^*(r_2) \frac{1}{r_{12}} \varphi_k(r_1) \varphi_l(r_2) \, dr_1 \, dr_2 \, .$$

(12.14)

Now, looking back at Eq. (12.12), the average value of the energy of a Slater determinant results from three groups of contributions. The first of these (first term in the right-hand side of Eq. (12.12)) is clearly a summation of the average values of the kinetic and nuclear-attraction energy of each electron, each described by the related ith orbital, of the form:

$$h_{ii} = \int \varphi_i^*(r) \left[-\frac{1}{2} \nabla^2 - \sum_a \frac{Z_a}{|r - r_a|} \right] \varphi_i(r) \, dr \, .$$

(12.15)

The second contribution results from a summation on all different couples (i, j) of terms of the form

$$V_{ijij} = \int \varphi_i^*(r_1) \varphi_j^*(r_2) \frac{1}{r_{12}} \varphi_i(r_1) \varphi_j(r_2) \, dr_1 \, dr_2$$

$$= \int |\varphi_i(r_1)|^2 \frac{1}{r_{12}} |\varphi_j(r_2)|^2 \, dr_1 \, dr_2 . \tag{12.16}$$

The reader will easily recognize in the second line of Eq. (12.16) the classical expression of the energy for two continuous charge distributions $|\varphi_i(r)|^2$ and $|\varphi_j(r)|^2$ interacting through a Coulomb potential. Thus a second contribution to the average value of the energy of a Slater determinant arises from the Coulomb repulsion between pairs of electrons whose spatial distribution is dictated by the shape of the related i and j orbitals.

Finally, the third contribution results from a summation on all different couples (i, j) of terms similar to those of Eq. (12.16), but obtained upon exchange of the last two indices ($V_{ijij} \to V_{ijji}$):

$$V_{ijji} = \int \varphi_i^*(r_1)\varphi_j^*(r_2)\frac{1}{r_{12}}\varphi_j(r_1)\varphi_i(r_2) \, dr_1 \, dr_2$$

$$= \int \varphi_i^*(r_1)\varphi_j(r_1)\frac{1}{r_{12}}\varphi_j^*(r_2)\varphi_i(r_2) \, dr_1 \, dr_2 . \tag{12.17}$$

This time, the expression in Eq. (12.17) has no classical counterpart and, looking at the second line of Eq. (12.17), seems to stem from a Coulombic interaction between 'mixed' charge distributions $\varphi_i^*(r)\varphi_j(r)$ and $\varphi_j^*(r)\varphi_i(r)$. Due to the above-mentioned reasons, the term V_{ijij} is called the 'Coulomb integral' and the term V_{ijji} is called the 'exchange integral'.

By noting that the Coulomb integral is equal to the exchange integral if $i = j$, and that both V_{ijij} and V_{ijji} give the same values if i and j are exchanged, we can replace the summation over all different couples in Eq. (12.12) with a double summation on i, j multiplied by $\frac{1}{2}$:

$$E = \sum_i h_{ii} + \frac{1}{2} \sum_{i,j} V_{ij[ij]} , \tag{12.18}$$

where $V_{ij[ij]} = V_{ijij} - V_{ijji}$.

We notice now that, if we rearrange Eq. (12.16) as follows:

$$V_{ijij} = \int \varphi_i^*(r) \left[\int \varphi_j^*(r')\frac{1}{|r - r'|}\varphi_j(r') \, dr' \right] \varphi_i(r) \, dr , \tag{12.19}$$

the term V_{ijij} can be seen as the average value of a one-electron operator, the so-called 'Coulomb' operator $\hat{j}^{(j)}$:

$$\hat{j}^{(j)} = \int \frac{|\varphi_j(r')|^2}{|r - r'|} \, dr' , \tag{12.20}$$

which, when acting on an orbital i, accounts for the Coulomb interaction with orbital j.

Analogously, the term V_{ijji} can be written as the average value of a one-electron operator by defining the one-electron 'exchange' operator $\hat{k}^{(j)}$ as:

$$\hat{k}^{(j)} \varphi_i(\boldsymbol{r}) = \left(\int \frac{\varphi_j^*(\boldsymbol{r}')\varphi_i(\boldsymbol{r}')}{|\boldsymbol{r} - \boldsymbol{r}'|} \, d\boldsymbol{r}' \right) \varphi_j(\boldsymbol{r}) , \qquad (12.21)$$

which acts on orbital i and accounts for the exchange interaction with orbital j, whereby

$$V_{ijji} = \int \varphi_i^*(\boldsymbol{r}) \hat{k}^{(j)} \varphi_i(\boldsymbol{r}) \, d\boldsymbol{r} . \qquad (12.22)$$

Note that the definition of $\hat{k}^{(j)}$ is somewhat more involved than that of $\hat{j}^{(j)}$, as for the exchange operator one cannot avoid including in its definition the orbital on which the operator acts.

If we now define the one-electron total Coulomb, $\hat{J}$, and total exchange, $\hat{K}$, operators:

$$\hat{J} = \sum_j \hat{j}^{(j)} \qquad (12.23)$$

$$\hat{K} = \sum_j \hat{k}^{(j)} , \qquad (12.24)$$

and introduce these into Eq. (12.18), then the energy of a Slater determinant can be written as a sum of average values of a one-electron operator $\hat{h} + \frac{1}{2}(\hat{J} - \hat{K})$:

$$E = \sum_i \langle \varphi_i | \hat{h} + \frac{1}{2}(\hat{J} - \hat{K}) | \varphi_i \rangle . \qquad (12.25)$$

In view of the above discussion, we have now some elements to infer which terms should enter the operator $\hat{F}$ of Eq. (12.7) for the one-electron problem that has to be solved in order to find the orbitals $\{\varphi_i\}$. Such an operator should certainly include the kinetic and nuclear-attraction term $\hat{h}$, but also account for the electron–electron interactions. If we limit ourselves to model these with only Coulombic terms, then we obtain the Hartree model, where the $\hat{F}$ operator reads

$$\hat{F}_i = \hat{h} + \hat{J}_i = \hat{h} + \sum_{\substack{j=1 \\ j \neq i}}^{N} \hat{j}^{(j)} . \qquad (12.26)$$

Note that this operator is different for each electron, as of course the self-repulsion term has to be excluded from the summation on the right-hand side of Eq. (12.26). Accordingly, within this model, one has to solve a set of N equations of the form

$$\hat{F}_i \varphi_i(\boldsymbol{r}) = \varepsilon_i \varphi_i(\boldsymbol{r}) \quad i = 1, \ldots, N , \qquad (12.27)$$

for each electron i.

On the other hand, the determinantal nature of the wavefunction – ultimately arising from the symmetry requirements dictated by the antisymmetry principle – forces us to also account for an exchange term in the electron–electron interactions, leading to an $\hat{F}$ operator of the form:

$$\hat{F}_i = \hat{h} + \hat{J}_i - \hat{K}_i = \hat{h} + \sum_{\substack{j=1 \\ j \neq i}}^{N} \hat{j}^{(j)} - \sum_{\substack{j=1 \\ j \neq i}}^{N} \hat{k}^{(j)} , \qquad (12.28)$$

where, again, the self-interaction terms are excluded from the summations. Now, due to the fact that the self-Coulomb and self-exchange interactions are equal and thus cancel out, there is actually no need to exclude the ith term from the summations in the operator of Eq. (12.28). As a consequence, despite the intricate and counterintuitive nature of the exchange interactions, its inclusion in the $\hat{F}$ operator curiously leads to the simplification that the same operator $\hat{F}$ can now be used for all the electrons. One has thus to solve a single one-electron eigenvalue problem that is the same for all electrons:

$$\hat{F} \varphi_i(\mathbf{r}) = \varepsilon_i \varphi_i(\mathbf{r}) , \qquad (12.29)$$

where

$$\hat{F} = \hat{h} + \hat{J} - \hat{K} \qquad (12.30)$$

is the so-called Fock operator.

Eqs. (12.29) and (12.30) are the final equations of the Hartree–Fock model. As can be evinced from these equations, the essence of the Hartree–Fock model is that each electron is described by a wavefunction φ_i that responds to the Fock operator, i.e., a Hamiltonian that describes the field generated by the nuclei and by all other electrons embodied by the respective orbitals. In other words, the motion of the electrons is uncorrelated and each electron feels the presence of the other electrons in an averaged way through their orbitals.

This also has the implication that, as is evident from Eqs. (12.20) and (12.21), the definition of the Fock operator includes the solutions of the eigenvalue problem (12.29), and thus the problem, as further discussed in Chapter 13, has to be solved through an iterative scheme. In practice, one starts with a guess for an initial set of orbitals, uses these orbitals to construct the Fock operator, and solves the Hartree–Fock equations to find a new set of orbitals. The lowest-energy N orbitals of this new set are 'occupied' with the electrons[1] (the remaining, so-called 'virtual' orbitals are discarded) and used to construct a new Fock operator leading to a new set of solutions. This procedure is iterated until the newly obtained set of orbitals is identical (within some convergence criterion) to the set of orbitals that has been used for generating them or, in

---

[1] Note that the determinantal nature of the wavefunction excludes that the same orbital can be used for more than one electron, otherwise the Slater determinant automatically becomes null.

other words, until the solutions are 'self-consistent' with the field that they generate in the Fock operator. Accordingly, the iterative procedure is called the self-consistent field.

A few additional remarks are due here before moving to the next sections. A first important note is that the Hartree–Fock Eqs. (12.30) and (12.29) can be rigorously derived using the variational principle stating, as already mentioned, that any trial wavefunction approximates the ground-state energy 'from above'. In particular, solving the mathematical problem of minimizing the energy of Eq. (12.25) with respect to variations in the orbitals leads exactly to the Hartree–Fock equations. In other words, the best approximation to the true wavefunction of an N-electron system that one can get by imposing the form of a Slater determinant, is that obtained by using orbitals that respond to the Fock operator and are solutions of the Hartree–Fock equations.

A further comment is that the energy associated to a wavefunction that is approximated by a Slater determinant is not equal to the sum of the energies of the orbitals that make up the determinant. In the energy of a given orbital, in fact, the Coulomb and exchange interaction with all the remaining electrons is included. When summing up the energies of all the orbitals the interaction between each electron pair would be counted twice (hence the appearance of a factor $\frac{1}{2}$ in (12.25)).

Finally, in the above discussion no mention has been made of the spin of the electrons. As is known, elementary particles (including the electrons) possess an intrinsic angular momentum (i.e., not related to their motion in space), the existence of which is inferred from experiment. While, as we will discuss in Chapter 17, the spin arises naturally within relativistic quantum mechanics, in the nonrelativistic framework (the one that we implicitly adopted so far) it is introduced ad hoc by labeling the electrons as either α or β (spin up or spin down). It is worth stressing now that the antisymmetry principle applies both to the exchange of the coordinates and of the spin properties of the electrons. An implication of this is the fact that electrons with different spin are allowed to have the same spatial distribution, i.e., occupy the same orbital. While a detailed account of these aspects (which can be found for instance in Ref. [210]) would stray us away from the focus of this book, it will be useful to summarize here some conclusions.

After noting that the $\hat{J}$ and $\hat{K}$ operators behave differently in relation to the spin, in the so-called 'unrestricted' Hartree–Fock model the spin can easily be included in the Hartree–Fock equations by setting up two different equations, one for the α electrons and one for the β electrons:

$$\hat{F}^{(\alpha)} = \hat{h} + \hat{J} - \hat{K}^{(\alpha)} \tag{12.31}$$

$$\hat{F}^{(\beta)} = \hat{h} + \hat{J} - \hat{K}^{(\beta)}, \tag{12.32}$$

where the $\hat{J}$ operator couples all the orbitals. Additionally, many molecular systems of interest in chemistry offer a further simplification due to the fact that they are so-called 'closed-shell' systems, i.e., the number of α electrons is equal to the number of β electrons ($N^{(\alpha)} = N^{(\beta)} = N/2$), as is also the case for the systems that appear in Chapters 14 and 16. In this case, adopting the so-called 'restricted' Hartree–Fock closed-shell approach, the two equations (12.31) and (12.32) can be reduced to the single equation:

$$\hat{F} = \hat{h} + 2\hat{J} - \hat{K}\,, \tag{12.33}$$

where the orbitals are now allowed to host two electrons and it is implied that the sums in the $\hat{J}$ and $\hat{K}$ operators run over the $N/2$ doubly occupied orbitals.

12.2 The electronic correlation

As just discussed, the Hartree–Fock model is inherently approximate as it is based on the unrealistic description of a molecular system in terms of uncorrelated electrons. Despite deriving from an approximate model, however, Hartree–Fock calculations are capable of estimating the total energy of a molecular system in a semiquantitative way, as in most cases the Hartree–Fock energy amounts to about 90% of the exact energy. This reflects the fact that the leading terms in the total energy of a molecule are those relating to the kinetic energy of the electrons and nuclear–electron attraction. The Hartree–Fock error with respect to the true total energy of a molecular system is clearly due to the neglect of correlation in the electron motion. Accordingly, the difference between the Hartree–Fock energy and the exact nonrelativistic energy of a molecular system is called the 'correlation energy', and the consequences of such a difference (e.g., changes in other molecular properties) are called 'correlation effects'.

While the Hartree–Fock model yields satisfactory results in the estimate of absolute energies, this unfortunately is of little use in chemistry. As has been partly anticipated in Section 3.3 and will be further elaborated in the following chapters, chemistry is, in fact, a science of differences, and the quantities of interest are often not the energies per se, but rather differences of energies, e.g., between different electronic states or between different nuclear configurations of a given electronic state (including the energy difference between a bound molecule and its unbound constituent fragments). Now, the order magnitude of these energy differences is unfortunately comparable with that of the correlation energy, which makes mandatory the inclusion of electronic correlation in any electronic-structure calculations aiming at an accurate description of chemical properties.

As already mentioned, the Hartree–Fock wavefunction is the best approximation that one can obtain using a single Slater determinant. As a consequence, in order to improve the description of the system, a natural choice is to add greater flexibility to the wavefunction by constructing it as a linear combination of multiple determinants resulting from electronic configurations in which one or more electrons have been promoted from the occupied to the virtual orbitals:

$$\psi = \sum_{\mu} c_{\mu} \Phi_{\mu} \, . \tag{12.34}$$

By introducing Eq. (12.34) into Eq. (12.1) and making use of the variational principle one obtains the following matrix eigenvalue problem

$$\boldsymbol{Hc} = \boldsymbol{Ec} \, , \tag{12.35}$$

where the eigenvectors $\boldsymbol{c}$ contain the expansion coefficients (which also ensure normalization of the wavefunction) and the elements of the $H_{\mu\mu'}$ matrix result from combinations of integrals h_{ij} and V_{ijkl} of the form of Eqs. (12.13) and (12.14). The solution of Eq. (12.35), if the expansion in Eq. (12.34) is extended to all possible determinants, would yield in principle the exact wavefunction and is at the core of the so-called full configuration interaction (FCI) method. While it turns out that $\boldsymbol{H}$ is a rather sparse matrix, it is nonetheless typically a huge one and this makes FCI calculations out of reach except for very small molecules.

Between the two extrema of the Hartree–Fock and FCI approaches, over the past decades several methods have been developed in order to cope with the electronic correlation. We limit ourselves here to a short overview of a few related aspects, while referring the reader to dedicated textbooks such as for instance Ref. [210] and [205] for a more detailed account.

We note first that two kinds of electronic correlation are traditionally recognized. In most cases, the electronic correlation is due to the instantaneous interaction between pairs of moving electrons. In this case, the Hartree–Fock determinant (obtained, as already mentioned, populating the lowest-energy orbitals solving the Hartree–Fock equations) is the leading term in the expansion of Eq. (12.34) and this type of electronic correlation is called a 'dynamical correlation'. In some other cases, the expansion of Eq. (12.34) features more than one leading term, with several determinants having similar weights because of near or exact degeneracy in frontier orbitals. In this second case, the error in the Hartree–Fock calculations does not derive from the fact that the Hartree–Fock model ignores the correlation in the motions of the electrons, but rather from the fact that a single determinant is inappropriate for the description of these systems. This second kind of correlation is called a 'nondynamical correlation'.

An appropriate scheme for treating systems featuring nondynamical correlation is the multiconfiguration–self consistent field, which involves a simultaneous optimization of the orbitals and of the expansion coefficients $\{c_{\mu}\}$ for a given set of determinants. Instead, dynamical correlation has to be addressed either through configuration-interaction methods based on a truncation of the summation in Eq. (12.34), or through perturbative approaches such as Møller–Plesset perturbation theory.

In alternative to the above-mentioned so-called post-Hartree–Fock methods, a powerful route to an efficient inclusion of electronic correlation in electronic-structure calculations is rooted outside of the domain of wavefunction-based methods. It falls instead in the domain of the so-called density-functional theory, which was developed in the 1960s and has over the years gained great importance in chemistry. For these reasons, and due to its contiguity with many aspects of Hartree–Fock calculations that we will address in Chapter 13, the basics of density-functional theory will be discussed in the next section.

12.3   Density-functional theory

According to the approach outlined so far, in order to calculate the physical properties pertaining to the electronic structure of an N molecular system one has to compute the electronic wavefunction $\psi(r_1, r_2, \ldots, r_N)$ and then manipulate it through some mathematical operations to obtain the relevant information. Focusing, for instance, on the energy E, which is one of the most important quantities in chemistry, this can be extracted from the wavefunction using the expression for its average value (as, for instance, in Eq. (12.11)). Now, a law that maps a function into a number (as that of Eq. (12.11)) is said to be a 'functional'. In our case, thus we can say that E is a functional of the wavefunction $\psi\{r_i\}$, and this relation is notated as $E[\psi(r_1, r_2, \ldots, r_N)]$.

On the other hand, as we had occasion to mention several times, the wavefunction is a rather complicated mathematical object, and one may wonder whether the full knowledge of the electronic wavefunction is really necessary to calculate the observable properties of a system or if some simpler mathematical object may suffice. Take, for instance, the more general version of the electronic Hamiltonian of Eq. (12.2):

$$\hat{H} = -\frac{1}{2}\sum_i \nabla_i^2 + V_{\text{ext}} + \sum_{i>j}\frac{1}{r_{ij}}, \tag{12.36}$$

where the first term and the last term on the right-hand side are the usual kinetic-energy and electron–electron repulsion terms, respectively, and V_{ext} is a generic external potential resulting from a summation of one electron operators:

$$V_{\text{ext}} = \sum_{i=1}^{N} v_{\text{ext}}(r_i). \tag{12.37}$$

These could, for instance, be (as is the case of the Hamiltonian of Eq. (12.2) that we have considered so far) the one-electron operators expressing the nuclear-attraction potential $v_{\text{ext}}(r_i) = -\sum_a \frac{Z_a}{|r_i - r_a|}$. Let us now focus on the average value of this external potential, which we know is given by the expression:

$$\langle V_{\text{ext}} \rangle = \sum_{i=1}^{N} \int v_{\text{ext}}(r_i)|\psi(r_1, r_2, \ldots, r_N)|^2 \, dr_1 \, dr_2 \ldots dr_N. \tag{12.38}$$

We notice here that, due to the fact that the electrons are indistinguishable particles, each electron must contribute to the average value in the same way. In

other words, the N terms of the sum in Eq. (12.38) must be equal, and this leads
to:

$$\langle V_{ext} \rangle = N \int v_{ext}(r)|\psi(r, r_2, \ldots, r_N)|^2 \, dr \, dr_2 \ldots dr_N . \qquad (12.39)$$

As the operator is a function only of r the integration on the remaining variables
can be carried out separately. However, this integration is exactly the one that we
have encountered in Section 3.3 of Chapter 3 when we introduced the concept of
electron density through Eq. (3.6). Accordingly, the average value of the energy
associated with the external potential V_{ext} can indeed be computed from the
simple electron density $\rho(r)$ as follows

$$\langle V_{ext} \rangle [\rho(r)] = \int v_{ext}(r)\rho(r) \, dr , \qquad (12.40)$$

where we explicitly notated that the average value for the energy associated with
the external potential V_{ext} is a functional of the electron density $\rho(r)$.

As opposed to the $(3N)$-dimensional wavefunction $\psi(\{r_i\})$, the electron
density $\rho(r)$ is a much simpler, albeit less detailed, three-dimensional function. While the wavefunction, in fact, when taken in square modulus returns the
probability density for finding electron i in r_i, conditional to electron j being
in r_j, electron k in r_k, etc., the electron density is more simply the probability
density for the position of an electron independently of the position of the others, multiplied by the number of electrons N. Accordingly, the electron density
represents the number of electrons per unit volume and its integration over the
whole space returns the number of electrons N:

$$\int \rho(r) \, dr = N . \qquad (12.41)$$

Incidentally, we note here that for a wavefunction expressed by a single
Slater determinant, the electron density (obtained by inserting Eq. (12.10) into
Eq. (3.6)), has a very simple and intuitive expression. It turns out, in fact, to be
made up of $N!$ terms equal in groups of $(N-1)!$, leading to:

$$\rho(r) = N \frac{1}{N!}(N-1)! \sum_{i=1}^{N} \varphi_i^*(r)\varphi_i(r) = \sum_{i=1}^{N} |\varphi_i(r)|^2 . \qquad (12.42)$$

As one might expect, for a system of uncorrelated electrons the electron density
is simply the sum of the probability densities described by each occupied orbital.

Now, as we know, the total energy of a system described by the Hamiltonian of Eq. (12.36) also contains a term associated with the kinetic energy
of the electrons and with the electron–electron repulsion. While the simplifications discussed above for the V_{ext} term do not apply for the kinetic and
electron–electron repulsion terms, it can nonetheless be shown that for calculating the average values of these terms a partial integration over most of the

variables of the wavefunction can similarly be performed, and that the average values can also in these cases be worked out from a function not as simple as the electron density but with considerably lower dimensionality than the full wavefunction. However, and this can be considered the cornerstone of density-functional theory, as demonstrated by Hohenberg and Kohn in 1964 [211], at least for evaluating the properties of the ground state of a system, the knowledge of the sole electron density must actually suffice, and thus the total energy E – and not only its V_{ext} term – must be a functional, $E[\rho(r)]$, of the electron density (hence the name of the theory).

If we consider a system described by a Hamiltonian of the form of Eq. (12.36), we note that the only 'parameters' that determine such a Hamiltonian differentiating one system from another are the number of electrons N (which is implied in the summations in Eq. (12.36)) and the particular form of the external potential $v_{ext}(r)$. Thus N and $v_{ext}(r)$ determine a particular Hamiltonian $\hat{H}$, which in turn determines the ground-state wavefunction $\psi(\{r_i\})$ – and, as a consequence, all the properties of the ground state of the system, including its electron density $\rho(r)$. Now, the starting point of density-functional theory is that actually also the opposite is true, i.e., the electron density $\rho(r)$ univocally determines the external potential $v_{ext}(r)$ (and implicitly also the number of electrons N, which is obtained by simple integration of $\rho(r)$), and thus also all the properties of the ground state of the system. This result, known as the first theorem of Hohenberg and Kohn, can be derived as follows via reductio ad absurdum.

Imagine that we have two systems with equal number of electrons N featuring a different external potential $v_{ext}(r)$ and $v'_{ext}(r)$, which give rise to two different Hamiltonians $\hat{H}$ and $\hat{H}'$, determining in turn two different ground-state wavefunctions ψ and ψ', but leading to the same electron density ρ. If we now use the wavefunction of a system as a trial wavefunction for the other system, and try to evaluate the average value for the energy of the two systems, leveraging on the variational principle we have:

$$E < \langle \psi'|\hat{H}|\psi'\rangle = E' + \langle \psi'|\hat{H} - \hat{H}'|\psi'\rangle$$
$$= E' + \int \rho(r)\left[v_{ext}(r) - v'_{ext}(r)\right] dr \qquad (12.43)$$

and

$$E' < \langle \psi|\hat{H}'|\psi\rangle = E + \langle \psi|\hat{H}' - \hat{H}|\psi\rangle$$
$$= E + \int \rho(r)\left[v'_{ext}(r) - v_{ext}(r)\right] dr, \qquad (12.44)$$

where we also made use of Eq. (12.40). Now, summing the two equations above leads to the absurd $E + E' < E + E'$, which proves that indeed, at least for the electronic ground state of a system, the electron density determines its observable properties and thus its knowledge would in principle suffice to access them.

While this result is certainly remarkable, it nonetheless leaves open two important questions. In light of the formalism reviewed so far, we would not know how to calculate the electron density $\rho(r)$ if not by integration of the wavefunction $\psi\{r_i\}$. Of course, if in order to obtain $\rho(r)$ one has first to calculate $\psi\{r_i\}$ no great simplification has been obtained with respect to a fully wavefunction-based approach. The first question is thus: how can we obtain $\rho(r)$ without passing through the wavefunction? The second question is equally challenging: provided that we determine an expression for $\rho(r)$ that does not involve the calculation of $\psi\{r_i\}$, what would be the expression for the energy functional $E[\rho(r)]$ or, in other words, which is the mathematical expression that would allow us to calculate the energy directly from the electron density?

These certainly are among the main topics that density-functional theory is concerned with. While the interested reader is referred to specialized textbooks such as, for instance, Ref. [212], we limit ourselves here to briefly review a powerful method, due to Kohn and Sham [213], for effectively putting density-functional theory into action.

The Kohn–Sham method builds on the consideration that to the true ground-state electron density of a system $\rho(r)$ we can formally associate a single determinant Φ resulting from a set of orbitals $\{\varphi_i(r)\}$ which, if inserted into Eq. (12.42), would return exactly the same electron density $\rho(r)$. This determinant would certainly differ from the one obtained via the Hartree–Fock equations, as the electron density of the true system of course includes the electronic correlation – which the Hartree–Fock model inherently excludes. In so doing, we have thus shifted the problem of finding the form of the electron density to that of finding the form of the orbitals that would make up a single determinant leading to the exact electron density of our system. This leads to the definition of a Kohn–Sham equation, similar to the Hartree–Fock one:

$$\left[-\frac{1}{2}\nabla^2 + \tilde{v}(r)\right]\varphi_i(r) = \varepsilon_i\varphi_i(r) \qquad (12.45)$$

for a hypothetical system of uncorrelated electrons that has nothing in common with the real one except for the electron density. In other words, the variational problem for the density of the real system is the same as that of a system of independent particles subject to a potential $\tilde{v}(r)$. This potential is given by:

$$\tilde{v}(r) = v_{\text{ext}}(r) + \int \frac{\rho(r')}{|r - r'|}\, dr' + v_{\text{xc}}(r)\,, \qquad (12.46)$$

and includes an external potential (such as, for instance, the one due to nuclear attraction), a Coulomb-repulsion potential, and a so-called 'exchange and correlation' potential $v_{\text{xc}}(r)$, which must now account not only for the exchange interaction (as in the Hartree–Fock equations) but also for the electronic correlation. In terms of the calculated orbitals and density, within the Kohn–Sham scheme the total energy of the system reads

$$E = -\frac{1}{2} \sum_{i=1}^{N} \langle \varphi_i | \nabla^2 | \varphi_i \rangle + \int v_{\text{ext}}(r) \rho(r) \, dr + \frac{1}{2} \int \frac{\rho(r)\rho(r')}{|r - r'|} \, dr \, dr' + E_{\text{xc}}[\rho],$$

$$(12.47)$$

or equivalently, in terms of the orbital energies,

$$E = \sum_{i=1}^{N} \varepsilon_i - \int v_{\text{xc}}(r) \rho(r) \, dr - \frac{1}{2} \int \frac{\rho(r)\rho(r')}{|r - r'|} \, dr \, dr' + E_{\text{xc}}[\rho]. \quad (12.48)$$

In Eqs. (12.47) and (12.48), the term $E_{\text{xc}}[\rho]$ is the so-called exchange and correlation functional, which is in charge of accounting for i) the exchange interaction, ii) the electron correlation, and iii) a correction term due to the fact that the kinetic energy of the electrons has been approximated by using the orbitals $\{\varphi(r_i)\}$.[2]

It is evident that the core problem of density-functional theory is that of finding a suitable and accurate expression for the exchange and correlation potential $v_{\text{xc}}(r)$ and functional $E_{\text{xc}}[\rho]$, and it is on this topic that much of the research in density-functional theory has focused so far. Nowadays, a wide range of exchange and correlation functionals have been derived, most of which result from a compromise between theoretical premises and calibration towards experimental data. It should be thus stressed that, unlike wavefunction-based methods, most of the current implementations of density-functional theory fall outside of the spectrum of the so-called 'ab initio' (exclusively based on first principles) methods and have instead a semiempirical nature.

In spite of that, the Kohn–Sham approach of density-functional theory offers a powerful way to calculate accurate estimates of the electronic energy (including electronic correlation) of a system using a scheme as conceptually simple and computationally cheap as the Hartree–Fock one. In fact, the same practical implementation of the Hartree–Fock model that we will soon illustrate in Chapter 13, can easily be adopted for Kohn–Sham calculations with minor modifications due to the presence of the exchange and correlation functional in lieu of the pure exchange term (this typically involves a numerical evaluation, for example by means of Becke's integration scheme [214], of the relevant integrals that, as described in Section 13.3, can be computed analytically in Hartree–Fock calculations).

---

[2] While the Kohn–Sham formalism has been here presented for the spin-free case, the Kohn–Sham equations can easily be extended to include the spin of the electrons in a similar way to the Hartree–Fock equations.

Chapter 13

From theory to computing: the Hartree–Fock model

13.1 The Roothaan–Hall equations

In Chapter 12, following the Hartree–Fock route, we arrived at a concise formulation of the equations that have to be solved (Eqs. (12.29) and (12.30)) in order to compute the orbitals entering the Slater determinant that approximates the electronic wavefunction of a given molecular system. This chapter is devoted to show how to effectively make those equations work.

The Hartree–Fock equations have no analytic solution. While they can in principle be solved through numerical techniques, a much more efficient method is the algebraic one devised independently by Roothaan [24] and Hall [25] in 1951. This method is based on an expansion of the orbitals in a basis set of analytic functions of known form, and the subsequent transformation of the Hartree–Fock equation in a matrix-diagonalization problem. As we will see, coupled to the use of the Gaussian-type basis functions introduced by Boys [26] in 1950, the Roothaan–Hall method provides a robust and efficient scheme that is currently widely adopted in many electronic-structure program packages.

In the Roothaan–Hall scheme, the orbitals are expressed as a linear combination of suitable analytic functions $\{\chi_p\}$ whose form is known a priori[1]:

$$\varphi_i(\boldsymbol{r}) = \sum_p c_{pi} \chi_p(\boldsymbol{r}). \tag{13.1}$$

In other words, rather than finding the 'shape' of the orbitals, we seek now the set of coefficients that express them within a given set of functions of fixed and known 'shape'. For the expansion of Eq. (13.1) to be effective, the form of the basis functions $\{\chi_p\}$ has to be carefully chosen (this will be discussed in Section 13.3) and enough flexibility has to be granted by expanding the summation to typically one or two orders of magnitude higher than the number of the electrons.

In contrast to the orbitals, the basis functions are not required to be orthonormal. The first evidence of the equivalence between the problem in terms of functions on the one hand, and in terms of vectors and matrices on the other

---

[1] As already mentioned in Section 1.2 of Chapter 1, indices i, j, k, l will be used for orbitals, while indices p, q, r, s will be used for basis functions.

Chemistry at the Frontier with Physics and Computer Science
https://doi.org/10.1016/B978-0-32-390865-8.00023-4

hand, can be seen in the orthonormality condition of the orbitals:

$$\int \varphi_i^*(\boldsymbol{r}) \varphi_j(\boldsymbol{r}) \, d\boldsymbol{r} = \int \left(\sum_p c_{pi}^* \chi_p^*(\boldsymbol{r}) \right) \left(\sum_q c_{qj} \chi_q(\boldsymbol{r}) \right) d\boldsymbol{r}$$

$$= \sum_{p,q} c_{pi}^* S_{pq} c_{qj} = \delta_{ij} , \tag{13.2}$$

where we introduced the elements of the 'overlap matrix' between the basis functions:

$$S_{pq} = \langle \chi_p | \chi_q \rangle = \int \chi_p^*(\boldsymbol{r}) \chi_q(\boldsymbol{r}) \, d\boldsymbol{r} . \tag{13.3}$$

The orthonormality condition for the orbitals Eq. (13.2) can in fact be expressed in matrix form as:

$$\boldsymbol{C}^\dagger \boldsymbol{S} \boldsymbol{C} = \boldsymbol{1} , \tag{13.4}$$

where the $\boldsymbol{C}$ matrix gathers in each column $\boldsymbol{c}_i$ the expansion coefficients of the ith orbital. Within this scheme, there is thus an equivalence, upon definition of a given basis set, between the orbital $\varphi_i(\boldsymbol{r})$ and the vector $\boldsymbol{c}_i$.

Analogously, the introduction of the expansion (13.1) in the Hartree–Fock equation (12.29)

$$\hat{F} \left(\sum_p c_{pi} \chi_p(\boldsymbol{r}) \right) = \varepsilon_i \left(\sum_p c_{pi} \chi_p(\boldsymbol{r}) \right) \tag{13.5}$$

leads to the algebraic equivalent:

$$\boldsymbol{F} \boldsymbol{c}_i = \varepsilon_i \boldsymbol{S} \boldsymbol{c}_i \tag{13.6}$$

$$\boldsymbol{F} \boldsymbol{C} = \boldsymbol{S} \boldsymbol{C} \boldsymbol{\varepsilon} , \tag{13.7}$$

i.e., a 'generalized eigenvalue problem' for a matrix $\boldsymbol{F}$ – the 'Fock matrix' – whose elements over the basis functions are defined as follows:

$$F_{pq} = \langle \chi_p | \hat{F} | \chi_q \rangle = \int \chi_p^*(\boldsymbol{r}) \hat{F} \chi_q(\boldsymbol{r}) \, d\boldsymbol{r} . \tag{13.8}$$

In Eq. (13.7), the matrix $\boldsymbol{\varepsilon}$ is a diagonal matrix whose diagonal elements are the orbital energies $\{\varepsilon_i\}$.

In other words, we need to find the particular orbital coefficients that make $\boldsymbol{C}$ diagonalize $\boldsymbol{F}$:

$$\boldsymbol{C}^\dagger \boldsymbol{F} \boldsymbol{C} = \boldsymbol{\varepsilon} \tag{13.9}$$

subject to the orthonormality condition of Eq. (13.4). Matrix diagonalization is a well-known algebra problem that can be solved through standard techniques using the procedures available in efficient computational libraries such as the already-mentioned (Section 8.3, Chapter 8) LAPACK (http://www.netlib.org/lapack/).

As in the analogous eigenvalue problem of Eq. (12.29), solving Eq. (13.7) means finding both the column vectors $\{c_i\}$ making up the C matrix – so-called eigenvectors, expressing the orbitals – and the eigenvalues $\{\varepsilon_i\}$. Thus provided that we are able to compute the overlap matrix elements between the chosen basis functions (Eq. (13.3)), all we need to do is set up the Fock matrix related to a given molecular system. As for the Fock operator, its matrix representation also can be split into terms relating to the one-electron part, $\hat{h}$, of $\hat{F}$, and to the Coulomb, $\hat{J}$, and exchange, $\hat{K}$ part:

$$F_{pq} = h_{pq} + J_{pq} - K_{pq}, \qquad (13.10)$$

where h_{pq}, J_{pq}, and K_{pq} are given by, respectively:

$$h_{pq} = \int \chi_p^*(r) \left[-\frac{1}{2}\nabla^2 - \sum_a \frac{Z_a}{|r - r_a|} \right] \chi_q(r)\, dr \qquad (13.11)$$

$$J_{pq} = \sum_i V_{piqi} = \sum_i \sum_{r,s} c_{ri}^* c_{si} V_{prqs} = \sum_{r,s} D_{sr} V_{prqs} \qquad (13.12)$$

$$K_{pq} = \sum_i V_{piiq} = \sum_i \sum_{r,s} c_{ri}^* c_{si} V_{prsq} = \sum_{r,s} D_{sr} V_{prsq}, \qquad (13.13)$$

where the sums on i run over the occupied orbitals. In Eqs. (13.12) and (13.13), the four-index elements V_{pqrs} are integrals of the form of Eq. (12.14)

$$V_{pqrs} = \int \chi_p^*(r)\chi_q^*(r') \frac{1}{|r - r'|} \chi_r(r)\chi_s(r')\, dr\, dr' \qquad (13.14)$$

now evaluated over the basis functions rather than the orbitals, and the co-efficients of the occupied orbitals can be gathered in the so-called 'density matrix' D:

$$D = \sum_i c_i c_i^\dagger = C_{occ} C_{occ}^\dagger, \qquad (13.15)$$

with C_{occ} being the rectangular block of C made up by the vectors of the occupied orbitals. The electron density, in terms of the density matrix and the basis functions, reads:

$$\rho(r) = \sum_i |\varphi_i(r)|^2 = \sum_{r,s} D_{sr} \chi_r^* \chi_s. \qquad (13.16)$$

The Fock matrix can be simply obtained by the h, J and K matrices:

$$F = h + J(D) - K(D), \qquad (13.17)$$

where the dependence of J and K on the density matrix D – which is the algebraic counterpart of the dependence of the Coulomb and exchange operators upon the occupied orbitals – has been made explicit. In terms of matrix element over the basis functions, the electronic energy reads

$$E = \frac{1}{2}\sum_i (h_{ii} + F_{ii}) = \frac{1}{2}\sum_i \sum_{p,q} c_{pi}^* c_{qi} (h_{pq} + F_{pq})$$

$$= \frac{1}{2}\sum_{p,q} D_{qp}(h_{pq} + F_{pq}) = \frac{1}{2}\mathrm{Tr}\,\mathbf{D}(\mathbf{h} + \mathbf{F}). \tag{13.18}$$

In the closed-shell case, for which some illustrative calculations are reported in Section 13.3, the matrix counterpart of the Fock operator of Eq. (12.33) is:

$$\mathbf{F} = \mathbf{h} + 2\mathbf{J}(\mathbf{D}) - \mathbf{K}(\mathbf{D}), \tag{13.19}$$

where it is implied that the density matrix results from the $N/2$ doubly occupied orbitals. The total energy is thus:

$$E = \sum_i (h_{ii} + F_{ii}) = \sum_i \sum_{p,q} c_{pi}^* c_{qi} (h_{pq} + F_{pq})$$

$$= \sum_{p,q} D_{qp}(h_{pq} + F_{pq}) = \mathrm{Tr}\,\mathbf{D}(\mathbf{h} + \mathbf{F}), \tag{13.20}$$

where the sum on i runs on the doubly occupied orbitals. The electron density can be computed from the density matrix as in Eq. (13.16) (with the sum on i now running over the doubly occupied orbitals) multiplied by an extra factor 2 accounting for the double occupation of the orbitals and ensuring the correct electron count.

Finally, it is worth stressing here that the above-outlined procedure can be entirely transposed to the Kohn–Sham approach of density-functional theory, after replacing the exchange matrix with an exchange-correlation matrix and adapting the expressions for the energy to Eq. (12.48) or its closed-shell counterpart (some of these aspects will be further elaborated in Chapter 17 when discussing relativistic density-functional theory calculations).

13.2 Self-consistent field procedure

As already mentioned, once the Hartree–Fock equation has been cast in matrix form, standard and efficient matrix-diagonalization techniques can be used to solve it. At this stage, we are thus left with two issues. The first issue is that, as already commented in Section 12.1, the Fock operator depends on the orbitals (or, equivalently, the Fock matrix depends on the orbital expansion coefficients through the density matrix), which in turn are the solution of the eigenvalue equation for the same operator, thus the solution of the Hartree–Fock problem must be sought through an iterative scheme called the SCF. The second issue is that we need to choose an appropriate basis set for the expansion of the orbitals.

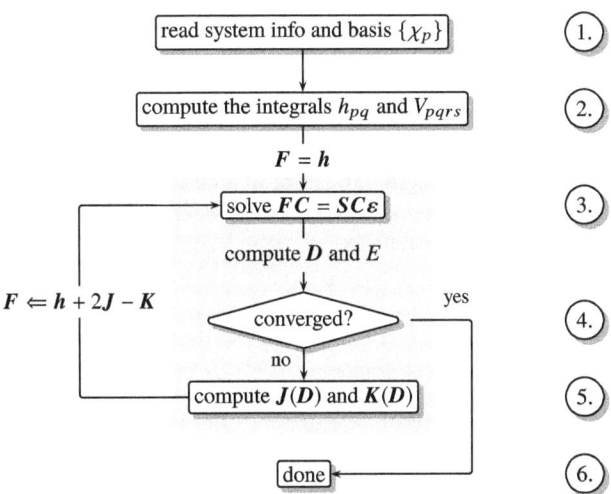

FIGURE 13.1   Flowchart for a SCF calculation. The iterative section is highlighted with a light blue (gray in print version) frame.

We will address this second issue in Section 13.3, while we will focus here on the iterative procedure.

As already discussed in Section 12.1, in a SCF calculation one has to start with a guess for the orbitals – or, in the Roothaan–Hall scheme, for the density matrix. A practical way of obtaining a reasonable initial density matrix is to solve the Roothaan–Hall equation (13.7) for an operator consisting of the sole one-electron part of $\hat{F}$, i.e., discarding the D-dependent part. This would correspond to a system of noninteracting electrons subject to an external potential due to the attraction exerted by the nuclei of the molecular system, and thus the associated density would approximately share with the true one at least its localization features.

Once a guess for the density matrix has been generated, one can start iterating the SCF procedure using the density matrix to construct a new Fock matrix through its Coulomb and exchange terms, and use this new Fock matrix to obtain a new set of orbitals and the related density matrix. The iterative procedure is stopped when convergence has been reached, e.g., the density matrix elements or the electronic energy differ from the previous cycle by less than a given threshold (see Ref. [215] for a review on the topic of convergence in SCF calculations). For the Hartree–Fock calculations reported in Section 13.3, a convergence criterion of 1×10^{-8} E_H on the electronic energy was used, coupled to a 'level-shifting' technique [216] to speed up the convergence.

Accordingly, the flowchart for a SCF restricted Hartree–Fock closed-shell calculation using the Roothaan–Hall method can be summarized as in Fig. 13.1. First, the system information (atomic numbers, molecular geometry, spin multiplicity, number of electrons) and the parameters defining the basis functions are read (step 1). Then the one-electron, h_{pq}, and two-electron, V_{pqrs} integrals over the basis functions are preliminarily calculated (step 2). At this point, the Fock matrix is set equal to the one-electron matrix h and one enters the iterative section (light blue (gray in print version) frame). The eigenvalue problem is solved (step 3) to obtain a first version of the density matrix. The convergence is checked in step 4 (at the first iteration the convergence will be checked against some 'previous' values purposely initialized so as to make the convergence check fail) and, if not reached, the density matrix is used to compute the two-electron part of the Fock matrix (step 5). A new Fock matrix is thus defined and the cycle is entered again at step 3 and iterated until the convergence criteria are met.

13.3 Basis functions and one- and two-electron integrals

At this point, the last aspect that needs to be addressed is the choice of an appropriate set of basis functions. As already mentioned, a number of basis functions much higher than the number of the electrons has to be used in order to grant enough flexibility to the expansion of the orbitals, and quite generally the larger the basis set the more accurate the results will be. On the other hand, the size of the Fock and overlap matrices involved in Hartree–Fock (or Kohn–Sham) SCF calculations scales with the second power of the number of basis functions, and the computational cost for the calculation of the two-electron integrals scales with the fourth power. It is thus important to choose an analytic form for the basis function that would grant the greatest accuracy while keeping as low as possible the computational requirements. The basis functions have thus to satisfy two criteria. They have to i) be adequate towards the function (the electronic orbitals) that they are called to represent, and ii) allow for an easy calculation of the one- and two- electron integrals h_{pq} and V_{pqrs}.

In trying to answer the question raised by the first of these criteria, i.e., which form to give to the basis functions, a starting point may be the consideration that the electron density of a molecular system may be reasonably thought of as a deformation of the (atomic) electron densities of the atoms that constitute the system. An appropriate choice may thus be to use basis functions whose form is related to the form of the atomic orbitals, and that are localized around the involved atoms. These functions, also referred to as atomic (as opposed to molecular) orbitals, are the so-called Slater-type orbitals (STO), which are given the expression:

$$\chi_{p \leftarrow \{n,l,m,\alpha,R_p\}}(r) = N_{\alpha,n} |r - R_p|^{n-1} e^{-\alpha|r-R_p|} Y_{lm}(\theta, \varphi), \qquad (13.21)$$

where n, l, and m are integer numbers, $N_{\alpha,n}$ is a normalization factor, $R_p = (X_p, Y_p, Z_p)$ centers the function in a given space point, $Y_{lm}(\theta, \varphi)$ is a spherical harmonic, and polar coordinates r, θ, φ have been used rather than the Cartesian x, y, z. A Slater-type orbital has thus long-range exponential decay and features a cusp at its origin, while its angular behavior is dictated by the l

and m integers that characterize the spherical harmonic $Y_{lm}(\theta, \varphi)$. The center of each basis function is typically chosen so as to coincide with the position of one of the nuclei of the system, $\boldsymbol{R}_p \equiv \boldsymbol{r}_a$.

However, and here we come to issues related to the second of the above-mentioned criteria, the calculation of the integrals – especially the two-electron integrals – over Slater-type orbitals is not straightforward. A great simplification in the calculation of these integrals can instead be achieved if use is made of Gaussian-type orbitals, where the radial part is represented by a Gaussian function and the angular part is expressed in Cartesian coordinates. The resulting functions, so-called Cartesian Gaussian-type orbitals (CGTO), are thus given the expression:

$$\chi_{p \leftarrow \{l,m,n,\alpha,\boldsymbol{R}_p\}}(\boldsymbol{r}) = N_p (x - X_p)^l (y - Y_p)^m (z - Z_p)^n e^{-\alpha |\boldsymbol{r} - \boldsymbol{R}_p|^2}, \quad (13.22)$$

where l, m, and n are integer numbers, and the normalization constant N_p

$$N_p = \frac{(4\alpha)^{(l+m+n)1/2}(2\alpha)^{3/4}}{[(2l - 1)!!(2m - 1)!!(2n - 1)!!]^{1/2}\pi^{3/4}} \quad (13.23)$$

ensures that

$$\int \chi_p^2(\boldsymbol{r}) \, d\boldsymbol{r} = 1. \quad (13.24)$$

The Gaussian exponent α dictates the behavior of the radial part and can have a wide range of values (from > 1000 for core functions to $\leq 10^{-3}$ for diffuse functions). Functions with the same value of $L = l + m + n$ are said to belong to a shell (s shell for $L = 0$, p shell for $L = 1$, d shell for $L = 2$, f shell for $L = 3$, and so on). Accordingly, each shell features $(L - 1)(L - 2)/2$ CGTOs. A graphical representation through so-called direct volume rendering (displaying a three-dimensional function as a cloud) for the complete set of χ_{lmn} of the s, p, d, and f shells with $\alpha = 1.0$ is given in Fig. 13.2. Blue (dark gray in print version) color is associated with positive values of the function, and red (gray in print version) color with negative values.

Besides being localized functions similarly to the Slater-type orbitals, the great advantage of Guassian-type orbitals is that the product of two Gaussian functions is another Gaussian function and this leads to a great simplification in the calculation of two electron integrals involving products of Gaussian functions. On the other hand, the radial behavior of atomic orbitals is better reproduced by Slater-type orbitals, as the Guassian-type orbitals of course miss the exponential decay and the cusp condition at the origin. In order to partly recover the radial behavior of a Slater-type orbital, use can be made of so-called 'contracted' CGTOs, which are linear combinations of (typically few) 'primitive' CGTOs (Eq. (13.22)) with the coefficients chosen so as to mimic the radial

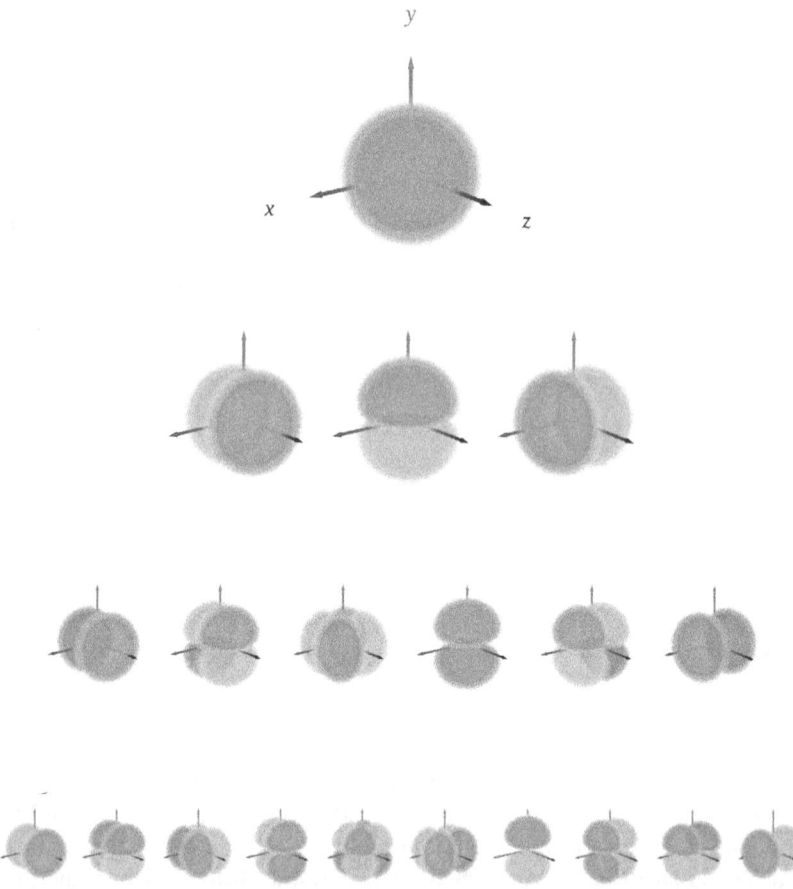

FIGURE 13.2   Graphical representation of the complete set of χ_{lmn} with $\alpha = 1.0$ for shells s (first row), p (second row), d (third row), and f (fourth row) listed in the so-called canonical order from left to right. The triples (lmn) are: first row: (000); second row: (100), (010), (001); third row: (200), (110), (101), (020), (011), (002); fourth row: (300), (210), (201), (120), (111), (102), (030), (021), (012), (003).

behavior of a Slater-type orbital. The contracted CGTO thus have the form:

$$\tilde{\chi}_{p \leftarrow \{l,m,n,\{\alpha_k\},\{c_k\}, R_p\}}(\boldsymbol{r}) = \tilde{N}_p (x - X_p)^l (y - Y_p)^m (z - Z_p)^n \sum_k c_k e^{-\alpha_k |\boldsymbol{r} - \boldsymbol{R}_p|^2}.$$

$$(13.25)$$

The introduction of contracted CGTOs completes now the hierarchy of functions ψ (electronic wavefunction), Φ_μ (Slater determinant), φ_i (orbital), $\tilde{\chi}_p$ (contracted basis function), χ_p (primitive basis function) that will serve us in the modeling of the electronic structure of molecular systems.

The one- and two-electron integrals over contracted basis functions are straightforwardly computed as combinations of the integrals over the relevant primitive basis functions.[2] The analytic expression for the overlap, kinetic-energy, nuclear-attraction, and electronic-repulsion integrals are given at the end of this section (the reader interested in their derivation is referred to Refs. [217] and [218]). It is worth adding here that the calculation of the integrals can be made much more efficient by exploiting recursive schemes such as the Obara–Saika [219] or the McMurchie–Davidson [220] schemes, which are routinely employed in most electronic-structure program packages.

As already mentioned, one typically makes use of several contracted basis functions for each electron, centering the basis function on the related atom. Over the years, a wide range of basis sets have been developed. The parameters determining the radial behavior of the basis functions, $\{\alpha_k\}$, together with the collection of the contraction coefficients, $\{c_k\}$, for these predetermined basis sets are made available through the Basis Set Exchange [221,222] portal (https://www.basissetexchange.org/). Once a basis set has been chosen and coupled to a particular electronic structure method (e.g., Hartree–Fock, multiconfiguration SCF, full configuration interaction (FCI), or, on another front, the particular exchange-correlation functional used for density-functional theory calculations), this makes a particular so-called 'model chemistry'.

It will be useful at this stage to add a note on the practical implementation of a Hartree–Fock program that anticipates a concept – that of 'object-based' programming – that will be further elaborated in Chapter 15. Besides the straightforward implementation of an iterative procedure such as that sketched in Fig. 13.1, the writing of a Hartree–Fock computer program involves the development of procedures for handling the basis functions and computing the one- and two-electron integrals. In an object-oriented programming paradigm (more on this will be given in Chapter 15) within the capabilities of modern Fortran, this could be efficiently achieved by setting up a module for the primitive CGTOs where a derived datatype with the following structure is defined:

```
TYPE, PUBLIC :: cgto
  PRIVATE
  INTEGER, DIMENSION(3) :: l
  REAL (KIND=wp), DIMENSION(3) :: r0
  REAL (KIND=wp) :: alpha, normfac
END TYPE cgto
```

where l is a vector of three integers storing the value of l, m, and n, r0 is a vector of three real numbers storing the values of X_p, Y_p, and Z_p, alpha and normfac are the Gaussian exponent α and the normalization factor N_p, respectively, and KIND=wp determines the numerical precision that will be adopted (the working precision, wp, will be typically defined elsewhere to be 'double precision'). The module would then contain procedures (functions or subroutines) to operate on

---

[2] The integrals over the Cartesian basis functions are then typically transformed into integrals over the related spherical functions to further reduce the dimension of the matrices, though this feature has not been used in the Hartree–Fock calculations reported later in this section.

TABLE 13.1 Input molecular geometry of NH_4^+ used in the Hartree–Fock calculations. Calculations on NH_3 were performed by removing the last H atom in the table (see Section 14.3 of Chapter 14 for further discussion).

	x/a_0	y/a_0	z/a_0
N	0.0000	0.0000	0.0000
H	0.0000	-1.8056	-0.6384
H	1.5637	0.9028	-0.6384
H	-1.5637	0.9028	-0.6384
H	0.0000	0.0000	1.9151

the CGTOs (e.g., setting the values of their attributes, computing overlap integrals between two of them or other one- and two-electron integrals) that would be made accessible to the main Hartree–Fock program, where an array of N_{bs} CGTOs would have been declared and used for the construction of the matrices of the integrals. A second module could then be analogously setup for the contracted basis functions, where the following derived datatype is defined:

```
TYPE, PUBLIC :: xgto
  PRIVATE
  REAL (KIND=wp), DIMENSION(:), POINTER :: c
  TYPE (cgto), DIMENSION(:), POINTER :: g
END TYPE xgto
```

Here the array c is the vector of the contraction coefficients and the array g is the vector of the primitive CGTOs. The calculations reported in the following have been performed with a similarly conceived object-based computer program, Waverley, written by the author and calculating the one- and two-electron integrals according to the formulae given at the end of this section.

For illustrative purposes, some details on the Hartree–Fock calculations performed to compute the electron density of systems NH_4^+ and NH_3 that will be discussed in Chapter 14 are given in the following. The molecular geometry used for the calculation on NH_4^+ is that reported in Table 13.1. For NH_3, calculations were performed using the same geometry after removing the H in the last line of Table 13.1 (see discussion in Chapter 14 for the reasons for such a choice). The so-called 6-31G* basis set was adopted for the H and the N atoms [223–225]. The related parameters, that as already mentioned can be retrieved through the Basis Set Exchange portal available at https://www.basissetexchange.org/, are listed in Tables 13.2 and 13.3. Accordingly, two contracted s-type functions were used for each H atom, one resulting from three primitive CGTOs and the other coinciding with a single primitive CGTO. For

TABLE 13.2 Parameters of the 6-31G* basis functions for atom H.

	α_k	c_k
s	0.1873113696E+02	0.3349460434E-01
	0.2825394365E+01	0.2347269535E+00
	0.6401216923E+00	0.8137573261E+00
s	0.1612777588E+00	1.0000000

TABLE 13.3 Parameters of the 6-31G* basis functions for atom N.

	α_k	c_k
s	0.4173511460E+04	0.1834772160E-02
	0.6274579110E+03	0.1399462700E-01
	0.1429020930E+03	0.6858655181E-01
	0.4023432930E+02	0.2322408730E+00
	0.1282021290E+02	0.4690699481E+00
	0.4390437010E+01	0.3604551991E+00
s	0.1162636186E+02	-0.1149611817E+00
	0.2716279807E+01	-0.1691174786E+00
	0.7722183966E+00	0.1145851947E+01
p	0.1162636186E+02	0.6757974388E-01
	0.2716279807E+01	0.3239072959E+00
	0.7722183966E+00	0.7408951398E+00
s	0.2120314975E+00	0.1000000000E+01
p	0.2120314975E+00	0.1000000000E+01
d	0.8000000000E+00	1.0000000

the N atom, instead, three contracted s-type functions were used alongside six contracted p-type functions and six d-type functions (note that all three possible $L = 1$ p and all six possible $L = 2$ d functions have to be included in the count).

For the NH_4^+ (10 electrons, 5 doubly occupied orbitals) this leads to a number of 44 CGTOs and 23 contracted CGTOs. The number of one-electron integrals over primitive CGTOs to be computed (which is lower than N_{bs}^2 for reasons of symmetry in the matrix elements) amounts to $N_{bs}(N_{bs} + 1)/2 = 990$. The number of two-electron integrals amounts to $(N_{bs}(N_{bs} + 1)/2)[(N_{bs}(N_{bs} + 1)/2) + 1]/2 = 490\,545$. Over the contracted CGTOs, instead, the one-electron integrals are 276, and the two-electron integrals 38 226. The memory required to store the full four-index matrix of the two-electron integrals amounts to $N_{bs}^4 \times 8/(1024 \times 1024) = 2.14$ MiB. The calculated Hartree–Fock energy is $E = -56.530771\ E_H$.

For NH_3 (10 electrons, 5 doubly occupied orbitals), one has 40 CGTOs and 21 contracted CGTOs. The one- and two-electron integrals over the primitive CGTOs are 820 and 336 610, respectively. Those over the contracted CGTOs are 231 and 26 796, respectively. The memory required to store the two-electron integrals amounts to 1.48 MiB. The calculated Hartree–Fock energy is $E =$ -56.183533 E_H.

Overlap integrals

The overlap integral between two CGTOs χ_1 and χ_2 is:

$$\langle \chi_1 | \chi_2 \rangle = \int \chi_{l_1, m_1, n_1, \alpha_1, R_1}(r) \chi_{l_2, m_2, n_2, \alpha_2, R_2}(r) \, dr \ . \tag{13.26}$$

The analytic expression of this integral reads:

$$\langle \chi_1 | \chi_2 \rangle = N_1 N_2 \left(\frac{\pi}{\gamma_p} \right)^{3/2} e^{\eta_p (R_1 - R_2)^2} \sum_{i_1, i_2, o} S_x \sum_{j_1, j_2, p} S_y \sum_{k_1, k_2, q} S_z \ , \tag{13.27}$$

where

$$\gamma_p = \alpha_1 + \alpha_2 \qquad \eta_p = \frac{\alpha_1 \alpha_2}{\gamma_p} \ , \tag{13.28}$$

and

$$\sum_{i_1, i_2, o} S_x = \frac{(-1)^{l_1} l_1! l_2!}{\gamma_p^{l_1 + l_2}} \sum_{i_1 = 0}^{[\frac{l_1}{2}]} \sum_{i_2 = 0}^{[\frac{l_2}{2}]} \sum_{o=0}^{[\frac{\Omega}{2}]} \frac{(-1)^o \Omega! \alpha_1^{l_2 - i_1 - 2i_2 - o} \alpha_2^{l_1 - 2i_1 - i_2 - o}}{4^{i_1 + i_2 + o} i_1! i_2! o!}$$
$$\times \frac{\gamma_p^{2(i_1 + i_2) + o} (X_1 - X_2)^{\Omega - 2o}}{(l_1 - 2i_1)! (l_2 - 2i_2)! (\Omega - 2o)!} \ , \tag{13.29}$$

with $\Omega = l_1 + l_2 - 2(i_1 + i_2)$. The S_y and S_z are similarly defined in terms of the y- and z-components. Note that in the summation notation $[k]$ denotes the largest integer $\leq k$.

Kinetic-energy integrals

The analytic expression for the kinetic-energy integral is a combination of overlap integrals:

$$\langle \chi_1 | -\frac{1}{2} \nabla^2 | \chi_2 \rangle = \frac{1}{2} N_1 N_2 \Big\{ \alpha_2 [4(l_2 + m_2 + n_2) + 6] \langle \chi_1 | \chi_2 \rangle$$
$$- 4\alpha_2^2 \Big[\langle \chi_1 | \chi_2, l_2 + 2 \rangle + \langle \chi_1 | \chi_2, m_2 + 2 \rangle + \langle \chi_1 | \chi_2, n_2 + 2 \rangle \Big]$$

$$- l_2(l_2 - 1)\langle \chi_1 | \chi_2, l_2 - 2 \rangle - m_2(m_2 - 1)\langle \chi_1 | \chi_2, m_2 - 2 \rangle$$
$$- n_2(n_2 - 1)\langle \chi_1 | \chi_2, n_2 - 2 \rangle \Big\} , \tag{13.30}$$

where the overlap is performed with unnormalized functions, and

$$|\chi_2, l_2 + 2\rangle = (x - X_2)^{l_2+2}(y - Y_2)^{m_2}(z - Z_2)^{n_2} e^{-\alpha_2(r - R_2)^2} , \tag{13.31}$$

and similarly for the other terms.

Nuclear-attraction integrals

The nuclear-attraction matrix elements are defined as follows:

$$\sum_a \langle \chi_1 | - \frac{Z_a}{r_a} | \chi_2 \rangle = - \sum_a Z_a \int \frac{\chi_1(r)\chi_2(r)}{|r - r_a|} \, dr . \tag{13.32}$$

The analytic expression for the ath nuclear attraction integral is:

$$\langle \chi_1 | - \frac{Z_a}{r_a} | \chi_2 \rangle = - \frac{Z_a N_1 N_2 \pi}{\gamma_p} e^{-\eta_p(R_1 - R_2)^2}$$
$$\times \sum_{\substack{i_1,i_2 \\ o_1,o_2 \\ r,u}} \mathcal{A}_x \sum_{\substack{j_1,j_2 \\ p_1,p_2 \\ s,v}} \mathcal{A}_y \sum_{\substack{k_1,k_2 \\ q_1,q_2 \\ t,w}} \mathcal{A}_z \, 2 F_v(\gamma_p(P - r_a)^2) , \tag{13.33}$$

where

$$\gamma_p = \alpha_1 + \alpha_2 \qquad \eta_p = \frac{\alpha_1 \alpha_2}{\gamma_p} \qquad P = \frac{1}{\gamma_p}(\alpha_1 R_1 + \alpha_2 R_2) , \tag{13.34}$$

$$\sum_{\substack{i_1,i_2 \\ o_1,o_2 \\ r,u}} \mathcal{A}_x = (-1)^{l_1+l_2} l_1! l_2! \sum_{i_1=0}^{[l_1/2]} \sum_{i_2=0}^{[l_2/2]} \sum_{o_1=0}^{l_1-2i_1} \sum_{o_2=0}^{l_2-2i_2} \sum_{r=0}^{[(o_1+o_2)/2]} \frac{(-1)^{o_2+r}(o_1 + o_2)!}{4^{i_1+i_2+r} i_1! i_2! o_1! o_2! r!}$$

$$\times \frac{\alpha_1^{o_2-i_1-r} \alpha_2^{o_1-i_2-r} (X_a - X_b)^{o_1+o_2-2r}}{(l_1 - 2i_1 - o_1)!(l_2 - 2i_2 - o_2)!(o_1 + o_2 - 2r)!}$$

$$\times \sum_{u=0}^{[\mu_x/2]} \frac{(-1)^u \mu_x!(X_P - x_a)^{\mu_x-2u}}{4^u u!(\mu_x - 2u)! \gamma_p^{o_1+o_2-r+u}} , \tag{13.35}$$

and

$$\mu_x = l_1 + l_2 - 2(i_1 + i_2) - (o_1 + o_2) \qquad v = \mu_x + \mu_y + \mu_z - (u + v + w) . \tag{13.36}$$

$\mathcal{A}_y$ and $\mathcal{A}_z$ are similarly defined in terms of the y- and z-components. The function $F_v(\gamma_p(P - r_a)^2)$ is evaluated as

$$F_\nu(u) = \int_0^1 dt^{2\nu} e^{-ut^2}$$

$$= \frac{(2\nu)!}{2\,\nu!} \left[\frac{\sqrt{\pi}}{4^\nu u^{\nu+1/2}} \operatorname{erf}\sqrt{u} - e^{-u} \sum_{k=0}^{\nu-1} \frac{(\nu-k)!}{4^k (2\nu - 2k)! u^{k+1}} \right]. \tag{13.37}$$

Electronic-repulsion integrals

The electronic-repulsion integrals are defined as:

$$\langle \chi_1 \chi_3 | \frac{1}{r} | \chi_2 \chi_4 \rangle = \int \frac{\chi_1(r)\chi_2(r)\chi_3(r')\chi_4(r')}{|r - r'|} \, dr\, dr'. \tag{13.38}$$

The analytic expression for the electronic-repulsion integral is:

$$\langle \chi_1 \chi_3 | \frac{1}{r} | \chi_2 \chi_4 \rangle = \frac{N_1 N_2 N_3 N_4 \pi^{5/2}}{\gamma_p \gamma_q \sqrt{\gamma_p + \gamma_q}} e^{-\eta_p (R_1 - R_2)^2} e^{-\eta_q (R_3 - R_4)^2}$$

$$\times \sum_{\substack{i_1,i_2,i_3,i_4 \\ o_1,o_2,o_3,o_4 \\ r_1,r_2,u}} \mathcal{J}_x \sum_{\substack{j_1,j_2,j_3,j_4 \\ p_1,p_2,p_3,p_4 \\ s_1,s_2,v}} \mathcal{J}_y \sum_{\substack{k_1,k_2,k_4,k_4 \\ q_1,q_2,q_3,q_4 \\ t_1,t_2,w}} \mathcal{J}_z \, 2F_\nu(\eta(P - Q)^2),$$

$$\tag{13.39}$$

where

$$\gamma_p = \alpha_1 + \alpha_2 \qquad \eta_p = \frac{\alpha_1 + \alpha_2}{\gamma_p} \qquad P = \frac{1}{\gamma_p}(\alpha_1 R_1 + \alpha_2 R_2), \tag{13.40}$$

$$\gamma_q = \alpha_3 + \alpha_4 \qquad \eta_q = \frac{\alpha_3 + \alpha_4}{\gamma_q} \qquad Q = \frac{1}{\gamma_q}(\alpha_3 R_3 + \alpha_4 R_4), \tag{13.41}$$

$$\eta = \frac{\gamma_p \gamma_q}{\gamma_p + \gamma_q}, \tag{13.42}$$

$$\sum_{\substack{i_1,i_2,i_3,i_4 \\ o_1,o_2,o_3,o_4 \\ r_1,r_2,u}} \mathcal{J}_x = (-1)^{l_1+l_2} \frac{l_1! l_2!}{\gamma_p^{l_1+l_2}}$$

$$\times \sum_{i_1=0}^{[l_1/2]} \sum_{i_2=0}^{[l_2/2]} \sum_{o_1=0}^{l_1-2i_1} \sum_{o_2=0}^{l_2-2i_2} \sum_{r_1=0}^{[(o_1+o_2)/2]} \frac{(-1)^{o_2+r_1}(o_1+o_2)!}{4^{i_1+i_2+r_1} i_1! i_2! o_1! o_2! r_1!}$$

$$\times \frac{\alpha_1^{o_2-i_1-r_1} \alpha_2^{o_1-i_2-r_1} \gamma_p^{2(i_1+i_2)+r_1}(X_a - X_b)^{o_1+o_2-2r_1}}{(l_1 - 2i_1 - o_1)!(l_2 - 2i_2 - o_2)!(o_1 + o_2 - 2r_1)!}$$

$$\times \frac{l_3! l_4!}{\gamma_q^{l_3+l_4}} \sum_{i_3=0}^{[l_3/2]} \sum_{i_4=0}^{[l_4/2]} \sum_{o_3=0}^{l_3-2i_3} \sum_{o_4=0}^{l_4-2i_4} \sum_{r_2=0}^{[(o_3+o_4)/2]} \frac{(-1)^{o_3+r_2}(o_3+o_4)!}{4^{i_3+i_4+r_2} i_3! i_4! o_3! o_4! r_2!}$$

$$\times \frac{\alpha_3^{o_4-i_3-r_2}\alpha_4^{o_3-i_4-r_2}\gamma_p^{2(i_3+i_4)+r_2}(X_c-X_d)^{o_3+o_4-2r_2}}{(l_3-2i_3-o_3)!(l_4-2i_4-o_4)!(o_3+o_4-2r_2)!}$$

$$\times \sum_{u=0}^{[\mu_x/2]} \frac{(-1)^u \mu_x! \eta^{\mu_x-u}(X_P-X_Q)^{\mu_x-2u}}{4^u u!(\mu_x-2u)!} , \tag{13.43}$$

and

$$\mu_x = l_1+l_2+l_3+l_4 - 2(i_1+i_2+i_3+i_4) - (o_1+o_2+o_3+o_4) \tag{13.44}$$

$$v = \mu_x+\mu_y+\mu_z - (u+v+w) . \tag{13.45}$$

$\mathcal{J}_y$ and $\mathcal{J}_z$ are similarly defined in terms of the y- and z-components.

Chapter 14

The atom and the bond

In Chapters 12 and 13 we have been concerned with the quantum-mechanical description of a molecular system. Within the Born–Oppenheimer approximation, a molecular system at fixed geometry is equated to its electronic wavefunction describing from a quantum-mechanical point of view a system of interacting electrons in a field generated by a set of nuclei fixed in space, and whose calculation involves a rather complex mathematical formalism and a great deal of computational resources. In that formalism, we must take note that there is no notion of simple chemical concepts such as the atom or the bond.

If we now look back at Lewis' figure (Fig. 3.1 in Section 3.3 of Chapter 3) of the $[H_3N–H]^+$ bond formation, we certainly 'see' the atoms and can easily conceptualize chemical bonding in terms of electron-pair sharing. This chapter deals with the problem of how to match the quantum-mechanical description of molecules with their 'chemical' interpretation, or in other words of where to seek the physical counterpart of simple chemical concepts bridging chemistry with physics.

For this purpose, we will analyze the above-mentioned concepts of the atom and the bond. The first concept, the atom, poses no significant issues if we are dealing with an ideally isolated atomic system. In this case, in fact, whether we think of the electrons as point particles belonging to that atom or as a charge distribution around its nucleus, we can safely agree on the fact that such system is an atom. The picture changes dramatically, however, if we consider an atom in a molecular context, such as that of NH_3 or NH_4^+ in Lewis's figure. To the extent that atoms can be considered the union of a nucleus and a set of electrons, it is clear that the central problem here is that of partitioning the electron density so as to determine which fraction of charge is ascribed to a given atom. In Section 14.1, by reviewing three popular partitioning schemes, we will be concerned with this topic.

The second concept, the chemical bond, is probably even more elusive and can be approached from different points of view. Over the past decades, several methods have been developed, some of which analyze the chemical bond from an energy point of view (e.g., symmetry-adapter perturbation theory, SAPT [226,227], or various energy decomposition analysis, EDA, schemes [228,229]), others from a wavefunction point of view (e.g., through the electron-localization function, ELF [230,231], or the electron-localizability indicator, ELI-D [232]), others from an orbital point of view (such as natural bond orbital, NBO, analysis [233]), and others from an electron-density point of view. It is this last class

Chemistry at the Frontier with Physics and Computer Science
https://doi.org/10.1016/B978-0-32-390865-8.00024-6

151

of methods that we will be considering here, where the focus is on the particular shape that the electron density of the atoms assumes in their molecular context. Accordingly, Section 14.2 will be devoted to an elegant and elaborate theory, Bader's quantum theory of atoms in molecules (QTAIM), providing a physics-based, consistent framework for finding the chemical concepts of atom and bond in the quantum-mechanical molecular electron density. In Section 14.3 we will face directly the electron-charge redistribution underlying Lewis' diagram and introduce a simple, yet powerful technique that will be used in Chapter 16 to address the more challenging chemical concepts of σ-donation and π-backdonation, and in Chapter 17 to probe the chemical character of the so-called superheavy elements.

14.1 Partitioning schemes

As mentioned above, the problem of identifying an atom in a molecule is tightly connected to the determination of the fraction of electron charge that can be ascribed to that atom – which, combined with the nucleus, can be considered to constitute the atom. Assigning a fraction of electron density to constituent atoms of a molecule falls in the topic traditionally referred to as 'population analysis' and has also wider implications in chemistry. Determining the fraction of electron charge pertaining to a given atom would in fact be the quantum-mechanical counterpart of the classical interpretation of partial charges, which is very popular in the context of organic and inorganic chemistry. On the one hand, this would also mean to provide these qualitative models with a physically grounded quantitative description. On the other hand, atomic charges in a molecule are not a quantum-mechanical observable and thus any attempt at their definition is inevitably arbitrary.

While the interested reader will find a comprehensive account on the topic of population analysis in dedicated reviews such as Ref. [234] and [235], we focus here on three schemes that are very different in spirit, and will provide the reader with a varied viewpoint on the problem. As we will see in the following subsections, the partition of the electron density can be performed either in physical regions of space that define the volume occupied by each atom in a molecule, or in the 'space' constituted by the set of atom-centered basis functions that, combined together through the orbital-expansion coefficient, make up the molecular electron density. A third approach, which is based neither on a clear-cut real-space partitioning nor on an orbital-space decomposition, makes use of fuzzy, overlapping atomic cells that bear the stamp of the atomic densities of the atoms constituting the molecule.

Voronoi tessellation

As anticipated above, a rather intuitive approach to defining an atom in a molecular system is to divide the space occupied by the molecule into regions that

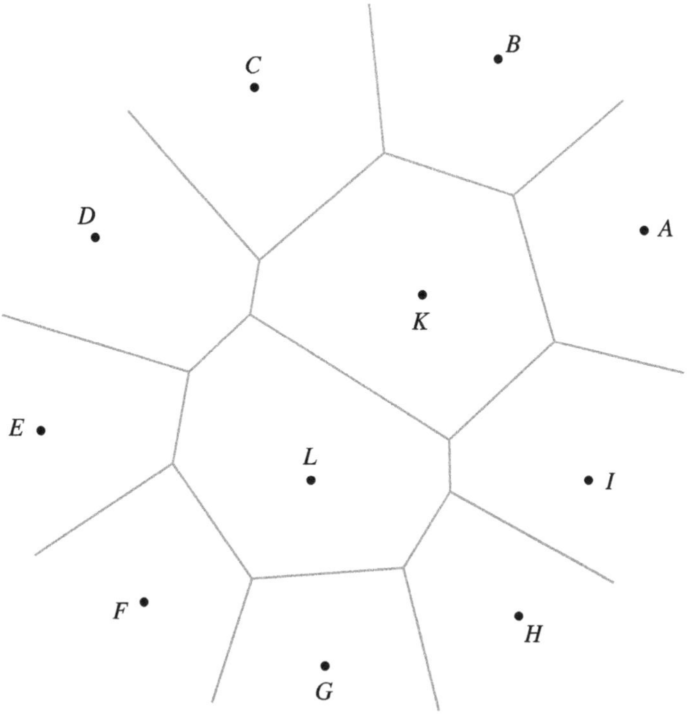

FIGURE 14.1 Voronoi tessellation for a set of points in a two-dimensional space. The regions delimited by the polygons are called Voronoi cells, while the line segments forming the polygons are sets of points in the plane that are equidistant to the two nearest sites, and the vertices are points equidistant to three or more sites.

can be assigned to each constituent atom. A simple way to do that is based on a purely geometric criterion according to which a given space point is assigned to an atom if the point is closer to that atom than to any other atom. The volume region 'belonging' to this atom would then be the set of points satisfying this condition. Such a partitioning scheme is known as 'Voronoi tessellation' and was originally proposed by the mathematician Georgy Voronoi in 1907 [236]. An illustration of the Voronoi tessellation for a set of points (referred to as 'sites') in a two-dimensional space is given in Fig. 14.1. The regions delimited by the polygons are called Voronoi cells, while the line segments forming the polygons are sets of points in the plane that are equidistant to the two nearest sites, and the vertices are points equidistant to three or more sites.

Within such a partitioning scheme, the number of electrons of an atom a is given by an integration of the molecular electron density over the volume of the associated Voronoi cell:

$$N_a = \int_a \rho(\boldsymbol{r}) \, \mathrm{d}\boldsymbol{r} \, , \qquad (14.1)$$

and the partial charge on atom a is given by the sum of its electronic charge and its atomic number:

$$Q_a = -N_a + Z_a .$$
(14.2)

In spite of its simplicity, the Voronoi scheme has indeed some popularity in computational chemistry and is adopted, with some refinements, in some electron density–partitioning methods (see, for instance, Refs. [237] and [238]). Before moving to the illustration of the second partitioning scheme, we anticipate here that in Section 14.2 we will encounter another scheme that makes use of the same Eqs. (14.1) and (14.2) but is based on more sophisticated, physically grounded space-partitioning criterion for determining the atomic volume regions.

Mulliken population analysis

A more elaborate scheme, which is rooted in the particular method that is often adopted to compute the electron density – the Roothaan–Hall method illustrated in Chapter 13 – is the popular so-called 'Mulliken population analysis', which was developed by Mulliken as early as 1955 [239]. In Mulliken's scheme, the partitioning, rather than being performed in real space, is made in the 'space' defined by the functions used for the orbital expansion.

In Section 13.1 we saw how in Hartree–Fock or Kohn–Sham calculations the molecular electron density can be expressed, upon definition of a set of basis functions, in terms of these basis functions and of the orbital expansion coefficients:

$$\rho(\mathbf{r}) = \sum_i |\varphi_i(\mathbf{r})|^2 = \sum_i \sum_{p,q} c_{pi}^* c_{qi} \chi_p^*(\mathbf{r}) \chi_q(\mathbf{r}) ,$$
(14.3)

where with $\varphi_i(\mathbf{r})$ we will indicate hereinafter the singly occupied spin-orbitals (in closed-shell calculations the set of spin-orbitals will result by duplication of the doubly occupied orbitals) and the sum i runs on the occupied spin-orbitals. Now, due to the fact that the basis functions – as discussed in Section 13.3 – are atom centered, the space of these functions already by itself responds to an atomic partition. We might thus seek the charge of a given atom directly in the coefficients multiplying the basis functions centered on that atom in Eq. (14.3).

Focusing on a single term of the first sum (the one running on i) of Eq. (14.3), the electron charge of an orbital $\varphi_i(\mathbf{r})$ can be decomposed in the following summation:

$$\int |\varphi_i(\mathbf{r})|^2 \, d\mathbf{r} = \sum_{p,q} c_{pi}^* c_{qi} S_{pq} = \sum_{a,b} N_{ab}^{(i)} ,$$
(14.4)

i.e., in a sum of charge fractions $N_{a,b}^{(i)}$ depending on pairs of atoms:

$$N_{a,b}^{(i)} = \sum_{\substack{p \in a \\ q \in b}} c_{pi}^* c_{qi} S_{pq} \, , \tag{14.5}$$

where $p \in a$ indicates the subset of basis functions centered on atom a and $q \in b$ indicates the subset of basis functions centered on atom b.

Now the diagonal terms $N_{aa}^{(i)}$ and $N_{bb}^{(i)}$ can be safely attributed to atoms a and b, respectively, while the off-diagonal terms $N_{ab}^{(i)} = N_{ba}^{(i)}$ can be grossly equally shared between the two atoms, e.g., assigning $N_{ab}^{(i)}$ to a and $N_{ba}^{(i)}$ to b. Accordingly, we can write the electron charge of the ith orbital on an atom a as:

$$N_a^{(i)} = \sum_b N_{ab}^{(i)} \, , \tag{14.6}$$

and the overall partial charge on that atom is:

$$Q_a = -\sum_i N_a^{(i)} + Z_a \, . \tag{14.7}$$

This approach was the most used in the early days and was further elaborated in successive years by Cusachs and Politzer in the so-called Löwdin population analysis [240]. Over the years, however, the method proved to be not always adequate, one of its major flaws being the strong dependence on the adopted basis set, and it has been nowadays superseded by more robust schemes, some of which will be reviewed in the following.

Hirshfeld partitioning scheme

The last partitioning scheme that we will illustrate in this section is rather different in spirit from the previous two, and was devised by Hirshfeld in 1977 [241] based on previous works of Politzer and Harris [242]. The idea behind Hirshfeld partitioning is to ground the partition of the electron density on the basis of a set of reference atomic electron densities as follows.

Within Hirshfeld's scheme, a so-called promolecular electron density is defined as the superposition of all spherically averaged atomic densities centered on the atoms constituting the molecular system:

$$\rho^{\text{pro}}(\boldsymbol{r}) = \sum_b^{N_{\text{atoms}}} \rho_b^{\text{at}}(\boldsymbol{r}) \, . \tag{14.8}$$

This promolecular density, which clearly has no physical meaning, thus refers to the density of the constituent atoms in an 'unrelaxed state' prior to molecular formation. Then, a share or weight function $w_a(r)$ is defined for each atom, returning at each point in space its relative share in the promolecular density:

$$w_a(r) = \frac{\rho_a^{at}(r)}{\rho^{pro}(r)} . \tag{14.9}$$

The several resulting overlapping weight functions are all positive and they sum up to one in every point of space. These weight functions, which are built on the basis of the shape of the atomic densities, represent an efficient and chemically grounded partitioning criterion and can be used to evaluate the share of atom a in other quantities such as the total electron density:

$$\rho_a(r) = w_a(r)\rho(r) . \tag{14.10}$$

The total electron density may thus be decomposed in atomic contributions as follows:

$$\rho(r) = \sum_a \rho_a(r) = \sum_a w_a(r)\rho(r) . \tag{14.11}$$

In other words, at each point in space the molecular density is divided among the atoms of the molecule in proportion to their respective contributions to the pro-molecule density. An overall molecular deformation density can also be defined as the density difference between the molecule and the promolecule:

$$\Delta\rho(r) = \rho(r) - \rho^{pro}(r) , \tag{14.12}$$

out of which the atomic deformation density can be extracted:

$$\Delta\rho_a(r) = w_a(r)\Delta\rho(r) . \tag{14.13}$$

Finally, the charge associated to the atom a can be obtained by the following simple integration over all the space:

$$Q_a = -\int w_a(r)\Delta\rho(r)\,dr = -\int w_a(r)\rho(r)\,dr + Z_a . \tag{14.14}$$

The Hirshfeld method is widely used in computational chemistry, among its advantages being the fact that the partitioning is grounded on the much more 'chemical' criterion of an atomic-density reference (rather than a purely mathematical one as in Voronoi's scheme) and that its results typically depend little on the quality of wavefunction.

14.2 The quantum theory of atoms in molecules

When illustrating the Hirshfeld partitioning scheme, we introduced the concept of promolecular electron density. As is evident, such a density is not a realistic one and its introduction is inevitably an arbitrary choice. One may wonder if it

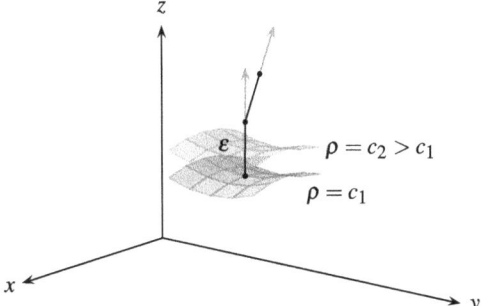

FIGURE 14.2 Schematic illustration of the gradient vector. The gradient is normal to the surface of constant value c_1 and it points towards a surface of a higher value c_2. The construction of the gradient path is also schematized (see text for discussion).

is really necessary to rely on an extra arbitrary reference density, or if chemical concepts such as the atom or the bond may stem directly from the only reliable physical quantity that we have discussed so far, the molecular electron density $\rho(\mathbf{r})$. This question led the physicist Richard Bader to the formulation of an elegant and elaborate theory, the QTAIM, providing a consistent framework in which chemistry (or some important facets of it) can be disclosed from the analysis of the mathematical properties of the three-dimensional function representing the molecular electron density (so-called 'topological analysis'). In the following, we shall illustrate the basics of this theory, while referring the reader to authoritative accounts such as Refs. [243–245] for further details and implications. Following Popelier's introduction to QTAIM [245], we will start by focusing on the following important concepts: the gradient, the gradient path, and gradient vector field.

The central quantity in Bader's theory is the gradient vector, or simply the gradient, of the electron density:

$$\nabla\rho = \left(\mathbf{u}_x \frac{\partial}{\partial x} + \mathbf{u}_y \frac{\partial}{\partial y} + \mathbf{u}_z \frac{\partial}{\partial z} \right)\rho = \begin{pmatrix} \frac{\partial\rho}{\partial x} \\ \frac{\partial\rho}{\partial y} \\ \frac{\partial\rho}{\partial z} \end{pmatrix}, \qquad (14.15)$$

where $\mathbf{u}_x$, $\mathbf{u}_y$, and $\mathbf{u}_z$ are the three unit vectors. Following from its definition, the gradient of the electron density in a given point of space is a vector pointing in the direction of greatest increase in ρ. It can, furthermore, be proven that the gradient vector in a given point in space is orthonormal to the surface defined by the equation $\rho = c$ containing that point. These properties are summarized in the schematic illustration of Fig. 14.2. In the figure, the gradient vector evaluated in a point of the surface $\rho = c_1$ is shown to be normal to the surface and to point in the direction of the greatest increase in ρ towards a surface $\rho = c_2$ with $c_2 > c_1$.

A second important concept, which can be built on top of the definition of the gradient, is that of 'gradient path'. The gradient path is formally defined as a curve to which the gradient vector is tangent at each of its points (where $\nabla\rho \neq 0$). This concept can be easily illustrated as follows. Consider the gradient evaluated at the point of the $\rho = c_1$ surface shown in Fig. 14.2, and imagine walking a distance ε along the vector gradient. If one now evaluates the gradient in this new position, one would obtain a second vector (see Fig. 14.2), pointing in a new direction that would follow the greatest increase in ρ. Imagine now walking again a distance ε along this second vector gradient, and iterate infinitely this procedure. The gradient path would be given by the generated 'trajectory' in the limit $\varepsilon \to 0$.

Now, the gradient paths have the following three important properties. First, they are orthogonal to isosurfaces of ρ. This can be easily rationalized on the basis of the consideration that a gradient path may be constructed in terms of infinite gradients, each orthonormal to the related isosurface, according to the procedure exemplified above. The second property of gradient paths is that a given point in space where $\nabla\rho \neq 0$ is 'threaded' by a single gradient path or, in other words, gradient paths never cross except when $\nabla\rho = 0$. The last property, resulting from the fact that a vector has an orientation, is that gradient paths have a beginning and an end.

A third important mathematical concept in QTAIM is the gradient vector field, which is simply an infinite collection of gradient paths. As we will note later, the gradient vector field is all that we need to analyze in order to obtain a good deal of chemical information within QTAIM. An efficient way of sampling the gradient vector field for a molecular electron density is to define a small sphere around the position of the nuclei, divide this sphere into evenly spaced points, and follow the gradient paths crossing these points. This procedure and the resulting gradient vector field are exemplified for a heteronuclear diatomic molecule in Fig. 14.3, where one of the possible infinite planes containing the two nuclei is shown (the three-dimensional picture can be easily mentally recovered by considering that this particular system has cylindrical symmetry).

In Fig. 14.3, a contour map of the molecular electron density is also shown (note that the gradient paths are orthonormal to the ρ isosurfaces, and that the depicted lines form a right angle with the isocontour lines at their mutual intersection). As can be expected, the electron density of the molecule has two peaks in coincidence of the position of the two nuclei, and decays when moving away from the nuclei, approaching zero at infinite distances. A first striking feature of the associated gradient vector field is thus that most of the sample gradient paths originate at infinity, walk along a path normal to the isosurfaces of ρ, and terminate on one of the two nuclei. Now, in the Hartree–Fock or Kohn–Sham calculations based on the expansion in Cartesian Gaussian-type orbitals (CGTOs) that we discussed in Chapter 13, the electron density results from a combination of CGTOs, and this has the implication that it will typically feature a maximum in the point coinciding with the position of the nuclei.

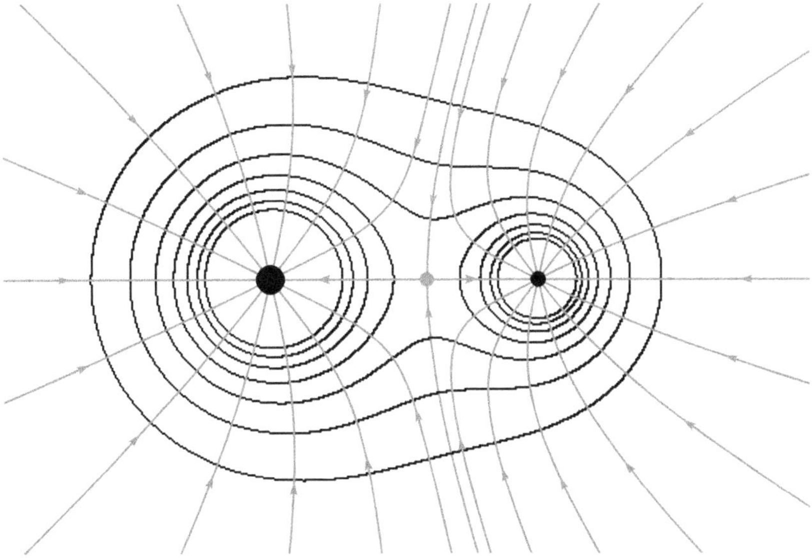

FIGURE 14.3 Scheme of the gradient vector field for a heteronuclear diatomic molecule on a plane containing the two nuclei, superimposed on a contour map of the molecular electron density.

We can at this point introduce the concept of 'critical points' in the gradient vector field, defined as the points where $\nabla\rho = 0$, and add the information that a gradient path starts at a critical point and terminates at a critical point. Critical points include points of maximum of ρ as those coinciding with the position of the nuclei, as well as saddle points, such as the one indicated by a blue (dark gray in print version) circle in Fig. 14.3 that lies on the straight line joining the nuclei. Considering that at infinity the electron density and its gradient approach zero ($|\nabla\rho| = 0$), also the infinite set of points at infinity are critical points.

If we look attentively at Fig. 14.3, we can distinguish three classes of gradient paths. The first class, comprising the vast majority of gradient paths, is represented by gradient paths that originate at infinity and are directed towards, or 'attracted by', one of the nuclei. Due to this feature, the site of a nucleus is also called a 'nuclear attractor'. The second class is represented by the two gradient paths connecting the saddle critical point (blue (dark gray in print version) circle) to the two nuclear attractors. Accordingly, these two gradient paths form a so-called 'bond path' and the saddle critical point between the nuclei is called the 'bond critical point'. The third and final class comprises the two gradient paths that originate at infinity and end on the bond critical point. Note that in the three-dimensional case there is an infinite bundle of gradient paths originating at infinity and terminating at the bond critical point. The collection of these infinite gradient paths makes up a so-called 'interatomic surface'.

Now, many chemical concepts can be, within QTAIM, defined in terms of these objects: the nuclear attractors, the bond paths, the bond critical points, and the interatomic surfaces. Consider, for instance, the infinite collection of gradient paths that originate at infinity and terminate at one nuclear attractor. The volume of space that this infinite collection of gradients paths defines is called an 'atomic basin', and the union of a nuclear attractor and the associated atomic basin – i.e., the region of space dominated by that nucleus – is what makes an atom in QTAIM. Looking back at the issue of determining the fraction N_a of electron charge belonging to an atom, and the associated partial charge Q_a, within QTAIM these can now be obtained by using Eqs. (14.1) and (14.2) where the integration volume is the atomic basin rather than the Voronoi cell.

In other words, as already mentioned, Bader's analysis leads to a space partitioning of a molecule in atomic regions based on the topological features of the molecular electron density, i.e., grounded on a rigorously defined physical criterion, rather than on the purely geometrical one of Voronoi's tessellation. As in Voronoi's scheme, also in Bader's analysis the atoms are objects possessing an awkward and rather counterintuitive shape, especially if compared to the shape they feature when they are taken as isolated. However, in QTAIM this shape, or equivalently the particular curvature of the interatomic surfaces, does now respond to the physics of the system through its electron density, which descends from a given Hamiltonian, rather turning into a polyhedron determined by a distance criterion.

14.3 Charge-redistribution analysis

As is apparent from the discussion in the previous section, the major strength of QTAIM is that it entirely follows from the physics of the system and needs no introduction of an external, artificial reference, such as that resulting from the superimposed atomic densities in Hirshfeld's partitioning. On the other hand, as mentioned several times, chemistry is a science of differences and many chemical phenomena can be understood and usefully rationalized through a comparison with an external reference. Think, for instance, of the deformation of the electron cloud upon an external perturbation such as the application of an electric field, or due to electronic excitation from the electronic ground state to an excited state, or due to the sharing of an electron pair – as in Fig. 3.1 – in chemical-bond formation. In this section we specifically target this last topic, i.e., the charge redistribution occurring when a chemical bond is formed.

The electron-charge redistribution taking place upon formation of a molecular adduct AB from two molecular fragments A and B can in principle be modeled through a comparison between the electron density $\rho^{(AB)}(r)$ of the adduct computed at its equilibrium geometry and the electron densities $\rho^{(A)}(r)$ and $\rho^{(B)}(r)$ of the two isolated fragments in their respective equilibrium geometries. For practical reasons, however, it is convenient to compute $\rho^{(A)}(r)$ and $\rho^{(B)}(r)$ using for A and B the same geometry that these fragments have in the adduct, thus leaving apart the deformation of the density associated with the geometrical distortion of the fragments and focusing only on the deformation resulting from the interaction.

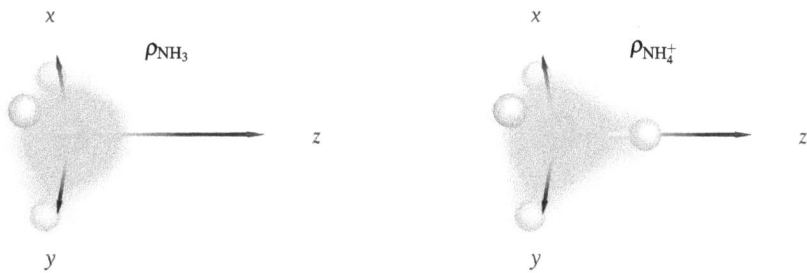

FIGURE 14.4   Electron density of NH_3 (left panel) and NH_4^+ (right panel).

In the simple case of $NH_3 + H^+ \rightarrow NH_4^+$ the only densities to compute would be those of the adduct NH_4^+ and of the fragment NH_3, as the fragment H^+ of course features no electrons. The electron densities of NH_3 and of NH_4^+ computed via the Hartree–Fock calculations described in Section 13.3 of Chapter 13 are shown as three-dimensional electron clouds in Fig. 14.4. As one might expect, the electron cloud of NH_3 (left panel) is thicker in the region of the electron-rich N atom, and extends smoothly over the whole molecular region to cover the three H atoms with a thinner layer. The accumulation of charge extending on the positive side of the z-axis, on the right of the N atom, reflects the existence of a lone pair on that atom. Moving to NH_4^+ (right panel), the electron density now has the tetrahedral symmetry of the compound. By comparison of the electron cloud of NH_4^+ with that of NH_3, one might easily guess that, as hypothesized by Lewis, upon formation of the $[H_3N–H]^+$ the fraction of charge relating to the electron lone pair on N has extended along the z-axis to envelope the newly bound atom.

This conjecture can be given an argument by computing the difference between the electron density of NH_4^+ and that of NH_3 for each point in space. In other words, we might define the electron-charge redistribution associated with formation of a chemical bond as the difference $\Delta\rho(x, y, z)$ between the electron density of AB and that of its two constituting, noninteracting fragments A and B frozen at their in-adduct geometries:

$$\Delta\rho = \rho^{(AB)} - \left[\rho^{(A)} + \rho^{(B)}\right] = \sum_i |\varphi_i^{(AB)}|^2 - \left[\sum_i |\varphi_i^{(A)}|^2 + \sum_i |\varphi_i^{(B)}|^2\right],$$

(14.16)

where we also explicitly given the expression of $\Delta\rho$ in terms of Hartree–Fock or Kohn–Sham spin-orbitals φ_i. As in Hirshfeld partitioning, the 'reference' density $\rho^{(A)} + \rho^{(B)}$ has no clear physical meaning but, as we shall see here and in Chapters 16 and 17, its introduction can help considerably in the analysis and interpretation of chemical bonding.

The $\Delta\rho(x, y, z)$ function computed from the densities of Fig. 14.4 associated with formation of the $[H_3N–H]^+$ bond is shown as a two-color cloud

FIGURE 14.5 Front view (left panel) and side view (right panel) of the charge-redistribution $\Delta\rho(x, y, z)$ upon formation of the $[H_3N-H]^+$ bond. Blue (dark gray in print version) regions are associated with positive values of the function (electron gain) and red (gray in print version) regions with negative values (electron loss).

in Fig. 14.5, with blue (dark gray in print version) regions being associated with positive values of the function (electron gain) and red (gray in print version) color with negative values (electron loss). A visual inspection of the figure clearly confirms that the lone-pair region on the right of the N atom has undergone an intense electron loss (thick red (gray in print version) cloud with a disk-like shape) to the benefit of the right-hand side H atom (surrounded by a blue (dark gray in print version) cloud). Furthermore, a visual analysis of the $\Delta\rho(x, y, z)$ function conveys the additional information that also the three H atoms of the ammonia fragment are involved in the overall charge redistribution following bond formation between the N and H atoms lying on the z-axis. In fact, an additional charge flow from these three H atoms to the N atom is seen to occur. The N atom itself, displays a red (gray in print version) electron cloud in its rear side with respect to the newly formed bond, indicating that an intrafragment charge redistribution is occurring.

The visual analysis of $\Delta\rho(x, y, z)$ often offers remarkable qualitative insight into the nature of a chemical bond. However, the analysis can be made even more stringent if formulated in a quantitative fashion. For this purpose, we note here that a chemical bond typically develops along a given direction, so that in most cases an 'interaction axis' can be defined. In the case of the $[H_3N-H]^+$, this can be easily recognized as the axis joining the N and H atoms involved in the formed bond. As implicitly already done in Figs. 14.4 and 14.5, we will conventionally adopt the z-axis as the interaction axis hereinafter.

We might thus transform the three-dimensional information contained in the $\Delta\rho(x, y, z)$ function into a more manageable one-dimensional information as a function of the interaction axis z. Considering the molecular electron density $\rho(x, y, z)$ of a given system, Brown and Shull [242,246] formulated two useful quantities, the planar density:

$$\rho(z) = \int_{-\infty}^{\infty} \int_{-\infty}^{\infty} \rho(x, y, z) \, \mathrm{d}x \, \mathrm{d}y , \qquad (14.17)$$

i.e., the electron density integrated over the xy-planes at each value of z, and the electron count function:

$$q(z) = \int_{-\infty}^{z} \rho\left(z'\right) \, dz' , \qquad (14.18)$$

obtained as a progressive integration of the planar density and yielding, for each z, the number of electrons contained in the volume at the left of a plane perpendicular to the z-axis through that point. If one adapts Brown and Harrison's formulae to the electron-density difference $\Delta\rho(x, y, z)$ rather than to a density itself, then one can formulate a planar density difference:

$$\Delta\rho(z) = \int_{-\infty}^{\infty} \int_{-\infty}^{\infty} \Delta\rho\left(x, y, z\right) \, dx \, dy , \qquad (14.19)$$

and an 'electron-gain' count function:

$$\Delta q(z) = \int_{-\infty}^{z} \Delta\rho\left(z'\right) \, dz' . \qquad (14.20)$$

This second function is the same 'charge-displacement' (CD) function introduced in Ref. [247] to study the chemical bond between gold and the noble gases:

$$\Delta q(z) = \int_{-\infty}^{z} dz' \int_{-\infty}^{\infty} \int_{-\infty}^{\infty} \Delta\rho\left(x, y, z'\right) \, dx \, dy . \qquad (14.21)$$

In the context of chemical-bond analysis, the CD function, besides being conceptually very simple, is a powerful tool as it quantifies, at any point z of the interaction axis, the exact amount of electron charge that, upon formation of the bond, has moved from right to left (or from left to right, for negative values of $\Delta q(z)$) across a plane perpendicular to the interaction axis through the z point. Accordingly, the CD function provides a clear-cut z-resolved quantitative picture of the charge transfer associated with chemical-bond formation. Fig. 14.6 shows a scheme that aids in visually grasping the interpretation of the CD function.

The planar density difference $\Delta\rho(z)$ and the CD function $\Delta q(z)$ associated with formation of the $[H_3N–H]^+$ bond are reported in Fig. 14.7. The planar-density difference exhibits, from left to right: i) a negative peak associated with electron depletion from the H atoms of the NH_3 fragment, ii) a positive peak associated with electron gain around the N atom, iii) a negative peak associated with electron loss from the N lone pair, and iv) a positive peak associated with electron gain around the newly added H atom. The CD function, whose value at each z equals the signed area bounded by $\Delta\rho(z)$ from $-\infty$ up to that z, is negative along the whole molecular region, indicating a z-resolved net flow of electrons always in the left-to-right direction. Considering the midpoint between the N and H atoms involved in the formed bond as a good indicator of

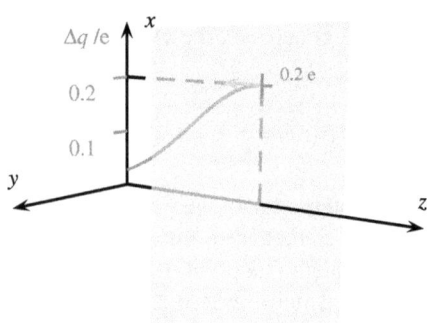

FIGURE 14.6 Scheme illustrating the interpretation of the charge-displacement function $\Delta q(z)$. At a given point z, $\Delta q(z)$ quantifies the fraction of electrons that, upon charge redistribution, has crossed in the direction of decreasing z an xy-plane perpendicular to the z-axis through that point.

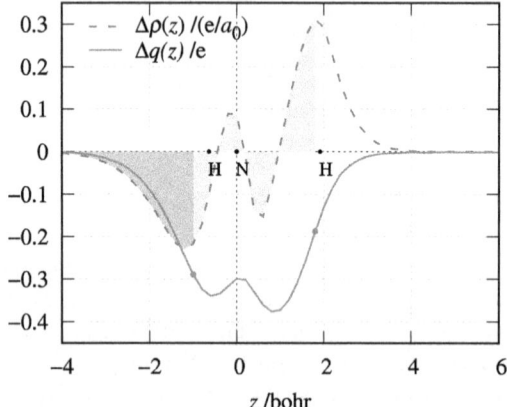

FIGURE 14.7 Planar density difference $\Delta\rho(z)$ (dashed curve) and charge-displacement function $\Delta q(z)$ (solid curve) for the formation of the $[H_3N\text{–}H]^+$ bond. The position of the nuclei along the interaction axis z is also indicated, with the three H atoms of the NH_3 fragment superimposed one on another. The first point indicated along the $\Delta q(z)$ curve quantifies the (signed) area bounded by $\Delta\rho(z)$ from $-\infty$ up to the end of the darker color, the second indicated point quantifies the signed area bounded by $\Delta\rho(z)$ from $-\infty$ up to the end of the lighter color.

the boundary between the fragments, then we can conclude (within the limits, of course, of the 'model chemistry' adopted for the density calculations) that formation of the $[H_3N\text{–}H]^+$ bond involves the transfer of about 0.4 e from NH_3 to H^+.

We conclude this section by mentioning that the CD analysis has been recently coupled to a Voronoi-tessellation technique on the one hand in order to generalize the techniques to the case of curvilinear interaction paths [248], and to an orbital-space partitioning idea on the other hand in order to obtain an orbital-resolved picture of the overall charge redistribution associated to chemical bonding. This last aspect deserves to be described in some detail, as the related methodology – natural orbitals for chemical valence/charge-displacement (NOCV-CD) analysis [249,250] – will be used for the chemical-bond characterization of the systems treated in Chapter 16.

The NOCV-CD analysis scheme combines an orbital-space and a real-space analysis of the electron-charge redistribution upon intermolecular interactions. As we shall see in Chapter 16, the technique is particularly successful in characterizing the coordination bond, where synergistic charge flows running in opposite directions occur and cannot be disentangled in ordinary CD analysis. In this context, NOCV-CD analysis provides a robust quantitative framework for singling out competitive charge flows as the σ-donation and π-backdonation (see the applications discussed in Chapter 16, or Ref. [251] for a comparison of quantitative analysis with qualitative models such as resonance structures and the octet rule, and Refs. [252–255] for application of the technique in other contexts).

As for the orbital-space part of the analysis, use is made of the so-called natural orbitals for chemical valence (NOCV) [256,257]. In NOCV theory, the charge rearrangement taking place upon formation of AB from fragments A and B is defined with respect to a slightly different reference density from the one considered so far. This reference density is no longer the simple superposition of the densities of A and B, but is rather obtained from the occupied orbitals of A and B mathematically manipulated so as to result as orthonormal to each other.[1] We shall refer to this new set of spin-orbitals as φ_i^0. The resulting density rearrangement

$$\Delta\rho = \sum_i |\varphi_i^{(AB)}|^2 - \sum_i |\varphi_i^0|^2 \tag{14.22}$$

can be brought into diagonal contributions in terms of NOCVs, i.e., the eigenfunctions $\tilde{\varphi}_{\pm k}$ of the so-called 'valence operator' of Nalewajski and Mrozek valence theory [261–263] that can be defined in terms of molecule $(\varphi_i^{(AB)})$ and reference (φ_i^0) occupied spin-orbitals, as

$$\hat{V} = \sum_i \left(|\varphi_i^{(AB)}\rangle\langle\varphi_i^{(AB)}| - |\varphi_i^0\rangle\langle\varphi_i^0| \right) . \tag{14.23}$$

Now, the NOCVs feature the peculiar and interesting property that they can be grouped in pairs of complementary orbitals $(\tilde{\varphi}_k, \tilde{\varphi}_{-k})$ corresponding to eigenvalues with the same absolute value but opposite sign (for an account on the algebraic properties of NOCVs, see Ref. [264]):

$$\hat{V}\tilde{\varphi}_{\pm k} = \pm v_k \tilde{\varphi}_{\pm k} \quad (v_k > 0) . \tag{14.24}$$

---

[1] For instance, using the symmetric orthonormalization procedure proposed by Löwdin [258,259] that guarantees that the orbitals of the new orthonormalized set are the closest to the original ones in a least-squares sense [260].

In terms of NOCV pairs, $\Delta\rho$ reads

$$\Delta\rho = \sum_k v_k \left(|\tilde{\varphi}_k|^2 - |\tilde{\varphi}_{-k}|^2 \right) = \sum_k \Delta\rho_k , \qquad (14.25)$$

where k ranges from 1 to the number of occupied spin-orbitals of the adduct. Eq. (14.25) now sheds light on the interpretation of the NOCVs: upon formation of AB from the promolecule, a fraction v_k of electrons is transferred from the $\tilde{\varphi}_{-k}$ to the $\tilde{\varphi}_k$ orbital. While, as already mentioned, k runs over the number of occupied spin-orbitals of the adduct (i.e., the number of electrons of the system), it turns out that only a small subset of the NOCV pairs in the sum of Eq. (14.25) actually contributes to the overall charge rearrangement $\Delta\rho$, because a large part of them presents values of v_k close to zero. As we shall see in Chapter 16, often these few relevant orbital-resolved $\Delta\rho_k$ contributions to the overall charge redistribution have a clear chemical meaning and, in the NOCV-CD analysis scheme, can be conveniently analyzed in real space through the quantitative tools of CD analysis.

It is worth adding here that the NOCV orbitals are the equivalent of the so-called 'most implicated natural orbitals' introduced in Ref. [265] for studying electronic excitations, and of the so-called 'electron deformation orbitals' used in Ref. [266] for analyzing the electric response of molecular conductors. The NOCV-CD method could thus be easily extended in these contexts to analyze phenomena related to a charge redistribution not necessarily arising from chemical bonding. We will see the NOCV-CD analysis scheme in action in Chapter 16 on different classes of coordination complexes. Before that, however, we will in Chapter 15 provide some details on practical aspects related to the computation and analysis of electron-charge redistributions.

Chapter 15

From theory to computing: analyzing the electron-charge redistribution

This chapter is devoted to computational aspects involved in the practical calculation of the electron-charge redistribution from molecular electron densities and in the application of charge-displacement (CD) analysis on it. While in Chapters 8 and 13, both of the same series *From theory to computing* as the present chapter, the focus was on procedural aspects summarized by the flowcharts in Figs. 8.1 and 13.1, here we will be mostly concerned with the design of the program and the organization of the involved data. In particular, in Section 15.1 we will further elaborate on the concept of object-based programming already anticipated in Section 13.3 (where customized variables were introduced for storing and operating on CGTO basis functions). In Section 15.2, we will discuss the representation and manipulation of electron densities inside the computer. Section 15.3 will provide final guidelines on the implementation of a CD-analysis program. As mentioned in Chapter 1, code examples will be based on the Fortran 95 syntax, though they can easily be transposed to other languages supporting object-oriented programming such as C++ or Python.

15.1 Object-based programming

In the previous paragraph, use was made of the expressions 'object based' and 'object oriented'. The two expressions are not equivalent and their use in the above paragraph was not casual. Thus a clarification of this matter is a due premise to our discussion.

Object-oriented programming (see Ref. [267] for an early account on this topic) is a computer-programming paradigm that organizes the design of a code around objects corresponding to the scientific notions of an application problem, rather than on the specific procedure that one has to implement to solve the problem. Objects are discrete entities that incorporate both data and behavior, and are usually 'instances' of a given abstract class that determines the data structure of the objects and offers computational procedures to operate on them.

As concisely summarized in Ref. [268], in order to be object-oriented a programming language must support these four features: i) 'identity', the organization of data in discrete, distinguishable entities called objects, ii) 'classification', the grouping of objects with the same structure and behavior into classes, iii) 'polymorphism', the differentiation of behavior of the same operation on different classes, and iv) 'inheritance', the sharing of structure and behavior among classes in a hierarchical relationship. A programming language that supports only identity and classification is said to be 'object based' rather than fully object oriented [269].

Now the Fortran standard that we will adopt, Fortran 95, supports identity, classification, and, only to some extent, polymorphism. It is thus an object-based programming language.[1] The reader interested in the object-oriented capabilities of Fortran can find more detailed information in Refs. [268] and [270], and in Ref. [271] where a comparison with C++ and other programming languages is made. In the following, we will actually adopt a clear and concise subset of Fortran 95, the so-called 'F language' [272] (see https://fortranwiki.org/fortran/show/F, and http://pages.swcp.com/~walt/F/F_bnf.html for a description of its syntax) which has been carefully crafted so as to reduce redundancies in the syntax while retaining all of the capabilities of Fortran 95, and to force programmers to conceive their programs in an object-based programming paradigm.

Coming to our problem, essentially what we want to do is to write a computer program that computes the CD function associated to bond formation between two molecular fragments A and B. In particular, given three input electron densities $\rho^{(AB)}$, $\rho^{(A)}$, and $\rho^{(B)}$, we want to write a program that implements Eqs. (14.16) and (14.21), which for our convenience we recast here in the following form:

$$\rho^{\text{ref}} = \rho^{(A)} + \rho^{(B)} \tag{15.1}$$

$$\Delta\rho = \rho^{(AB)} - \rho^{\text{ref}} \tag{15.2}$$

$$\Delta q(z) = \int_{-\infty}^{z} dz' \int_{-\infty}^{\infty} \int_{-\infty}^{\infty} \Delta\rho\left(x, y, z'\right) dx \, dy . \tag{15.3}$$

We will soon see in Section 15.2 that electron densities are represented in the computer in the form of so-called 'cube' files. Thinking in terms of the scientific notions of our problem, we want thus to define an abstract class of electron densities, or more generally of molecular–three dimensional functions, and be able to manipulate objects belonging to this class. In particular, we want to be able do the sum and the difference of two objects, and to perform the partial progressive integration along a given axis (Eq. (15.3)). In Fortran 95, a class is encoded in a 'module'. A module contains the definition of the data structure of the class (the so-called 'derived datatype') and a list of procedures (functions or subroutines) for operating on the objects of the class. As the computed electron densities are stored in the form of cube files, we will name our module `cubes` and our derived datatype `cube`.

---

[1] Full support for object-oriented programming is introduced in the Fortran 2003 standard.

Accordingly, we want to write a computer program that, closely following the logic and the notions of our scientific problem, looks as simple as:

```
1   PROGRAM cda
2     USE kinds, ONLY: wp => dp
3     USE cubes
4     IMPLICIT NONE
5     TYPE (cube) :: rho_ab, rho_a, rho_b, rho_ref, drho
6     REAL (KIND=wp), DIMENSION(:), ALLOCATABLE :: cdz
7     ...
8     rho_ref = rho_a + rho_b
9     drho = rho_ab - rho_ref
10    cdz = cube_cdz(drho)
11    ...
12  END PROGRAM cda
```

where we omitted the input/output operations. In the above code, wp => dp (line 2) instructs the computer to use double precision for real numbers. The kinds module, in fact, will contain two parameters, sp and dp, relating to single and double precision, respectively, and will consist of the following few lines:

```
MODULE kinds
  IMPLICIT NONE
  INTEGER, PARAMETER, PUBLIC :: sp = SELECTED_REAL_KIND (p=6, r=37)
  INTEGER, PARAMETER, PUBLIC :: dp = SELECTED_REAL_KIND (p=13, r=300)
END MODULE kinds
```

The variables that contain rho in their names are declared of type cube. The sum and the difference between cube variables (lines 8 and 9) storing the electron densities are performed simply using the operators + and -, exactly as in Eqs. (15.1) and (15.2), and cube_cdz (line 10) is a Fortran function taking in input a cube variable and returning a one-dimensional array containing the values of Δq. The nature and behavior of the cube objects will entirely be 'encapsulated' in the cubes module (residing, as is typical, in an external file), which will contain the core procedures for performing the relevant mathematical operations, and that will be described in the following sections.

15.2 Working with discretized electron densities

As we learnt in Chapters 12 and 13, in most electronic-structure approaches molecular–three dimensional functions such as orbitals or electron densities are expressed in terms of a set of a coefficients reflecting their expansion in a basis set. Their natural representation in the computer is thus the matrix of the orbital coefficients. For visualization purposes, however, these functions are usually discretized in the form of so-called volumetric data, i.e., a set of samples

> **TABLE 15.1** Simplified structure of a 'cube' data file (see https://gaussian.com/cubegen/ for more information).
>
> (string)
>
> (string)
>
> $N_{atoms}, x_{min}, y_{min}, z_{min}$
>
> $N_x, \Delta x, 0.0, 0.0$
>
> $N_y, 0.0, \Delta y, 0.0$
>
> $N_z, 0.0, 0.0, \Delta z$
>
> $Z_1, charge_1, x_1, y_1, z_1$
>
> ...
>
> $Z_{N_{atoms}}, charge_{N_{atoms}}, x_{N_{atoms}}, y_{N_{atoms}}, z_{N_{atoms}}$
>
> (array of the $N_x \times N_y \times N_z$ values of the density at each point of the grid)

(x, y, z, f) representing the value f of a given function at the points (x, y, z), e.g., those of a regular (x, y, z) grid.

A popular file format for storing such discrete representation in the context of molecular electronic structure is the 'cube' file format of the Gaussian program package [273], which is supported by most quantum-chemistry visualization programs. Besides visualization purposes, cube files are also useful for analyzing the quantities that they represent, for instance through the analysis schemes illustrated in Chapter 14, and are indeed the standard choice for performing CD analysis.

A simplified structure of a cube file is summarized in Table 15.1. The first two lines are character strings (typically containing information on the content of the cube file). The third line contains an integer number indicating the number of atoms making up the molecular system, and three real numbers indicating the origin of the discretization grid. The next three lines contain the number of grid points N_x, N_y, and N_z and the discretization steps Δx, Δy, and Δz along the three directions x, y, and z, respectively.[2] There then follows a list of the atomic number, charge, and x, y, and z coordinate of each atom, and, after that, the entire array of $N_x \times N_y \times N_z$ sampled values of the three-dimensional function. These are listed in an order that reflects the loop structure given in the following pseudocode:

$i = 0$

DO $i_x = 1, N_x$

    DO $i_y = 1, N_y$

        DO $i_z = 1, N_z$

            $i = i + 1$

---

[2] The simplified cube file structure illustrated in Table 15.1, whereby some zeros appear in lines 4–6, imposes that the discretization grid is aligned with the references axes, which is not required by the cube file format itself. For the purposes of CD analysis, it is thus assumed that the user has orientated the system so as to feature the interaction axis along z prior to running the electronic-structure calculation.

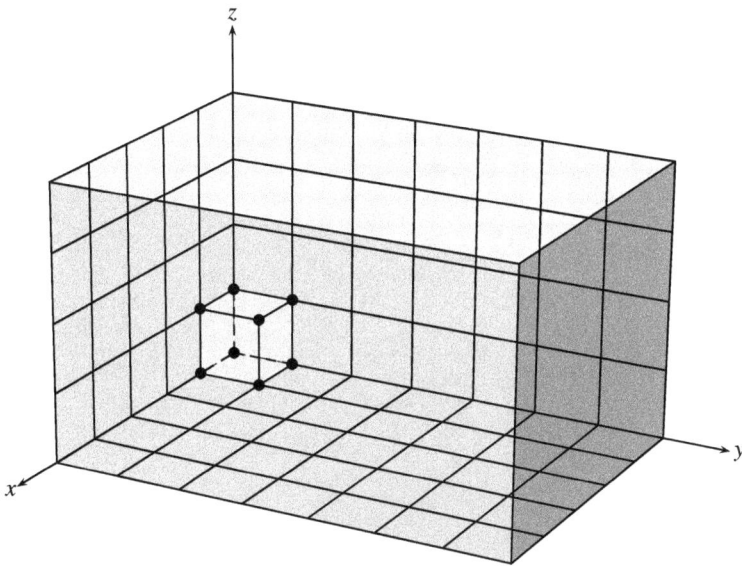

FIGURE 15.1 Illustration of the discretization box adopted in 'cube' files. The box is designed around the molecular region so as to contain the whole electron density, which is sampled at the points of the associated three-dimensional grid, eight of which are explicitly shown.

```
      WRITE array(i)
    ENDDO
  ENDDO
ENDDO
```

In practice, for a given molecule, one defines a discretization box more or less centered on the molecule and large enough to contain the whole electron density, with a discretization grain that is appropriate for visualization or analysis purposes. A graphical illustration of a discretization grid is given in Fig. 15.1. The electron density is sampled at the points defined by the grid, eight of which are explicitly shown in the figure.

For illustrative purposes the 'header' and the first six sampled values for the cube file of the electron density of NH_4^+ that has been used for generating the electron-density figure in the right panel of Fig. 14.4 is reported in Table 15.2. For the purposes of CD analysis the parameters of the discretization grid can reasonably be set as follows: adding a margin of 5 a_0 to each of the minimum and maximum values of x, y, and z in the molecular geometry and using a discretization step of 0.2 a_0 for all directions (a finer grid may be defined along the interaction axis z, e.g., using $\Delta z = 0.1 \ a_0$).

Going back to our problem, once the molecular electron densities have been discretized in the form of cube files, the sum and the difference in Eqs. (15.1) and (15.2) – provided that the discretization grid is the same for all the involved

TABLE 15.2 Header and the first six sampled values for the cube file of the electron density of NH_4^+ that has been used for generating the electron-density figure in the right panel of Fig. 14.4.

```
rho(NH4+)
Generated by Waverley, HF/6-31G*
   5 -5.6000000E+00 -6.0000000E+00 -4.8000000E+00
  57  2.0000000E-01  0.0000000E+00  0.0000000E+00
  56  0.0000000E+00  2.0000000E-01  0.0000000E+00
  56  0.0000000E+00  0.0000000E+00  2.0000000E-01
   7  0.0000000E+00  0.0000000E+00  0.0000000E+00  0.0000000E+00
   1  0.0000000E+00  0.0000000E+00 -1.8055658E+00 -6.3836394E-01
   1  0.0000000E+00  1.5636659E+00  9.0278292E-01 -6.3836392E-01
   1  0.0000000E+00 -1.5636659E+00  9.0278292E-01 -6.3836392E-01
   1  0.0000000E+00  0.0000000E+00  0.0000000E+00  1.9150918E+00
1.2681206E-013
2.1470767E-013
3.5434762E-013
5.7004402E-013
8.9389171E-013
1.3663387E-012
...
```

densities – reduces to a point-by-point sum or difference, respectively, of the relevant arrays, while the discrete analog of Eq. (15.3) becomes:

$$\Delta q(i_z) = \sum_{i_{z'}=1}^{i_z} \Delta z \sum_{i_x=1}^{N_x} \sum_{i_y=1}^{N_y} \Delta\rho(x_{i_x}, y_{i_y}, z_{i_{z'}}) \Delta x \Delta y . \qquad (15.4)$$

15.3 Implementation notes

This section gives the final details on the missing piece in the program cda discussed in Section 15.1, i.e., the Fortran module cubes that, as already mentioned, has to encapsulate the definition of the cube derived data type and the procedures for manipulating cube objects.

Code excerpts for a cubes module are shown in Table 15.3. The module contains the definition of the cube-derived data type (lines 6–16) in terms of simple data types that map onto the information contained in a cube file and summarized in Table 15.1. The module also contains a set of procedures that are declared 'public' (i.e., accessible to programs using the cubes module) in line 5. Here, the name of all public procedures of the module will be listed, though the code in Table 15.1 only shows three of them, cube_get_natom, cube_add, and cube_sub, while the procedures themselves will be listed after line 25.

TABLE 15.3 Code excerpts for a cubes module.

```
1    MODULE cubes
2      USE kinds, ONLY: wp => dp
3      IMPLICIT NONE
4
5      PUBLIC :: ..., cube_get_natom, cube_add, cube_sub, ...
6
7      TYPE, PUBLIC :: cube
8        PRIVATE
9        CHARACTER (LEN=72) :: str1
10       CHARACTER (LEN=72) :: str2
11       REAL (KIND=wp) :: xmin, ymin, zmin, dx, dy, dz
12       INTEGER :: nx, ny, nz, natom
13       INTEGER, DIMENSION(:), POINTER :: zahl
14       REAL (KIND=wp), DIMENSION(:), POINTER :: chrg, x, y, z
15       REAL (KIND=wp), DIMENSION(:), POINTER :: array
16     END TYPE cube
17
18     INTERFACE OPERATOR(+)
19       MODULE PROCEDURE cube_add
20     END INTERFACE
21     INTERFACE OPERATOR(-)
22       MODULE PROCEDURE cube_sub
23     END INTERFACE
24
25     CONTAINS
26
27     ...
28
29     FUNCTION cube_get_natom(mycube)
30       INTEGER :: cube_get_natom
31       TYPE (cube), INTENT(IN) :: mycube
32       cube_get_natom = mycube%natom
33     END FUNCTION cube_get_natom
34
35     ...
36
37   END MODULE cubes
```

Immediately below the definition of the cubes-derived data type, the code section of lines 18–23 performs the so-called 'overloading of the operators', whereby the + and − operators are overloaded so as to become meaningful when performing the respective operation on variables of type cube. This is achieved by interfacing the operators with the module procedures cube_add and cube_sub, respectively, which are functions (not shown in the code) that take as arguments two cube variables and return a new cube variable whose array results from the sum or the difference, respectively, of the two cubes in the arguments. Particular care will be taken in the crafting of these functions, as a cube file is not simply the array of the sampled density but also the associated header and one has to define what the sum and difference operations mean with regard to the quantities in the header of the cube. For instance, when summing two cube variables, it is reasonable that the resulting cube variable contains the atoms of both the original cubes. On the contrary, when subtracting the reference density to the molecular density, it is more appropriate to replicate in the final cube only the geometry of one of the two original cubes (which of the two does not matter, as they both relate to the same molecular system).

Finally, lines 29–33 list the simple procedure `cube_get_natom` (a function that takes as its argument a cube variable and returns the number of atoms associated with that cube), which shows how to access the attributes of a cube through the % symbol. This should complete the information necessary for a student with intermediate-level experience in scientific programming to complete the module and write a program performing CD analysis on electron densities given in the form of cube files similar to the one – the CUBES library and toolkit [274] (http://www.srampino.com/code.html#Cubes) – that has been used for calculating the CD function of Fig. 14.7.

Chapter 16

Application: donation and backdonation in coordination chemistry

This chapter is meant to show how elusive chemical concepts such as σ-donation and π-backdonation in coordination bonding can be modeled from a quantum-mechanical point of view. Based on the bond-analysis technique reviewed in Section 14.3 of Chapter 14, in Section 16.1 we will first focus on the paradigmatic case of the metal–carbonyl bond and show how σ-donation and π-backdonation charge flows can effectively be given a clear quantitative picture rooted in quantum-mechanical calculations. Then, in Section 16.2 we will address the relation between detailed bond properties (such as the extent of the σ-donation and π-backdonation charge transfer) and structural properties of the complexes (such as the carbonyl bond distance and stretching frequency) that can be measured experimentally and, thus used as probes of the bond properties. Finally, in Section 16.3 we will illustrate a strategy for selectively probing the σ-donor and π-acceptor power of a given class of ligands through measurements of carbonyl-stretching frequencies by synthesizing complexes with opportune coordination geometry.

16.1 The metal–carbonyl coordination bond

Since their introduction in the early 1950s, the concepts of σ-donation and π-backdonation deriving from the Dewar–Chatt–Duncanson (DCD) model [275,276] of the η^2 coordination of ethene to a coinage-metal atom have provided a popular and successful framework for the analysis and rationalization of the electronic properties of the interaction between a ligand and a metal, which is of fundamental importance in organometallic chemistry. In the field of catalytic reactions, in fact, ligands featuring specific electronic properties may be purposely designed so as to control the activation of substrates and drive the outcome and efficiency of the reactions [277].

The chemical bond M–CO between carbon monoxide CO and a transition metal M is commonly explained through the synergistic interplay between a σ-donation and a π-backdonation charge flow. In particular, as sketched in Fig. 16.1, the σ-donation charge flow involves a charge transfer from the CO lone pair on the carbon side (which is hosted by the CO highest occupied–

FIGURE 16.1   Scheme illustrating the σ-donation and π-backdonation concepts in metal–carbonyl complexes. In σ-donation (left panel), a fraction of electron charge is transferred from a donor orbital (red (gray in print version) color) to an acceptor orbital (blue (dark gray in print version) color) of σ symmetry. In π-backdonation (right panel), the electron charge transfer occurs in the opposite direction between orbitals of π symmetry. The frequency associated with the stretching of the carbonyl bond, which will be of use in Sections 16.2 and 16.3, is also highlighted.

molecular orbital, HOMO) to empty orbitals of σ symmetry of the metal. On the contrary, the π-backdonation charge flow involves a charge transfer from the filled d orbitals of π symmetry of the metal to the lowest unoccupied–molecular orbital (LUMO) of CO having π symmetry and antibonding character.

Now, a legitimate question for the computational chemist is: can σ-donation and π-backdonation be computed? For certain, σ-donation and π-backdonation are not quantum-mechanical observables and an unambiguous definition of these concepts is thus out of the question. However, as anticipated in Section 14.3, the natural orbitals for the chemical valence/charge-displacement (NOCV-CD) analysis scheme offers a robust framework rooted in quantum-mechanical calculations for a quantitative representation of these concepts.

Let us consider the simple case of the metal–carbonyl bond in two basic complexes, $CuCO^+$ and FCuCO. As detailed in Ref. [278], an NOCV-CD analysis reveals that in both cases the overall charge redistribution upon formation of the copper–carbonyl bond results mainly from the first three NOCV pairs (see Eq. (14.25)). The associated charge-redistribution components $\Delta\rho_k$, $k = 1$, 2 and 3, for the first of these two complexes, $CuCO^+$, are reported in Fig. 16.2. A visual inspection of these components reveals that the most important one ($k = 1$ and weight $v_1 = 0.40$, top panel) can be identified with the σ-donation component of the interaction sketched in Fig. 16.1, $\Delta\rho_1 \equiv \Delta\rho_{\sigma-\text{don}}$. The charge depletion from the carbon lone-pair region is readily visible (red (gray in print version) cloud in a disk-like shape), and so is the charge accumulation (blue (dark gray in print version) cloud) around the copper atom. Moreover, the red (gray in print version) cloud on the rear side of copper reveals that and intrametal reorganization of the electron cloud is occurring in response to the interaction.[1]

The other two components ($k = 2$ and $k = 3$, bottom panel), which feature equal weight ($v_2 = v_3 = 0.23$), can equally well be identified with π-backdonation. In the associated $\Delta\rho_2 \equiv \Delta\rho_{\pi_y-\text{back}}$ and $\Delta\rho_3 \equiv \Delta\rho_{\pi_x-\text{back}}$, one can clearly discern the involvement of the d_{yz} and d_{xz} orbitals of the metal on the one hand, and of the π^* orbital of CO on the other hand, corresponding to those sketched in the right panel of Fig. 16.1. More precisely, the charge-

---

[1] The reader is referred to Ref. [278] for a more detailed discussion on the nature of this charge reorganization.

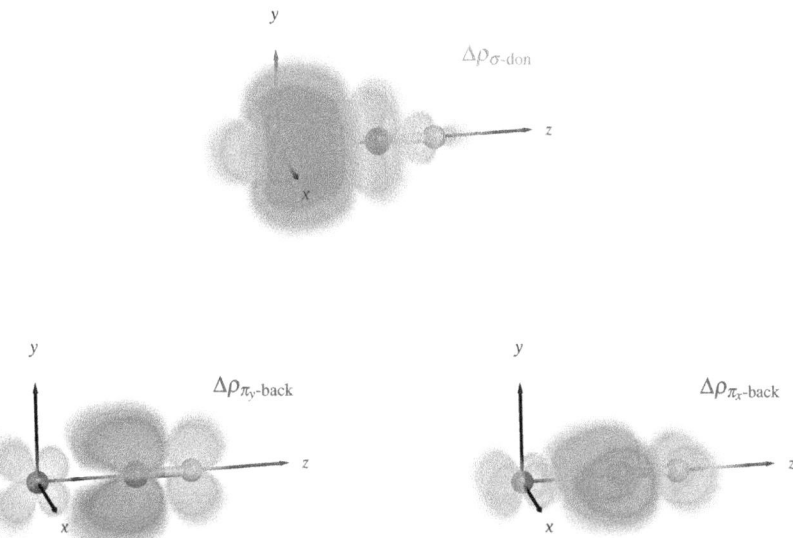

FIGURE 16.2 Electron-charge redistribution associated with the σ-donation, $\Delta\rho_{\sigma-\text{don}}$, and with the two degenerate π-backdonation, $\Delta\rho_{\pi_y-\text{back}}$ and $\Delta\rho_{\pi_x-\text{back}}$, components of the interaction.

redistribution plots associated with the two π-backdonation components show that there is a π-symmetry charge transfer from the metal to the carbon atom, but also that a simultaneous polarization of the π electron cloud on CO in the C$\leftarrow$O direction is occurring, indicated by the fact that the cloud in the regions of the O lobes of the π^* CO orbitals is red (gray in print version), while that in the regions of the C lobes is blue (dark gray in print version).

These findings can be given quantitative support by computing the CD function associated with $\Delta\rho_{\sigma-\text{don}}$, $\Delta\rho_{\pi_y-\text{back}}$, and $\Delta\rho_{\pi_x-\text{back}}$. CD functions associated with the σ-donation (red (gray in print version) line) and the degenerate π-backdonation components (blue (dark gray in print version) line and dots) in the metal–carbonyl bond in CuCO$^+$ are shown as full-color curves in Fig. 16.3. Before commenting on these curves, it will be useful to recall that the CD function quantifies at each z the exact amount of electron charge that, upon formation of the bond, is transferred from right to left across a plane perpendicular to the z-axis through that point. Accordingly, negative values of the CD function correspond to charge flow in the opposite direction, i.e., form left to right.

As is evident, the CD curve associated with σ-donation is almost always positive in the molecular region, thus indicating a charge flow from right to left, with the only exception of a small portion in the negative z-axis, at the rear side of the copper, which is due to the above-mentioned reorganization of the metal electron cloud. Now, a reasonable estimate of the charge transfer (CT) between the two fragments may be obtained by taking the value of the CD function at the middle of the bond (solid gray line in Fig. 16.3), which would quantify the

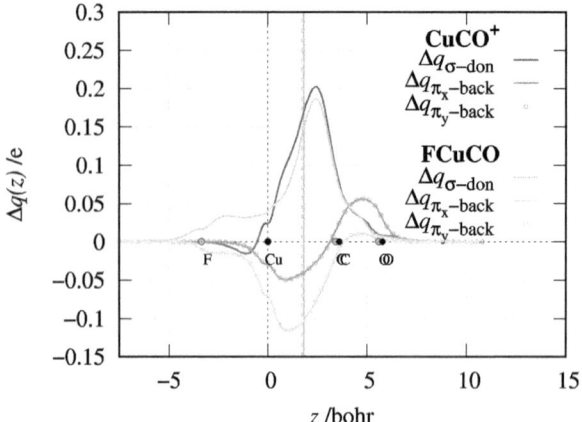

FIGURE 16.3 Charge-displacement function associated with the σ-donation and the two degenerate π-backdonation components of the overall charge redistribution upon coordination bonding between Cu and CO in CuCO$^+$ and FCuCO. Reprinted with permission from Ref. [278]. ©2018 Wiley Periodicals, Inc.

number of electrons that has crossed a plane halfway between the interacting fragments. Accordingly, the σ-donation charge transfer may be estimated to amount to 0.16 e.

As to the degenerate π-backdonation components, the associated CD curves show negative values in the metal–carbon region. As expected, this corresponds to a charge flow in the direction from left to right, i.e., from the metal to the carbonyl group. At the same time, the curve shows a positive peak in the region of the carbon–oxygen bond, thus indicating a polarization of the CO electrons in the C $\leftarrow$ O direction that, as shown in Ref. [279], is mainly due to the positive charge on the copper atom. The charge transfer associated with π-backdonation, modeled as the value of the CD function halfway from Cu to C, amounts to 0.04 e.

This picture undergoes significant changes if we now consider the second complex, FCuCO, where a fluorine ion is added as a ligand to copper. The related CD functions for this complex are shown as light-color curves in Fig. 16.3. As is apparent from the figure, the absence of a positive charge at the metal center and the presence of an electron-rich ligand enhances significantly the π-backdonation component of the interaction while almost nullifying the C $\leftarrow$ O polarization of the π CO orbitals. The σ-donation component is also affected by these changes, featuring a slightly smaller amount of M$\leftarrow$CO charge transfer and a larger spatial extension of the charge received by the metal, which propagates far into the rear side of fluorine.

16.2 Bond properties and experimental observables

The above discussion shows that NOCV-CD analysis is capable of providing a neat and compact description of the σ-donation and π-backdonation components of the interaction. Through the calculation of the CD function of these components, a z-resolved quantitative picture of the respective charge flow can be obtained. From the CD functions, estimates of the σ-donation and π-backdonation charge transfers can be extracted by taking the value of the CD function at an opportune z point serving as the boundary between the fragments.

Equipped with these tools, one may now wonder if there is any relation between the detailed properties of the coordination bond on the one hand, and structural properties of the complex, such as for instance the strength of the carbonyl bond, on the other hand. In fact, if this is the case, then simple experimental measures – e.g., the bond distance r_{CO} or the carbonyl-stretching frequency ν_{CO} – could be used as probes of the electronic properties of the metal–carbonyl bond.

Consider a generic $[(L)_n M(CO)]^m$ complex. If we focus on the M(CO) moiety of a complex of this class, the following three resonance structures can be written:

$$\text{a) } {}^-\text{M-C}\equiv\text{O}^+ \longleftrightarrow \text{b) } \text{M}=\text{C}=\text{O} \longleftrightarrow \text{c) } {}^+\text{M}\equiv\text{C-O}^- . \tag{16.1}$$

Structure a) results from simple σ-donation from the C lone pair of ${}^-\!:\text{C}\equiv\text{O}^+$. Structures b) and c) result from metal-to-ligand π-backdonation of the available electrons of the metal to the π^* orbitals of CO. In going from a) to b) and c), where one has zero, one and two π^* orbitals of CO engaged in backbonding, the CO bond multiplicity goes from three to two to one, suggesting that π-backdonation might have a bond-weakening effect leading to a stretching of the CO bond and to a shift of the carbonyl-stretching frequency towards lower values. Of course, the relative weight of each of the a), b) and c) structure will depend upon the π donor properties of the specific $[(L)_n M]^m$ fragment, and differences – as shown by the comparison between $CuCO^+$ and $FCuCO$ in the previous section – can be significant.

In Ref. [279], the relation between the detailed properties of the M–CO bond and the CO distance and frequency shifts with respect to free CO (Δr_{CO} and $\Delta \nu_{CO}$, respectively) for a wide class of carbonyl complexes with varied ligands and metal centers is addressed. In particular, twenty-three linear gold(I) complexes of general formula $[(L)Au(CO)]^{0/+}$ and ten – so-called 'homoleptic' – complexes of formula $[(CO)_n M(CO)]^m$ are considered. The net charge transfer, CT_{net}, along with its σ-donation $CT_{\sigma-\text{don}}$ and π-backdonation $CT_{\pi-\text{back}}$ components, are evaluated via NOCV-CD analysis of charge redistributions obtained through density-functional theory.[2] Leveraging on the high versatility of the CD function, the polarization of the CO electron cloud upon M–CO bond formation also can be quantified. In particular, the net charge transfer $CT_{r_{CO}/2}$

---

[2] In Ref. [279], the charge transfers are evaluated in a more elaborate way than we did in Section 16.1. In particular, rather that taking the midpoint between M and C, the boundary between the fragments is taken as the z point where equal-valued isodensity surfaces of the fragments become tangent.

TABLE 16.1 Calculated $\Delta\nu_{CO}$ (in cm^{-1}), Δr_{CO} (in Å) and charge transfers (in e) for a series of [(L)Au(CO)]$^{0/+}$ complexes. Where available (rows in boldface), experimental data are reported in parentheses along with the related reference in square brackets. Data from Ref. [279].

	$\Delta\nu_{CO}$ (exp. $\Delta\nu_{CO}$)	Δr_{CO}	CT$_{net}$	CT$_{\sigma-don}$	CT$_{\pi-back}$	CT$_{r_{CO}/2}$	CT$^{\sigma}_{r_{CO}/2}$	CT$^{\pi}_{r_{CO}/2}$
'Nonclassical' behavior								
[(CO)Au(CO)]$^+$	72 (74 [280])	-0.010	0.08	0.21	-0.13	0.15	0.06	0.09
[(PF$_3$)Au(CO)]$^+$	80	-0.009	0.09	0.22	-0.13	0.15	0.06	0.09
[Au(CO)]$^+$	**75 (94 [281])**	**-0.008**	**0.16**	**0.34**	**-0.18**	**0.16**	**0.07**	**0.09**
[(Ne)Au(CO)]$^+$	76	-0.008	0.15	0.33	-0.18	0.16	0.07	0.09
[(C$_2$H$_4$)Au(CO)]$^+$	63	-0.007	0.08	0.24	-0.16	0.13	0.06	0.07
[(PH$_3$)Au(CO)]$^+$	60	-0.007	0.06	0.22	-0.16	0.12	0.05	0.07
[(C$_2$H$_2$)Au(CO)]$^+$	63	-0.007	0.06	0.23	-0.17	0.13	0.06	0.07
[(Xe)Au(CO)]$^+$	62	-0.006	0.08	0.27	-0.19	0.13	0.06	0.07
[(P(CH$_3$)$_3$)Au(CO)]$^+$	38	-0.004	0.05	0.22	-0.17	0.10	0.05	0.05
[(NHC)Au(CO)]$^+$	39	-0.004	0.00	0.20	-0.20	0.10	0.05	0.05
[(C$_5$H$_5$N)Au(CO)]$^+$	46	-0.003	0.01	0.23	-0.22	0.10	0.05	0.05
[(Sidipp)Au(CO)]$^+$	**23 (54 [282])**	**-0.002**	**-0.01**	**0.20**	**-0.21**	**0.08**	**0.05**	**0.03**
[(Idipp)Au(CO)]$^+$	**17 (49 [282])**	**-0.001**	**-0.02**	**0.20**	**-0.22**	**0.08**	**0.05**	**0.02**
'Classical' behavior								
[(CF$_3$)Au(CO)]	**-2 (51 [283])**	**0.002**	**-0.02**	**0.22**	**-0.24**	**0.05**	**0.05**	**0.00**
[(CN)Au(CO)]	-6	0.003	-0.06	0.20	-0.26	0.05	0.05	0.00
[(H)Au(CO)]	-23	0.004	-0.07	0.19	-0.26	0.02	0.04	-0.02
[(CH$_3$)Au(CO)]	-42	0.007	-0.07	0.21	-0.28	0.01	0.04	-0.03
[(C$_6$H$_5$)Au(CO)]	-45	0.007	-0.06	0.22	-0.28	0.02	0.04	-0.02
[(I)Au(CO)]	-45	0.008	-0.08	0.24	-0.32	0.02	0.05	-0.03
[(Cl)Au(CO)]	**-29 (13 [284])**	**0.008**	**-0.11**	**0.23**	**-0.33**	**0.02**	**0.05**	**-0.03**
[(Br)Au(CO)]	**-40 (10 [284])**	**0.008**	**-0.09**	**0.24**	**-0.33**	**0.02**	**0.05**	**-0.03**
[(F)Au(CO)]	-23	0.009	-0.13	0.22	-0.35	0.02	0.05	-0.04
[(DPCb)Au(CO)]$^+$	**-76 (-30 [285])**	**0.010**	**-0.06**	**0.26**	**-0.32**	**0.03**	**0.05**	**-0.02**

between C and O, along with its σ- and π-symmetry components CT$^{\sigma}_{r_{CO}/2}$ and CT$^{\pi}_{r_{CO}/2}$, can be estimated as the value of the CD function at a z point halfway between C and O.

The considered complexes, together with the calculated frequency and distance shifts, $\Delta\nu_{CO}$ and Δr_{CO}, and the values of CT$_{net}$, CT$_{\sigma-don}$, CT$_{\pi-back}$, CT$_{r_{CO}/2}$, CT$^{\sigma}_{r_{CO}/2}$, CT$^{\pi}_{r_{CO}/2}$ are reported in Tables 16.1 and 16.2. For the gold complexes, where available, the experimental frequency shifts are also given. Reference values for free CO are $\nu_{free-CO} = 2106$ cm^{-1} (experimental: 2143 cm^{-1}), and $r_{free-CO} = 1.137$ Å. For the homoleptic carbonyls, only the distance shifts are reported, as the nonuniform influence of vibrational mode coupling, and the more complicated CO vibration modes in the homoleptic carbonyls, make $\Delta\nu_{CO}$ a less reliable parameter than Δr_{CO} for a quantitative analysis of its relation with the M–CO bond characteristics.

TABLE 16.2 Calculated Δr_{CO} (in Å) and charge transfers (in e) for a series of homoleptic carbonyl complexes $[(CO)_n M(CO)]^m$. Data from Ref. [279].

	Δr_{CO}	CT_{net}	$CT_{\sigma-don}$	$CT_{\pi-back}$	$CT_{r_{CO}/2}$	$CT^{\sigma}_{r_{CO}/2}$	$CT^{\pi}_{r_{CO}/2}$
'Nonclassical' behavior							
$Hg(CO)_2^{2+}$	-0.018	0.29	0.31	-0.02	0.13	0.08	0.21
$Ir(CO)_6^{3+}$	-0.015	0.17	0.31	-0.13	0.08	0.10	0.18
'Classical' behavior							
$Ni(CO)_4$	0.012	-0.16	0.16	-0.32	-0.01	0.04	-0.05
$Fe(CO)_5(ax.)$	0.014	-0.18	0.23	-0.41	-0.02	0.06	-0.08
$Cr(CO)_6$	0.016	-0.21	0.17	-0.37	-0.02	0.05	-0.07
$Mo(CO)_6$	0.016	-0.23	0.14	-0.37	-0.02	0.04	-0.06
$Fe(CO)_5(eq.)$	0.018	-0.20	0.20	-0.40	-0.03	0.05	-0.08
$Co(CO)_4^-$	0.038	-0.36	0.16	-0.52	-0.15	0.03	-0.18
$Ir(CO)_4^-$	0.039	-0.28	0.30	-0.58	-0.14	0.04	-0.18
$Ru(CO)_4^{2-}$	0.066	-0.42	0.25	-0.67	-0.24	0.03	-0.27
$Fe(CO)_4^{2-}$	0.069	-0.55	0.16	-0.71	-0.28	0.02	-0.30

The complexes are listed in order of increasing Δr_{CO} (decreasing CO bond strength) and are divided according to their 'nonclassical' (featuring a strengthened bond with respect to free CO) or 'classical' (featuring a weakened bond with respect to free CO) behavior [286]. While the interested reader is referred to Ref. [279] for an extensive analysis of these data, we limit ourselves here to the following considerations:

i) Looking at the values reported in Tables 16.1 and 16.2, while σ-donation remains almost constant across the complexes, π-backdonation features a wide variability. Variations in the extent of π-backdonation ($0.02 < |CT_{\pi-back}| < 0.71$ e) exactly triple those in the extent of σ-donation ($0.11 < CT_{\sigma-don} < 0.34$ e).

ii) Negligible backdonation is found for $Hg(CO)_2^{2+}$, while $CT_{\pi-back}$ for $[Fe(CO)_4]^{2-}$ is as high as 0.71 e. This picture is completely consistent with the above discussion of the resonance structures (Eq. (16.1)), in that we move from a purely σ M–CO bond (structure a, $Hg(CO)_2^{2+}$) to a situation in which all π^* CO orbitals are engaged in backbonding (structure c, $[Fe(CO)_4]^{2-}$).

iii) As the rows of the tables are descended, the values of $CT_{\pi-back}$ and $CT^{\pi}_{r_{CO}/2}$ are seen to decrease, while no such trend is evident in the values of $CT_{\sigma-don}$. This suggests a selective correlation between π-backdonation (and also the interrelated π-polarization of the CO electron cloud) and the CO interatomic distance in the complexes r_{CO}. This last conclusion is given a quantitative basis by plotting the values of $CT_{\sigma-don}$ and $CT_{\pi-back}$ as a function of r_{CO} in Fig. 16.4. While, as expected, no correlation is found for $CT_{\sigma-don}$

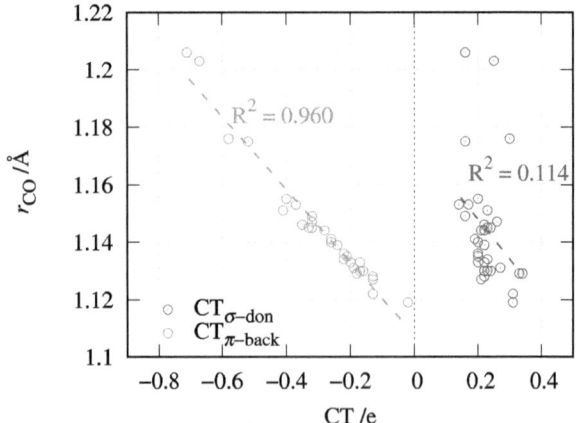

FIGURE 16.4 Correlation between the calculated r_{CO} and the σ-donation and π-backdonation charge transfers in the series of complexes of Tables 16.1 and 16.2.

values, a strict correlation ($R^2 = 0.960$) is instead found between $CT_{\pi-back}$ and r_{CO} (and thus also Δr_{CO}).[3]

We can thus confidently conclude that in carbonyl complexes the CO stretching responds selectively and quantitatively to the sole π-backdonation component of the interaction, and that experimental measurements of r_{CO} in the complexes can indeed be used as a selective probe for the π-acceptor power of the ligand or the π-backdonor ability of the metal.

16.3 Selectively probing σ-donation and π-backdonation

Consider now the more challenging coordination bond of PP to Ni in nickel dicarbonyl complexes of formula [Ni(CO)₂(PP)]:

$$(16.2)$$

where PP is a chelating bidentate phosphine. Chelating diphosphines represent a popular and versatile class of ligands in homogeneous catalysis [287], as, upon coordination at the metal center, the chelating structure is able to impose a relatively rigid environment, tuning the electron richness of the metal and, in turn, the reactivity of the coordinated complex.

In Ref. [288], a joint theoretical and experimental study on a class of nickel dicarbonyl complexes with six different bidentate phosphines whose structure is

---

[3] In Ref. [279] it is further shown that an equally good correlation exists between r_{CO} and the midpoint CO π-polarization values $CT^{\pi}_{r_{CO}/2}$. Furthermore, the sign of this polarization also seems to control the sign of the CO stretching with respect to free CO Δr_{CO}.

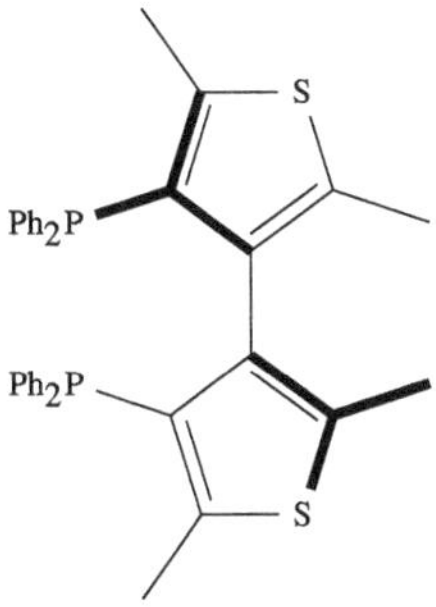

bitiop

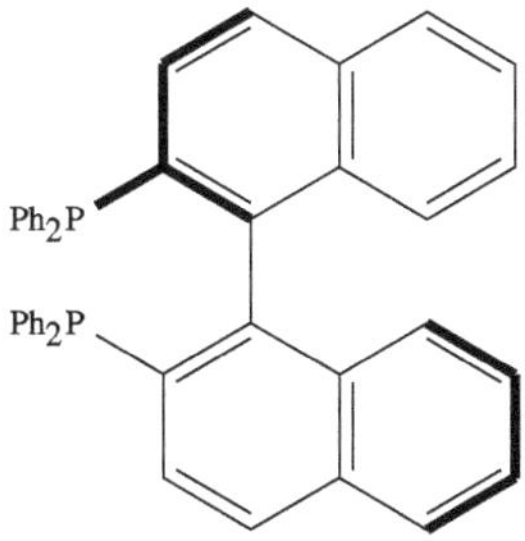

binap

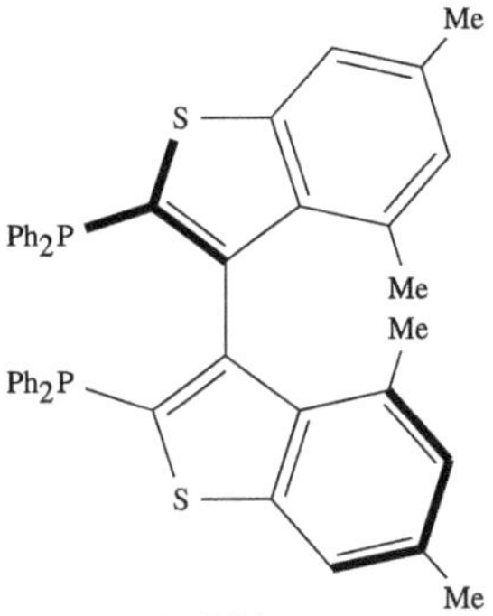

tmbitianp

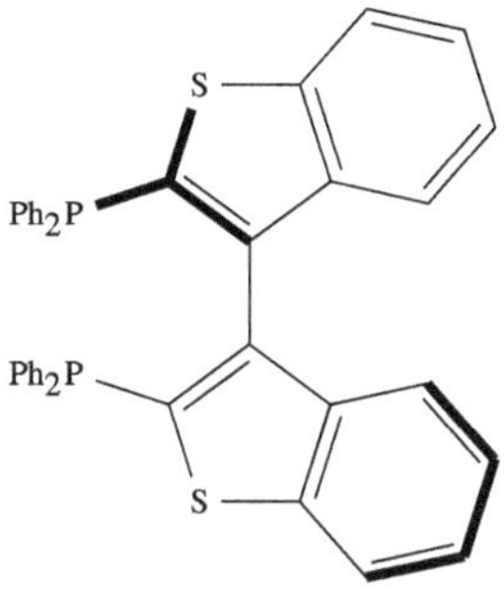

bitianp

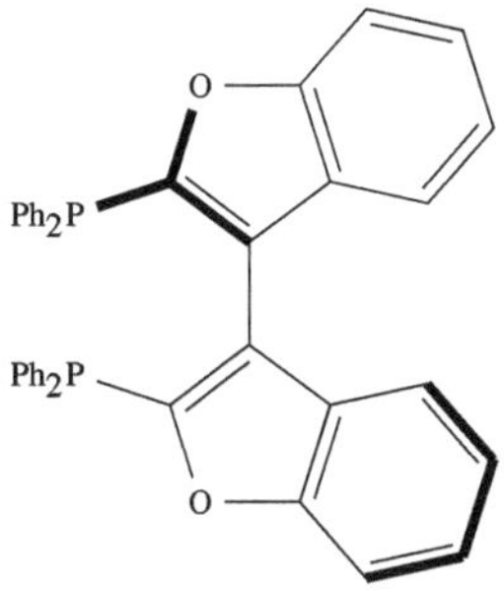

bifurp

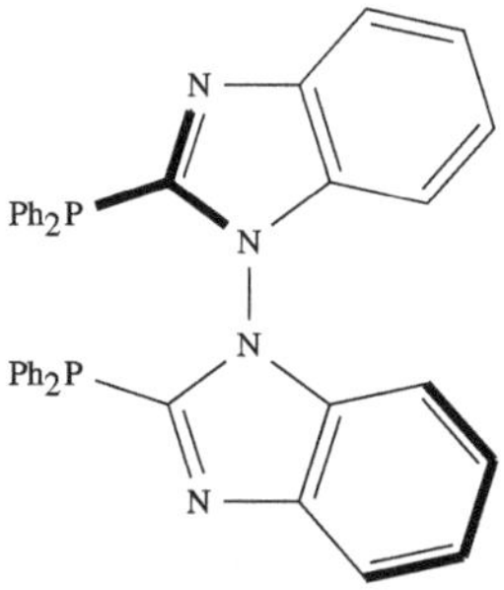

bimip

FIGURE 16.5 Chemical structures and abbreviated names for the six chelating diphosphine ligands.

reported in Fig. 16.5, shows that both the basicity and the oxidative potentials of the free ligands are strongly correlated to the observed carbonyl-stretching frequency, thus suggesting that in this class of complexes the carbonyl stretching responds selectively to the donor ability of the PP ligands. In the same work, this hypothesis is given a solid foundation by performing NOCV-CD analy-

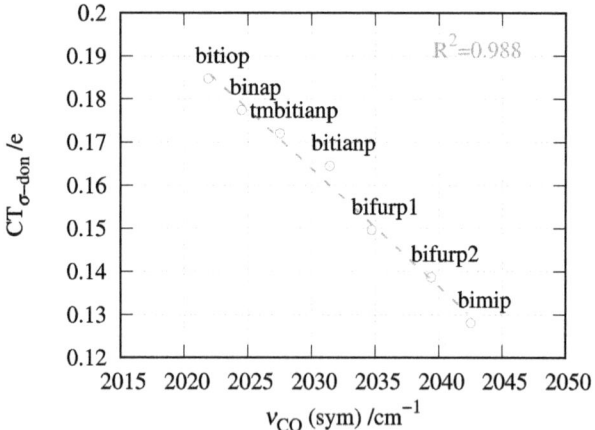

FIGURE 16.6 Correlation between the calculated CO stretching frequency and the σ-donation charge transfer in the nickel dicarbonyl complexes with the chelating diphosphine ligands of Fig. 16.5. Republished with permission of the Owner Societies from Ref. [288]. Permission conveyed through Copyright Clearance Center, Inc. ©the Owner Societies 2017.

sis of the charge redistribution upon coordination of the bidentate phosphine to the nickel dicarbonyl moiety obtained by density-functional theory. The results, which are summarized in Fig. 16.6, show that indeed there is tight correlation ($R^2 = 0.988$) between the carbonyl-stretching frequency and the charge transfer $CT_{\sigma-\text{don}}$ associated with σ-donation from the P lone pairs to the metal center, while no correlation at all is found for the other charge-redistribution components. For a better appreciation of the features of the σ-donation component of the charge redistribution upon coordination of PP to Ni in the nickel–dicarbonyl complexes, the related $\Delta\rho_{\sigma-\text{don}}$ is shown in Fig. 16.7 for the exemplary case of PP = bimip. The figure clearly shows two red (gray in print version) clouds indicating electron depletion in the region of each phosphorus facing the metal center, and the subsequent electron gain at the metallic moiety (in the form of a spread out blue (dark gray in print version) cloud).

Now, we saw in Section 16.1 how in the series of carbonyl complexes therein considered, it was exclusively the electron movement associated with π-backdonation from the metal that drove the carbonyl-stretching response. Here instead, for tetrahedral nickel–carbonyl complexes with chelating diphosphines, it is now the PP $\rightarrow$ M σ-donation component that exclusively controls the carbonyl-stretching shift upon formation of the metal–ligand bond. As can be verified by an attentive inspection of Fig. 16.7, the rationale of these apparently puzzling results is that due to the tetrahedral coordination geometry of the metal center, it is the charge flow resulting from each P $\rightarrow$ Ni σ donation that pushes electron charge into the carbonyl LUMO of π^* symmetry with respect to the C–O bond axis, thus weakening this bond and lowering the associated stretching frequency.

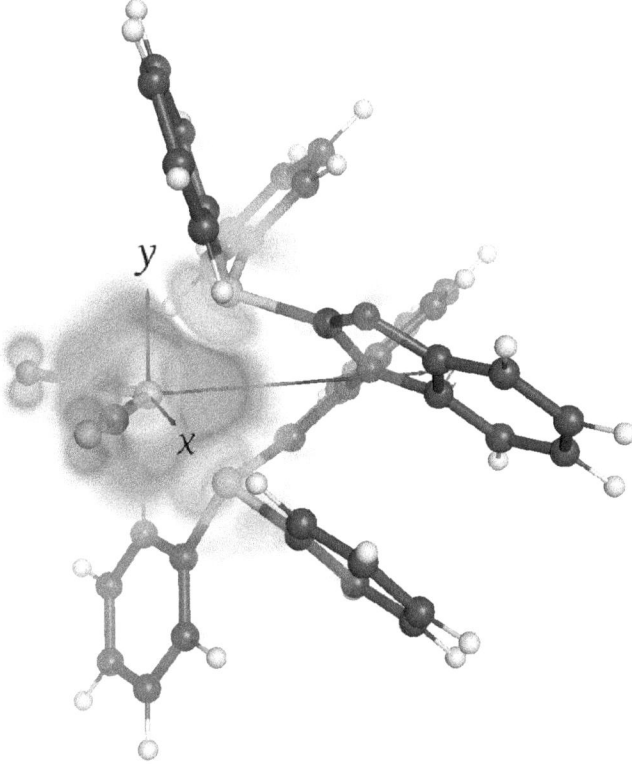

FIGURE 16.7 Charge redistribution $\Delta\rho_{\sigma-\mathrm{don}}$ associated with the P $\rightarrow$ Ni σ-donation in the [Ni(CO)$_2$(PP)] complex with PP = bimip.

FIGURE 16.8 Metal–ligand charge-flow component pushing (left)/drawing (right) charge into/from the carbonyl LUMO in tetrahedral/square-planar [M(CO)(L')(L)$_2$] complexes. Republished with permission of The Royal Society of Chemistry from Ref. [289]. Permission conveyed through Copyright Clearance Center, Inc. ©The Royal Society of Chemistry 2018.

This triggers the idea that, if one switches to a coordination geometry where the out-of-plane π-backdonation charge flow correlates with the symmetry of the carbonyl LUMO, such as for instance in square-planar complexes, then the carbonyl-stretching frequency should be selectively influenced by this bond component. This idea is concisely sketched in the scheme of Fig. 16.8, where the charge movements and the symmetry of the involved orbitals for a generic tetrahedral (left panel) or square-planar (right panel) complex of formula [M(CO)(L')(L)$_2$] are highlighted.

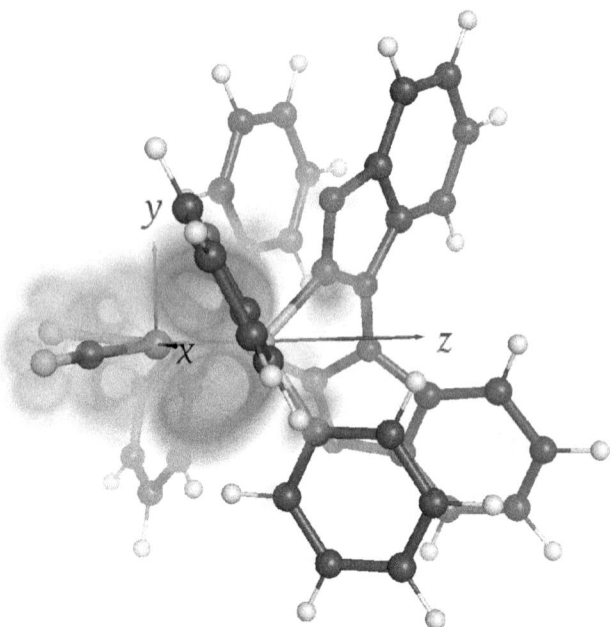

FIGURE 16.9 Charge redistribution $\Delta\rho_{\pi-\text{back}}$ associated with the Rh $\rightarrow$ P out-of-plane π-backdonation in the [Rh(CO)(Cl)(PP)] complex with PP = bimip.

In Ref. [289], square-planar analogs of the nickel complexes considered so far, are formulated as rhodium-based complexes of formula [Rh(CO)(Cl)(PP)]. The charge-redistribution component associated with Rh $\rightarrow$ P out-of-plane π-backdonation, disentangled from the overall charge-redistribution by NOCV-CD analysis, is shown in Fig. 16.9 for the case of PP = bimip, for a comparison with the analogous Fig. 16.7 showing the P $\rightarrow$ Ni σ-donation in a tetrahedral context. Fig. 16.9 clearly shows a charge flow of π symmetry with respect to the plane of the metallic moiety involving electron depletion not only from the metal but also from the π^* orbitals of CO.

In Ref. [289], for the purpose of assessing whether and to what extent the scheme of Fig. 16.8 is a valid model, an extensive theoretical analysis for the general case of ligand pairs ((L)$_2$, as in the scheme of Fig. 16.8) in lieu of chelating diphosphines is reported. The analysis encompasses a wide and varied class of thirty-four ligands, resulting in a total of sixty-eight tetrahedral nickel-based and square-planar rhodium-based complexes. The results of this analysis are summarized in Figs. 16.10 and 16.11, where also the complete set of the considered ligands is listed. Figs. 16.10 and 16.11 display the CD functions associated with the L $\rightarrow$ Ni σ-donation in tetrahedral [Ni(CO)$_2$(L)$_2$] complexes and with the Rh $\rightarrow$ L out-of-plane π-backdonation in square-planar [Rh(CO)(Cl)(L)$_2$] complexes, respectively, both evaluated along the z-axis bisecting the $\widehat{(L)M(L)}$ angle. The complete set of ligands is listed in the key of each plot (the interested

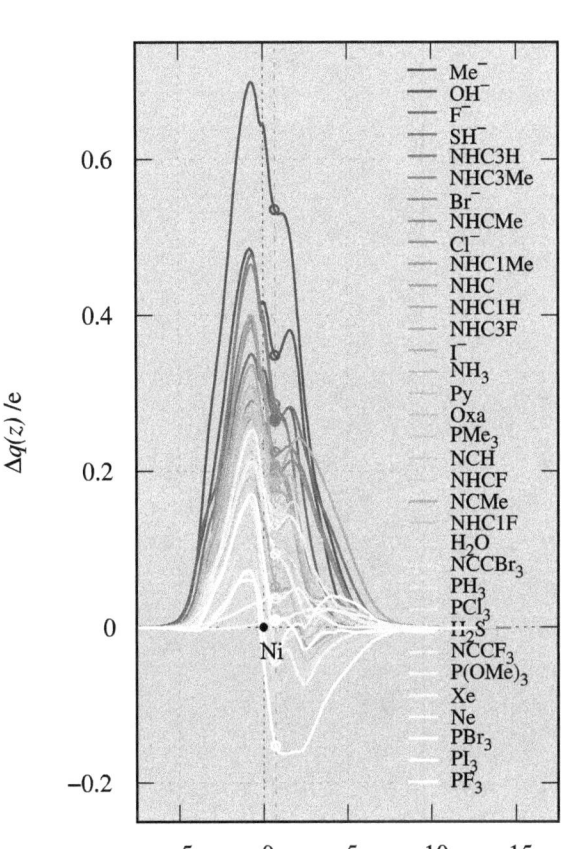

FIGURE 16.10 CD functions associated with L $\rightarrow$ Ni σ-donation in tetrahedral [Ni(CO)$_2$(L)$_2$] complexes evaluated along the z-axis bisecting the $\widehat{\text{(L)M(L)}}$ angle. Republished with permission of The Royal Society of Chemistry from Ref. [289]. Permission conveyed through Copyright Clearance Center, Inc. ©The Royal Society of Chemistry 2018.

reader is referred to Ref. [289] for explanation of the abbreviated names) in ascending order of computed ν_{CO}(sym) for Fig. 16.10, and in descending order of computed ν_{CO} for Fig. 16.11. The CD functions are drawn using a color scale that reflects this order. As is apparent from the two figures, and perfectly in line with our hypothesis, the extent of the displaced charge is in both cases proportional to the carbonyl-stretching frequency (see the curves getting darker as the extent of the related charge flow grows). More specifically, larger σ-donation in the tetrahedral complexes corresponds to lower frequencies, while larger out-of-plane π-backdonation in the square-planar complexes corresponds to higher frequencies.

This finding is put on a more quantitative foundation by extracting CT estimates from the related CD functions at $z = 0.7$ bohr and correlating them to

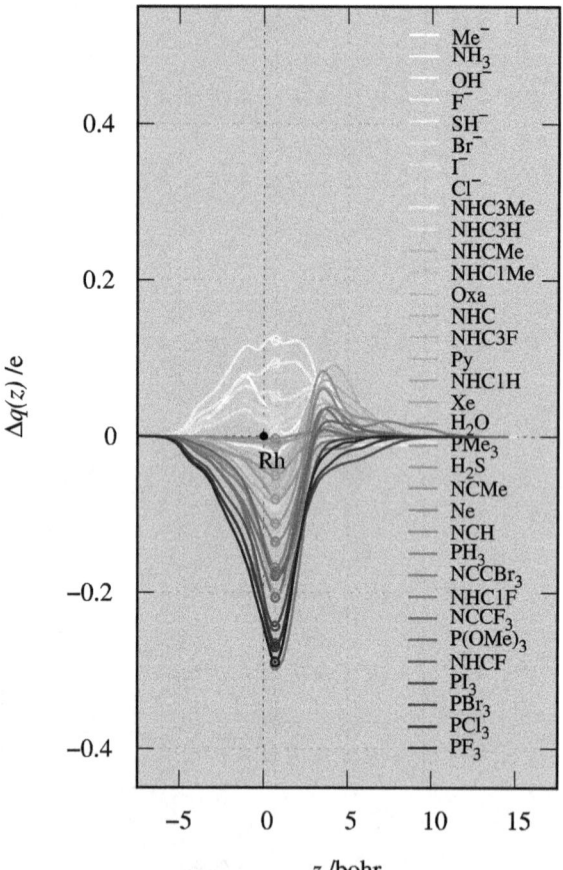

FIGURE 16.11 CD functions associated with Rh → L out-of-plane π-backdonation in square-planar [Rh(CO)(Cl)(L)$_2$] complexes evaluated along the z-axis bisecting the $\widehat{(L)M(L)}$ angle. Republished with permission of The Royal Society of Chemistry from Ref. [289]. Permission conveyed through Copyright Clearance Center, Inc. ©The Royal Society of Chemistry 2018.

the related carbonyl-stretching frequencies, as in Figs. 16.12 and 16.13. Considered the great variety of the examined complexes (a range of about 300 cm^{-1} is encompassed by the calculated frequencies, for a charge-transfer range of 0.7 e), the results are rather satisfactory, with $R^2 = 0.848$ for the correlation of CT$_{\sigma\text{-don}}$ versus ν_{CO}(sym) in the tetrahedral complexes, and $R^2 = 0.725$ for the correlation of CT$_{\pi\text{-back}}$ versus ν_{CO} in the square-planar complexes. This correlation considerably improves on restricting the analysis to a narrower series of homologous ligands such as OH$^-$, F$^-$, SH$^-$, Br$^-$, Cl$^-$, I$^-$ (see the filled circles in Figs. 16.12 and 16.13 and the related best-fitting solid lines), yielding $R^2 = 0.920$ for the data in Fig. 16.12 and $R^2 = 0.983$ for the data in Fig. 16.13.

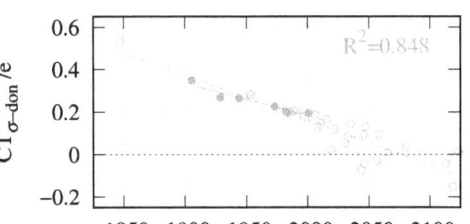

FIGURE 16.12 Correlation between the (symmetric) carbonyl-stretching frequencies and the σ-donation charge transfers evaluated at $z = 0.7\ a_0$ in the Ni complexes. Republished with permission of The Royal Society of Chemistry from Ref. [289]. Permission conveyed through Copyright Clearance Center, Inc. ©The Royal Society of Chemistry 2018.

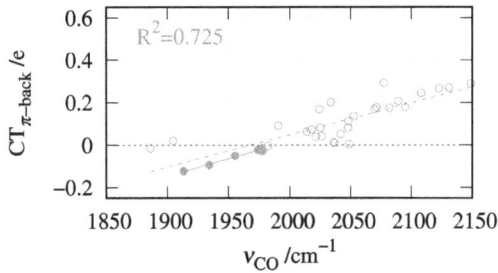

FIGURE 16.13 Correlation between the carbonyl-stretching frequencies and the out-of-plane π-backdonation charge transfers evaluated at $z = 0.7\ a_0$ in the Rh complexes. Republished with permission of The Royal Society of Chemistry from Ref. [289]. Permission conveyed through Copyright Clearance Center, Inc. ©The Royal Society of Chemistry 2018.

Going back to the set of our bidentate phosphine ligands, following the same procedure Ref. [289] shows that also in the case of [Rh(CO)(Cl)(PP)] complexes, the extent of the out-of-plane π-backdonation charge flow is directly proportional to the carbonyl-stretching frequency and the correlation between this quantity and the extracted CTs at $z = 0.7\ a_0$ is impressive ($R^2 = 0.974$).

The above-discussed results thus show how coordination geometry can be effectively exploited to switch the carbonyl-stretching frequency into a selective probe of the σ-donation and π-backdonation components of the interaction in coordination complexes. Based on the above-evidenced correlations, model equations of striking simplicity:

$$CT_{\sigma\text{-don}} = m\nu_{CO}^{Ni}(\text{sym}) + q \quad CT_{\pi\text{-back}} = n\nu_{CO}^{Rh} + p, \quad (16.3)$$

with parameters m, q, n and p to be determined via best fitting, can be set up and used to quantitatively predict electronic properties of the coordination bond from spectroscopic data on complexes with opportune coordination geometry.

Chapter 17

Relativity and chemistry

"We live in a world with a finite speed of light, c. Yet, most of the existing theoretical chemistry refers to an imaginary world where c would be infinite." We might start this chapter with this lapidary incipit from an important paper by Pekka Pyykkö and Jean-Paul Desclaux dated 1979 [290]. Indeed, through Chapters 1 to 16 of this book we moved within the boundaries of physical laws that implicitly assume an infinite speed of light. In this conclusive chapter of Part III we cope with the realization that the light has a finite speed – which, as is known, led Einstein to the formulation of the theory of relativity – and address the consequences on chemistry of the inclusion of relativity in quantum mechanics.

In Section 17.1, basic theoretical aspects are reviewed, the Dirac equation for the relativistic electron is introduced, and an overview on the so-called relativistic effects in chemistry is given. Section 17.2 shows how the Kohn–Sham approach of density-functional theory presented in Section 12.3 of Chapter 12 can be extended to the relativistic framework resulting in the so-called Dirac–Kohn–Sham approach. In Section 17.3, this approach is shown in action through a study of the interaction of superheavy elements with gold by means of the bond-analysis techniques illustrated in Section 14.3 of Chapter 14, and the inclusion of relativity in the quantum equations governing the atomic and molecular electronic structure is shown to have a strong impact on the periodic system of the elements.

17.1 Relativistic quantum chemistry

Einstein's special relativity follows from two postulates of striking simplicity: i) the laws of physics are invariant in all inertial systems, and ii) the speed of light in a vacuum is constant regardless of motion. Accounting for these two principles in classical mechanics, leads to the following expression of the energy for a free particle of mass m and linear momentum p:

$$E^2 = m^2 c^4 + c^2 p^2 \,, \tag{17.1}$$

which differs considerably from its nonrelativistic variant

$$E_{\text{n.r.}} = \frac{p^2}{2m} \,, \tag{17.2}$$

Chemistry at the Frontier with Physics and Computer Science
https://doi.org/10.1016/B978-0-32-390865-8.00027-1

and states a mass–energy equivalence that is alien to nonrelativistic physics. As concisely pointed out by Saue in Ref. [291], the two expressions can be connected by a Taylor expansion of the positive-energy branch of the relativistic energies,

$$E = mc^2\sqrt{1 + \frac{p^2}{m^2c^2}} = mc^2 + E_{n.r.} - \frac{p^4}{8m^3c^2} + \dots . \tag{17.3}$$

In Eq. (17.3), the first term of the expansion is the rest mass of the particle, the second term is identical to the nonrelativistic energy, and the remaining terms are relativistic corrections that, as expected, vanish in the limit $c \to \infty$.

As already partly mentioned in Chapter 12, in the early days of quantum mechanics the so-called 'duplexity' phenomena had led to the formulation of the spinning-electron model, which was 'fitted' into the nonrelativistic quantum equations by Pauli. In 1928, for the purpose of providing a rigorous explanation of these phenomena, after previous independent attempts by Gordon and Klein,[1] Dirac derived a wave equation for an electron in an external field by introducing Einstein's relativity into quantum mechanics [292]. The resulting equation is known as the Dirac equation and has the following form:

$$\left[c\boldsymbol{\alpha} \cdot \boldsymbol{p} + \beta m_e c^2 + eV(r) \right] \psi = E\psi , \tag{17.4}$$

where $\boldsymbol{\alpha} = (\alpha_x, \alpha_y, \alpha_z)$ and $\alpha_x, \alpha_y, \alpha_z$, and β are 4×4 matrices:

$$\alpha_x = \begin{pmatrix} 0 & 0 & 0 & 1 \\ 0 & 0 & 1 & 0 \\ 0 & 1 & 0 & 0 \\ 1 & 0 & 0 & 0 \end{pmatrix} \tag{17.5}$$

$$\alpha_y = \begin{pmatrix} 0 & 0 & 0 & -i \\ 0 & 0 & i & 0 \\ 0 & -i & 0 & 0 \\ i & 0 & 0 & 0 \end{pmatrix} \tag{17.6}$$

$$\alpha_z = \begin{pmatrix} 0 & 0 & 1 & 0 \\ 0 & 0 & 0 & -1 \\ 1 & 0 & 0 & 0 \\ 0 & -1 & 0 & 0 \end{pmatrix} \tag{17.7}$$

$$\beta = \begin{pmatrix} 1 & 0 & 0 & 0 \\ 0 & 1 & 0 & 0 \\ 0 & 0 & -1 & 0 \\ 0 & 0 & 0 & -1 \end{pmatrix} . \tag{17.8}$$

---

[1] The interested reader can find further details in Dirac's original article, Ref. [292].

This new equation, which bears some similarity with the nonrelativistic Schrödinger equation but also the remarkable difference that its solutions are four-component vectors, gave birth to a new theory that inherently accounted for the spin of the electrons and naturally explained the observed duplexity phenomena.[2] One year later, in a work that for different reasons has already been cited in Chapter 3 [12], the same founder of this theory wrote the following famous, peremptory paragraph:

The general theory of quantum mechanics is now almost complete, the imperfections that still remain being in connection with the exact fitting of the theory with relativity ideas. These give rise to difficulties only when high-speed particles are involved, and are therefore of no importance in the consideration of atomic and molecular structure and ordinary chemical reactions in which it is, indeed, usually sufficiently accurate if one neglects relativity variation of mass and velocity and assumes only Coulomb forces between the various electrons and atomic nuclei.

Now, despite the fact that at the end of the 1930s the theoretical toolset of quantum mechanics in both its nonrelativistic and relativistic flavor had already been developed and the related working equations were present, it was not until 1979, i.e., fifty years after the above quote from Dirac – and probably also due to the influence of such an authoritative viewpoint – that scientists became aware of the importance of the inclusion of relativity in the description of chemical systems containing heavy elements. In an important work entitled 'Relativity and the periodic system of elements' [290] which we already had the occasion to cite in the opening of this chapter, the Finnish chemist Pekka Pyykkö and the French physicist Jean-Paul Desclaux, after a focused investigation of relativistic effects on the electronic structure of some heavy elements via relativistic–self consistent field (SCF) calculations based on Dirac's equation (so-called Dirac–Hartree–Fock calculations), could write the following concluding remarks:

The chemical difference between the fifth row and the sixth row seems to contain large, if not dominant, relativistic contributions which, however, enter in an individualistic manner for the various columns and their various oxidation states, explaining, for example, both the inertness of Hg and the stability of Hg_2^{2+}. These relativistic effects are particularly strong around gold. A detailed understanding of the interplay between relativistic and shell-structure effects will form the impact of relativity on chemistry.

---

[2] The relativistic quantum equation for hydrogen-like atoms also led to orbitals whose shapes remarkably differ from the usual s, p, d, etc. orbitals of nonrelativistic quantum mechanics. Well before the advent of the scientific-visualization techniques that will be discussed in Chapter 19, a curious graphical representation of the electron cloud of the relativistic orbitals of hydrogen-like atoms showing unusual features in their spatial properties was given as early as in 1931 by White, who produced photograph-like pictures of the relativistic orbitals based on an ingenious mechanical device [293].

In the same work, Pyykkö and Desclaux pointed out the three main relativistic effects (defined as the difference between relativistic and nonrelativistic results) on the atomic electronic structure of the elements. The starting point of their discussion is the consideration that high-speed particles of rest mass m_0 undergo a mass increase in the relativistic framework:

$$m = m_0/\sqrt{1 - (v/c)^2} \, . \tag{17.9}$$

For an electron of the $1s$ shell of Hg ($Z = 80$, sixth period of the periodic table), for instance, whose average radial velocity in atomic units can be estimated approximately as $\langle v_r \rangle = Z$, the average ratio v/c would be about $80/137 = 0.58$, leading to a relativistic mass increase of $m \sim 1.2 m_0$.

Given this premise, the three main relativistic effects can be summarized as follows:

i) **'Relativistic contraction'**: due to the high speed reached by electrons of the s and p shells in the vicinity of the nuclei, these electrons undergo a relativistic mass increase. As the mass appears at the denominator in the expression of the Bohr radius

$$a_0 = 4\pi \epsilon_0 \hbar^2 / m Z e^2 \tag{17.10}$$

the average radius of a relativistic electron turns out to be smaller than the nonrelativistic one (for the $1s$ electron of Hg, the average radius is 20% smaller than the nonrelativistic one). This means that the orbitals undergo a radial contraction (and a subsequent energy stabilization). This is especially valid for the electrons of the s shell and of the $p_{1/2}$ subshell that results from effect ii).

ii) **'Spin-orbit splitting'**: due to the spin-orbit coupling, orbitals with orbital angular momentum higher than zero lose some degree of degeneracy (e.g., the degenerate orbitals of the nonrelativistic p shell split into one stabilized orbital of subshell $p_{1/2}$ and two degenerate destabilized orbitals of subshell $p_{3/2}$), and the splitting between the energy levels may reach a few eV for the valence electrons of the heaviest elements.

iii) **'Relativistic self-consistent expansion'**: due to the relativistic contraction of the s and p orbitals, the high-angular momentum d and f electrons, which mostly 'orbit' in the peripheral regions of the atom, feel the nuclei in a more screened way and thus as an indirect effect of the contraction of the s and p orbitals, undergo a radial expansion (and subsequent energy destabilization).

These effects, become nonnegligible starting from the fifth period of the periodic table.[3] Besides being responsible, in fact, for some striking macroscopic anomalies showing up in the properties and behavior of some elements (such as

---

[3] Note that in the calculations reported in Chapter 16 on coordination complexes featuring heavy elements, though we did not explicitly mention it, while working within a nonrelativistic framework relativistic effects were partly accounted for in an approximate way by using relativistic pseudopotentials for the core electrons. The reader is referred to the original works cited in Chapter 16 for further details.

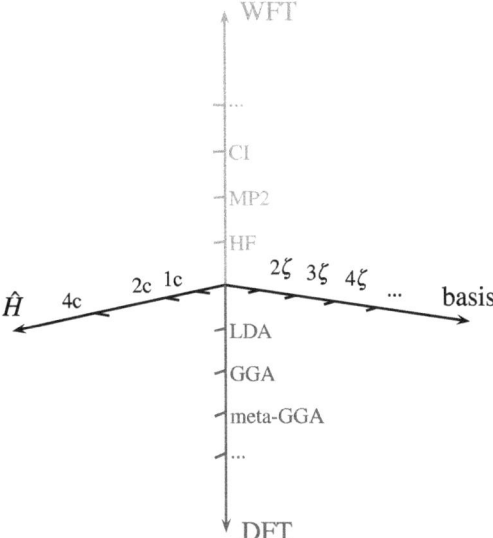

FIGURE 17.1 The three degrees of freedom defining a given 'model chemistry' in terms of the adopted Hamiltonian, basis set and level of theory in the domain of either wavefunction theory (WFT) or density-functional theory (DFT).

the yellow color and noble character of gold, the difference between its chemistry and that of silver, or the liquid state of mercury), as we shall see later in Section 17.3, they actually alter the electronic structure of the heaviest elements in such a way as to bring into question the validity itself of the periodic system in the bottom-right corner of the periodic table.

17.2 Dirac–Kohn–Sham calculations

As partly discussed in the previous section, the Dirac equation may form the basis for building relativistic molecular–electronic structure methods analogous to those developed in the nonrelativistic framework. The most fundamental one, the Dirac–Hartree–Fock, is in fact the one used by Pyykkö and Desclaux in the above-discussed Ref. [290]. As illustrated in Fig. 17.1, the possibility of adopting a relativistic Hamiltonian adds now an additional degree of freedom in the landscape of the possible so-called 'model chemistries'. In fact, besides the conventional degrees of freedom already discussed in Section 13.3 of Chapter 13, i.e., the quality of the basis set (e.g., the so-called double-ζ, triple-ζ, etc., abbreviated as 2ζ, 3ζ, etc., in the figure) and the specific level of theory within either wavefunction-theory (abbreviated as WFT in the figure) methods or density-functional theory (DFT) methods, one can also choose between Hamiltonians of different complexity, ranging from the one-component (1c) nonrelativistic one to the approximate-relativistic two-component (2c) and the full-relativistic four-component (4c) one.

To provide the reader with an idea of the additional complexity involved in relativistic electronic-structure calculations with respect to the nonrelativistic framework, we shall in the following illustrate with some detail the relativistic version of the Kohn–Sham approach discussed in Section 12.3 of Chapter 12, known as the Dirac–Kohn–Sham approach. In particular, we will focus on the formalism developed by Grant and Quiney and based on the use of the relativistic analog of the Gaussian-type basis functions already discussed in Section 13.3 of Chapter 13, the so-called G-spinors, and on a relativistic generalization of the McMurchie–Davidson recursive scheme for the evaluation of the integrals over the G-spinor basis functions [220,294,295]. This formalism is the basis of the relativistic Dirac–Kohn–Sham computer program BERTHA [294–300], named after the British scientist Bertha Swirles who in 1935 set up the SCF equations in the Dirac relativistic framework [301].

By combining the Kohn–Sham method with Dirac's Hamiltonian including only the so-called longitudinal interactions, one obtains the so-called Dirac–Kohn–Sham equation, which in atomic units reads:

$$\left[c \begin{pmatrix} 0 & \boldsymbol{\sigma} \\ \boldsymbol{\sigma} & 0 \end{pmatrix} \boldsymbol{p} + \begin{pmatrix} \boldsymbol{I} & 0 \\ 0 & -\boldsymbol{I} \end{pmatrix} c^2 + \tilde{v}^{(l)}(\boldsymbol{r}) \right] \boldsymbol{\Psi}_i(\boldsymbol{r}) = \varepsilon_i \boldsymbol{\Psi}_i(\boldsymbol{r}), \qquad (17.11)$$

where c is the speed of light in vacuum, $\boldsymbol{p}$ is the electron four-momentum, and the $\boldsymbol{\alpha}$ and $\boldsymbol{\beta}$ matrices of Eq. (17.4) have been recast in terms of the matrices $\boldsymbol{\sigma}$ and $\boldsymbol{I}$, where $\boldsymbol{\sigma} = (\sigma_x, \sigma_y, \sigma_z)$ with σ_q being a 2×2 Pauli spin matrix and $\boldsymbol{I}$ is a 2×2 identity matrix. As in its nonrelativistic counterpart, the diagonal potential operator consists of a nuclear potential term $v_N(\boldsymbol{r})$, a Coulomb-interaction term $v_H^{(l)}[\rho(\boldsymbol{r})]$ and an exchange-correlation term $v_{xc}^{(l)}[\rho(\boldsymbol{r})]$:

$$\tilde{v}^{(l)}(\boldsymbol{r}) = v_N(\boldsymbol{r}) + v_H^{(l)}[\rho(\boldsymbol{r})] + v_{xc}^{(l)}[\rho(\boldsymbol{r})]$$

$$= \sum_a \int \frac{\rho_a(\boldsymbol{r}')}{|\boldsymbol{r} - \boldsymbol{r}'|} \, d\boldsymbol{r}' + \int \frac{\rho(\boldsymbol{r}')}{|\boldsymbol{r} - \boldsymbol{r}'|} \, d\boldsymbol{r}' + v_{xc}^{(l)}[\rho(\boldsymbol{r})], \qquad (17.12)$$

the last two terms being a functional of the relativistic electron density $\rho(\boldsymbol{r})$.[4] Note that while the nuclear attraction term in Eq. (17.12) has the same nature as in nonrelativistic calculations, in the relativistic context the nuclear charges are modeled by spherically symmetric Gaussian-charge distributions $\rho_a(\boldsymbol{r})$, rather than point charges [297].

---

[4] We note here that genuine relativistic exchange-correlation functionals should depend on the relativistic four-current [302–304]. However, research activity on these functionals is still at an early stage and mostly limited to the nonrelativistic framework.

The solutions of the Dirac–Kohn–Sham equation are so-called four-spinors of the form:

$$\Psi_i(r) = \begin{pmatrix} \psi_i^{(1)}(r) \\ \psi_i^{(2)}(r) \\ \psi_i^{(3)}(r) \\ \psi_i^{(4)}(r) \end{pmatrix},$$

(17.13)

where the first two components define the so-called 'large' (L) component and the last two the 'small' (S) component. In terms of the spinors, the relativistic electron density reads:

$$\rho(r) = \sum_i \Psi_i^\dagger(r) \cdot \Psi_i(r) = \sum_i \sum_{k=1}^{4} \psi_i^{(k)*}(r)\psi_i^{(k)}(r),$$

(17.14)

and the total energy is:

$$E = \sum_i \varepsilon_i - \int v_{\text{xc}}^{(l)}[\rho(r)]\rho(r)\,\mathrm{d}r - \frac{1}{2}\int v_{\text{H}}^{(l)}[\rho(r)]\rho(r)\,\mathrm{d}r + E_{\text{xc}}^{(l)}[\rho(r)],$$

(17.15)

where the sums in Eqs. (17.14) and (17.15) extend over the occupied positive-energy spinors.

Following a similar approach to that leading to the Roothaan–Hall equations discussed in Section 13.1, the four-spinor solution of Eq. (17.11) can be expanded in a set of $2N$ G-spinor basis functions $M_\mu^{(T)}(r)$ with T being either L or S

$$\Psi_i(r) = \begin{bmatrix} \sum_{\mu=1}^{N} c_{\mu i}^{(L)} M_\mu^{(L)}(r) \\ i\sum_{\mu=1}^{N} c_{\mu i}^{(S)} M_\mu^{(S)}(r) \end{bmatrix},$$

(17.16)

where L and S identify the already-mentioned large and small component, respectively, and $c_{\mu i}^{(T)}$ are the coefficients to be determined. The G-spinors $M_\mu^{(T)}(r)$ – the analogs of the Cartesian Gaussian-type orbitals described in Section 13.3 – are Gaussian-based two-component objects labeled by the collective index μ that maps univocally onto the set of parameters (Gaussian center and exponent, fine-structure quantum number and magnetic quantum number) necessary to completely characterize the functions (see Ref. [297] for details on these basis functions).

As in the nonrelativistic context, the spinor expansion leads to the formulation of Eq. (17.11) in matrix form, i.e., to a matrix-diagonalization problem:

$$H_{\text{DKS}} \begin{bmatrix} c^{(L)} \\ c^{(S)} \end{bmatrix} = E \begin{bmatrix} S^{(LL)} & 0 \\ 0 & S^{(SS)} \end{bmatrix} \begin{bmatrix} c^{(L)} \\ c^{(S)} \end{bmatrix},$$

(17.17)

where $c^{(T)}$ are the spinor expansion vectors of Eq. (17.16). The matrix representation H_{DKS} of the Dirac–Kohn–Sham operator in the G-spinor basis is

$$\begin{bmatrix} v^{(LL)} + J^{(LL)} + K^{(LL)} + mc^2 S^{(LL)} & c\Pi^{(LS)} \\ c\Pi^{(SL)} & v^{(SS)} + J^{(SS)} + K^{(SS)} - mc^2 S^{(SS)} \end{bmatrix},$$
(17.18)

where the v, J, K, S and Π matrices are the basis representations of the nuclear, Coulomb, and exchange-correlation potentials, the overlap matrix and the matrix of the kinetic-energy operator, respectively (where the T superscript is dropped, the whole matrix composed of the LL, LS, SL and SS blocks is referred to). Their matrix elements are defined by

$$v_{\mu\nu}^{(TT)} = \int v_N(r)\rho_{\mu\nu}^{TT}(r)\,dr$$
(17.19)

$$J_{\mu\nu}^{(TT)} = \int v_H^{(l)}[\rho(r)]\rho_{\mu\nu}^{TT}(r)\,dr$$
(17.20)

$$K_{\mu\nu}^{(TT)} = \int v_{xc}^{(l)}[\rho(r)]\rho_{\mu\nu}^{TT}(r)\,dr$$
(17.21)

$$S_{\mu\nu}^{(TT)} = \int \rho_{\mu\nu}^{TT}(r)\,dr$$
(17.22)

$$\Pi_{\mu\nu}^{(TT')} = \int M_{\mu}^{(T)\dagger}(r)(\sigma \cdot p)M_{\nu}^{(T')}(r)\,dr,$$
(17.23)

with $\rho_{\mu\nu}^{(TT)}(r)$ being G-spinor overlap densities.

The relativistic electron density $\rho(r)$ (figuring in Eqs. (17.20) and (17.21)), in terms of the basis functions is

$$\rho(r) = \sum_T \sum_{\mu\nu} D_{\mu\nu}^{(TT)} \rho_{\mu\nu}^{(TT)}(r),$$
(17.24)

with

$$D_{\mu\nu}^{(TT')} = \sum_i c_{\mu i}^{(T)*} c_{\nu i}^{(T')},$$
(17.25)

where the sum runs over the occupied positive-energy states. The matrix H_{DKS} depends, because of J and K, on the spinor-orbitals produced by its diagonalization, so that the solution c must be obtained recursively to self-consistency. As in the nonrelativistic context, once a guessed density has been provided the problem formally reduces to the evaluation of the integrals in Eqs. (17.19)–(17.23) for the assembling of H_{DKS} (Eq. (17.18)), and the iterative solution of the eigenvalue problem (Eq. (17.17)) with up-to-date J and K integrals at each cycle, in an analogous scheme to the flowchart of Fig. 13.1.

As is evident from a comparison with the analogous equations of Chapter 13, four-component relativistic Dirac–Kohn–Sham calculations are far more expensive than their nonrelativistic counterpart from a computational point of view. This arises from the complex representation required for the involved matrices and the intrinsically larger basis sets resulting from the four-component structure of the Dirac equation. While these aspects will be analyzed in greater detail in Chapter 19 when addressing high-performance computing, the above-sketched formalism is shown in action in the next section. In particular, we shall see how Dirac–Kohn–Sham calculations can shed light onto the chemical properties of the rather exotic so-called superheavy elements and contribute to the discussion on the validity of the periodic table in its bottom-right corner.

17.3 Relativity and the periodic table

No branch of science has probably an icon so popular and universally recognized as has chemistry in the periodic table of the elements. While being a symbol of the whole discipline and a summary of the organizing principles of matter, it is also a useful work tool in the everyday life of a chemist that can be used to predict and interpret the properties and the behavior of the elements.

Early attempts to classify and organize matter date back to the work of Lavoisier (1789). However, the foundations for the modern periodic table were laid by the Russian scientist Mendeleev who in 1869 discovered the law of periodicity, according to which, at regular intervals of atomic weight, the elements featured similar properties and behavior [305]. Mendeleev's table was perfected and given a formal theoretical justification with the advent of quantum mechanics and of atomic and molecular physics. Nowadays, in fact, the two 'old' organizing criteria of atomic weight and chemical behavior are replaced by, respectively, the atomic number and the ground-state electronic configuration, rooted in quantum–electronic structure theory and resulting from the Madelung, Hund, and Pauli rules, and the few related exceptions.

The recent discovery of the elements with atomic number $Z = 117$ [306] and 118 [307] has extended the periodic table to the end of the seventh period. These elements had temporary IUPAC names Uus and Uuo until 2015, when they were named, respectively, Tennessine (Ts) and Oganesson (Og). Now, the periodic table, in the form that we commonly know it today, is based on an essentially nonrelativistic picture of the electronic structure that inherently ignores the relativistic orbital contraction/expansion and spin-orbit splitting discussed in Section 17.1. However, as we learnt there, relativistic effects become significant in the electronic structure of elements around the fifth and sixth rows, and are expected to be even stronger for the so-called 'superheavy' elements of the seventh period. This has a deep impact on the legitimacy itself of the periodic table in its bottom-right corner. In fact, due to relativistic effects, periodicity trends may break along the 7th period [308], with consequences that might even impact the naming of the new elements [309] and cause predictions based on traditionally recognized group trends to fall short.

The problem was first given evidence by Pitzer in 1975, [310] whose early atomic calculations on Cn ($Z = 112$, featuring a relativistically contracted and stabilized $7s$ outer shell) and Fl ($Z = 114$, featuring a contracted and stabilized quasiclosed outer shell $7p_{1/2}$ arising due to spin-orbit splitting) suggested that these two elements should reveal a noble-gas character of volatility and relatively high inertness. Since then, much debate followed among experimentalists and theoreticians on whether Cn first and Fl later behave more like metals, as do lighter homologs of their respective group, or noble gases [311–316]. Conversely, Og (featuring a relativistically expanded and destabilized outer shell $7p_{3/2}$), is expected to be rather reactive compared to lighter noble gases [317].

Now, investigating the physical and chemical properties of superheavy elements is not a walk in the park. Superheavy elements, also known as transactinides ($Z \geq 104$), are among the unstable ($Z > 83$, beyond bismuth) and nonnatural ($Z > 94$, beyond plutonium) elements. As a consequence, these elements have to be produced on an atom-at-a-minute/atom-at-a-month scale in heavy ion–induced fusion reactions [318,319]. Due to the fact that they feature very rapid decay times (half-lives in the range of seconds), little is known of their chemistry [320–323]. The few available experimental data result from sophisticated experiments based on gas-phase thermocromatography measurements of volatility through adsorption on gold surfaces [324]. Theory, thus besides being the only source of information on several chemical properties, plays also a fundamental role in providing experimentalists with valuable information for planning and tuning costly experiments.

In the following, we shall illustrate the results of a study [325] on the interaction of superheavy elements Cn, Fl and Og with gold based on the analysis (in terms of the techniques discussed in Section 14.3) of relativistic electron densities obtained through Dirac–Kohn–Sham calculations. The calculations, carried out with a recently implemented full-parallel version of the already-mentioned program BERTHA [300], aimed at a comparative chemical characterization of the above-mentioned superheavy elements – and of their lighter homologs Hg, Pb, Rn and Xe – in their interaction with a gold fragment, mimicking the gold surface used in the experiments. Three gold fragments (a gold atom, a seven-atom planar cluster, and a twenty-atom pyramidal cluster) of increasing complexity were chosen with the purpose of assessing the relative reactivity trends with increasing cluster size. In particular, the computational campaign involved the diatomics Au–E (E = Hg, Cn, Pb, Fl, Xe, Rn, Og) and, limited to those elements whose experimental adsorption enthalpies had been determined, the gold-cluster adducts Au_7– and Au_{20}–E (E = Hg, Cn, Pb, Fl, Rn). For the reader's convenience, the periodic table reproduced in Fig. 17.2 highlights in light red (light gray in print version) color the general collocation of the superheavy elements, in dark red (gray in print version) color the considered E elements, and in blue (dark gray in print version) color the Au element.

The interaction was characterized both in terms of the energetics of the complexes and of the charge redistribution accompanying their formation. As for the energetics, the interaction energies for Au–E (E = Hg, Cn, Pb, Fl, Xe, Rn, Og), Au_7– and Au_{20}–E (E = Hg, Cn, Pb, Fl, Rn) complexes are reported in Fig. 17.3

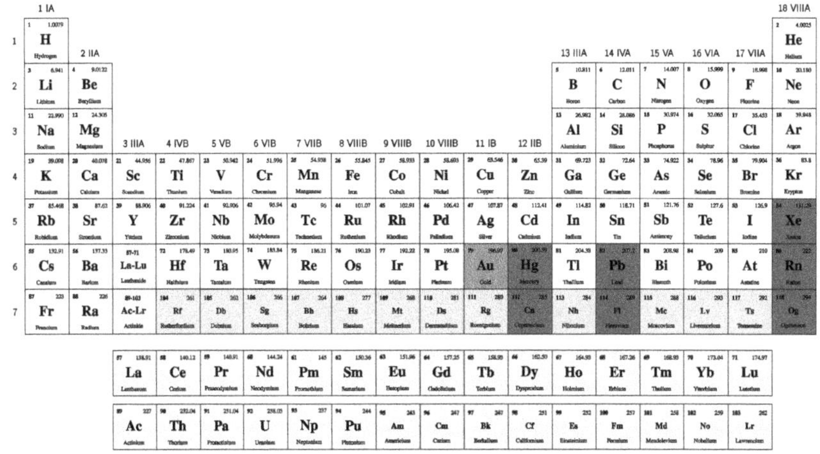

FIGURE 17.2 Collocation of the superheavy elements Cn, Fl, and Og, and some of their lighter homologs on the periodic table. The Au element is also highlighted as the experimental characterization of superheavy elements targets their interaction with gold surfaces.

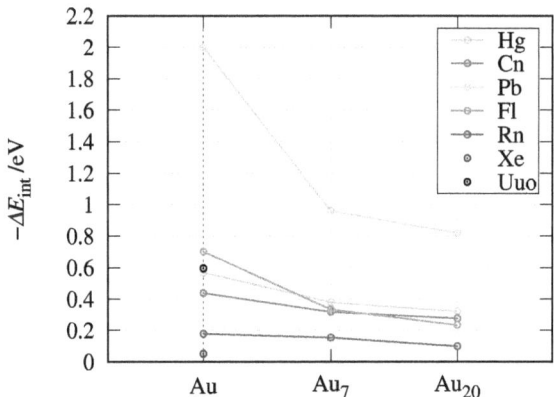

FIGURE 17.3 Interaction energies for Au–E (E = Hg, Cn, Pb, Fl, Xe, Rn, Og), Au$_7$– and Au$_{20}$–E (E = Hg, Cn, Pb, Fl, Rn) as a function of the gold-fragment size. Note that in the original figure, here reproduced as is, Og had not yet been given its final name and features here with its temporary IUPAC name. Reprinted from Ref. [325], with the permission of AIP Publishing. ©2015 AIP Publishing LLC.

as a function of the gold-fragment size. Focusing on diatomics, Cn with its contracted $7s$ orbital, and more markedly Fl with its quasiclosed shell $7p_{1/2}$ outer shell, are less reactive than their lighter homologs Hg and Pb, while an opposite trend is observed for the considered elements of group 18 (Xe < Rn ≪ Og) with Og (featuring highly destabilized $7p_{3/2}$ electrons) resulting even slightly more reactive than Hg. This picture is fully consistent with the above-discussed predictions based on the consideration of relativistic effects on the electronic structure of these elements.

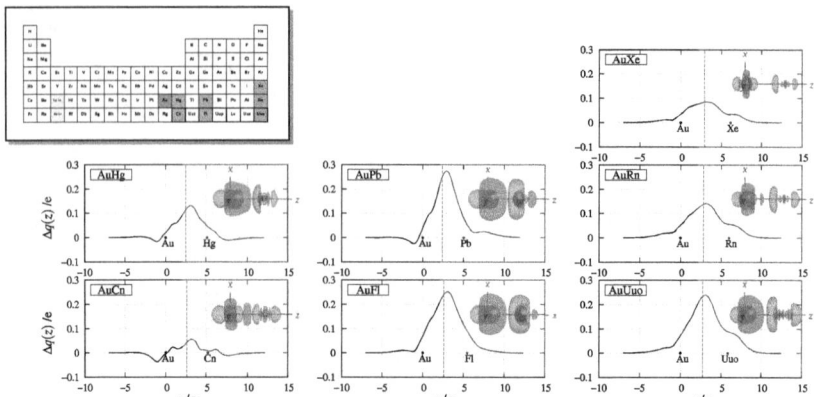

FIGURE 17.4 Charge-displacement (CD) functions for diatomics Au–E (E = Hg, Cn, Pb, Fl, Xe, Rn, Og). Labeled black dots mark the z position of the atoms. A vertical dashed line marks the boundary between fragments Au and E. In the top-right corner of each plot area, the related $\Delta\rho(x, y, z)$ isodensity surface (± 0.0015 e/a_0^3) is shown (blue (dark gray in print version) lobes indicate charge accumulation, red (gray in print version) lobes indicate charge depletion). Plots are arranged according to the position of E in the periodic table sketched at the top-left corner for the reader's convenience. Note that in the original figure, here reproduced as is, Og had not yet been given its final name and features here, together with some other superheavy elements, with its temporary IUPAC name. Reprinted from Ref. [325], with the permission of AIP Publishing. ©2015 AIP Publishing LLC.

Moving on to the cluster complexes, the interaction energies of elements Hg, Cn, Fl, Pb, and Rn with the Au$_7$ and Au$_{20}$ show a decreasing trend with increasing cluster size, with such a trend being more pronounced for elements of group 14 Pb and Fl. As a result, Fl (which in the interaction with a gold atom is modestly more reactive than Hg and Cn) becomes less reactive than Hg in the interaction with Au$_7$ and also of Cn in the interaction with Au$_{20}$. As shown in Ref. [325], due to these inversions, the order of the experimental adsorption enthalpies (Rn < Cn ≈ Fl < Hg ≪ Pb), which is missed in diatomics calculations, is finally reproduced in the cluster calculations.

Now, the interaction energy is one of the criteria on the basis of which a chemical interaction can be characterized and, as we learnt in Chapter 14, complementary information can be gained by the analysis of the charge redistribution upon the interaction between the fragments. CD functions $\Delta q(z)$ for diatomics Au–E (E = Hg, Cn, Pb, Fl, Xe, Rn, Og) are shown in Fig. 17.4. Labeled black dots mark the z position of the atoms. A vertical dashed line marks the boundary between fragments Au and E.[5] Density-difference maps (isodensity surfaces of $\Delta\rho(x, y, z)$ at ± 0.0015 e/a_0^3) are also shown in the top-right corner of the related plot area. For clearer readability, plots are arranged ac-

---

[5] This is defined, as in Section 16.2, as the z point where equal-valued isodensity surfaces of the fragments become tangent.

cording to the positions of E in the periodic table that is sketched at the top-left corner of Fig. 17.4 for the reader's convenience.

A first qualitative insight on the nature of the interaction may be gained by the visual inspection of the density-difference maps. Focusing on that of AuHg, a fraction of charge is seen to flow from Hg (red (gray in print version) lobes in its region indicate charge depletion) towards the partially empty $6s$ outer shell of gold (see the half-sphere-like blue (dark gray in print version) lobe centered on the latter). Charge depletion on the rear side of gold opposite to the bond region is also observed, indicating an intrametal charge reorganization similar to that observed in Section 16.1 of Chapter 16. Density-difference maps of AuPb, AuFl and AuOg, especially in the gold region, closely resemble that of AuHg. A markedly different charge rearrangement is instead exhibited by AuCn that, at least qualitatively, resembles that of the gold–noble gases Xe and Rn compounds.

A more quantitative picture of the charge rearrangement occurring upon bond formation is provided by the CD function. All CD curves shown in Fig. 17.4 are positive in the region between the atoms, measuring the flow of electrons from the E to the Au region of the system. The already-discussed charge depletion in the rear side of gold causes some of the curves (precisely those of Hg, Cn and Pb), to be negative in a short range of $z < 0$. In line with the above discussion of the contour plots, the CD curve of AuCn is markedly different from that of its homolog compound AuHg, the former showing a charge displacement smaller than 0.1e everywhere along z, only comparable to that of AuXe. On the contrary, the compounds of the elements of group 14, AuPb and AuFl, exhibit a very similar CD curve, despite their interaction energies being very different. The CD curves of compounds of elements of group 18, on the other side, show a charge flow all over the molecular region growing rapidly with the atomic number of E, in line with the trend of the corresponding interaction energies (Xe < Rn ≪ Og).

The CD analysis, thus returns a 'chemical' picture of Cn and Og that is consistent with predictions based on the consideration of relativistic effects, and with the results of the analysis of the energetics, according to which Cn – compared to Hg – features a rather inert character and Og – compared to Rn and Xe – features an enhanced reactivity. On the other hand, the differences in the chemistry of Fl and Pb, which is evident in the much larger interaction strength of Pb with respect to Fl due to the quasiclosed shell nature of the latter, are curiously transparent to CD analysis. Preliminary unpublished calculations by the author of this book based on a relativistic extension of the NOCV-CD scheme described in Section 14.3 of Chapter 14, show that the net CD functions of these two systems result from two additive charge-flow components running in opposite directions and very similar to the σ-donation and π-backdonation charge flows of coordination chemistry discussed and analyzed in Chapter 16. Now, these partial charge flows are both much more pronounced (contributing synergistically to a stronger bond) in AuPb than in AuFl, but as they have opposite sign the effect of this simultaneous increase of both flows does cancel in the net CD function, and this explains why the net CD functions for AuPb and AuFl displayed in Fig. 17.4 are very similar despite the rather different interaction strength in the two diatomics.

The topics covered in this chapter, through the discussion of the relativistic theory of the electron, led us back to the beginning of our journey, i.e., to aspects of the fundamental theory governing the properties and behavior of matter in terms of its constituting particles at the 'chemical' level of detail, the nuclei and the electrons. Having so far explored several aspects of the relation between chemistry and physics, we can now proceed on another front in the landscape of the connections between chemistry and computer science, which in part have already tangibly emerged in some of the previous chapters.

Part IV

Chemistry and computer science

Chapter 18

Scientific computing

This last part of the book is devoted to illustrating the connections of chemistry with computer science, especially in relation to the themes treated in the previous parts. We shall start our survey from the most fundamental aspect, that of scientific computing (with which we already had the chance of getting involved in some of the previous chapters), and move on in the next chapters to illustrate challenges and new perspectives offered by recent advances in the field of virtual-reality technology and data-driven science, and by the emergence of open-science paradigms.

Accordingly, this chapter deals with the 'computerization' of a scientific problem. In particular, in Section 18.1 the human/computer communication is addressed and the basics of scientific programming are reviewed in the conventional context of serial computing. In Section 18.2, an overview on parallel-computing architectures and programming models enabling the management of increasingly larger problems is given, with a focus on so-called high-performance and high-throughput computing paradigms. Some of the ideas presented in this section are then applied to a concrete case of a computationally demanding scientific problem in Section 18.3, where the strategy and performances of the parallelization of the Dirac–Kohn–Sham program described in Section 17.2 of the previous chapter are discussed.

18.1 Scientific programming

As we have seen throughout the book, and specifically in Chapters 8, 13 and 15, scientific problems relating to the properties and behavior of molecular systems involve a rather intricate mathematical treatment and, as a matter of fact, due to the involved large number of operations to be performed and data to be handled, seldom can they be solved without the aid of a computer. As already mentioned in Chapter 1, these reasons have made scientific computing a constitutive aspect of theoretical chemistry, so that the discipline is more often than not referred to as theoretical and computational chemistry. Conversely, the progress and the needs of scientific research in physics, chemistry, and related fields, have often represented the driving force in the development of several areas of computer science.

The computerization of a scientific problem involves essentially breaking the problem into a discrete series of instructions to be executed by one or more computers in a predetermined order. It thus in the first place involves a com-

Chemistry at the Frontier with Physics and Computer Science
https://doi.org/10.1016/B978-0-32-390865-8.00029-5

munication between humans and the computer. This communication occurs in the first stage through so-called 'high-level' programming languages, that are languages *tout court* (albeit very peculiar ones, with a limited vocabulary and a grammar polarized towards imperative statements) and act as a communication interface between the programmer and the computer. The instructions are then, in a second stage, translated into 'low-level' binary code instructions that are ready to be processed by the computer.

Depending on the specific adopted programming language, instructions can either be 'interpreted' by the computer on the fly (i.e., the translation into low-level binary code is performed in run time, one instruction after another), or they can be translated into binary code in a preliminary stage, so that the translation does not burden the execution. In this case, the preliminary translation is performed by specific computer programs known as compilers, and the translation process is referred to as compilation. Popular so-called 'interpreted languages' include Java and Python, while 'compiled languages' include C, C++, and Fortran.

The set of instructions required to solve a scientific problem is organized in a computer program, or in portions of code that can be (re)used by one or more computer programs, such as modules (see Table 15.3 for an example of a Fortran module) or procedures made available through external libraries (such as the LAPACK and FFTW mentioned in Chapter 8). To provide the unfamiliar reader with basic principles in writing a scientific program, the anatomy and basic functioning of a computer program will be briefly illustrated in the following, adopting, as already done in Chapters 13 and 15, the terminology of the Fortran 95 programming language.

The basic ingredients of a computer program are variables, operators, and expressions. Variables are abstract storage locations, referenced through a symbolic name, that contain the value of some information. This information, and thus the related variable, can be of different type. Adopting the terminology of Fortran 95, variables can represent integer, real or complex numbers (INTEGER, REAL, COMPLEX, respectively), character strings (CHARACTER), or logical values (LOGICAL, with possible values either .TRUE. or .FALSE.). Additionally, as already discussed in Chapters 13 and 15, values can also take the form of user-defined so-called 'derived data types' (TYPE), i.e., composite data structures resulting from the aggregation of simple (or other derived) data types. Operators can be arithmetic ones (+, -, *, /, and **, being the addition, subtraction, multiplication, division and exponentiation operator), relational ones (==, /=, <, <=, >, and >=, corresponding to, respectively: equal to, not equal to, less than, less than or equal to, greater than, greater than or equal to), or logical ones (such as .AND., .OR., .NOT., .EQV., and .NEQV.). Expressions, as in their mathematical equivalent, combine together variables and operators according to particular rules of precedence and association.

In a computer program, the instructions are coded in 'statements', which are written line after line in the program and that, according to the specific syntax of the adopted language, combine variables, operators, expressions, and language-specific directives such as those related to input/output operations. The value of a variable (which can also be the result of an expression) is (re)set through a so-called 'assignment' statement, which in Fortran 95 is performed through the = operator.

The typical structure of a computer program can be easily illustrated by commenting the code example given in Section 15.1 of Chapter 15 (PROGRAM cda). The first statement marks the beginning of the program and also specifies its name. There then follows a section (lines 2 and 3) where external modules (if any) that are referenced in successive lines of the computer program are listed. The next section of code (lines 4–6) is dedicated to the so-called variable declaration, where all the variables referenced in the program are listed and 'declared' to be of a given type. The main body of the program, comprised between the variable-declaration section and end of the last 'end program' statement (line 12) contains the actual operational statements encoding the operations that have to be executed at run time (lines 7–11).

The statements of a computer are typically executed one after the other according to a top-to-bottom flow. However, this flow can be altered if special instructions organized in the form of 'cycle' or 'alternative' constructs are encountered, or if an instruction is encountered referencing an external procedure (such as those contained in module cubes reported in Table 15.3 of Chapter 15, of which the function cube_get_natom is explicitly shown). In this case, the external procedure is entered and the contained portion of code is executed before returning to the main program.

With no pretense of completeness, the following two simple examples will give an idea of how the execution flux is altered by 'cycle' or 'alternative' constructs. In the following basic 'cycle' construct (where variables isum and i would have been preliminarily declared of integer type):

```
1   isum = 0
2   DO i = 1, 10
3      isum = isum + i
4      ...
5   ENDDO
```

first isum is assigned the value 0 (line 1). Then, an iterative section is reached (lines 2–5), where the indented block of instructions is repeated ten times with the variable i taking the values from 1 to 10 with a default increment of 1. During each iteration, isum is assigned the result of the expression isum + i. Accordingly, at the end of the cycle the value of isum will be 55.

In the following basic 'alternative' construct:

```
1  IF ( i > 0) THEN
2     block A
3     ELSEIF ( i < 0) THEN
4     block B
5     ELSE
6     block C
7  ENDIF
```

different blocks of instructions are executed according to the result of the logical expressions $i > 0$ (line 1) and $i < 0$ (line 3). In particular, the instructions of *block A* are executed if $i > 0$, those of *block B* if $i < 0$, those of *block C* if $i = 0$.

The above-sketched account mainly refers to a so-called 'procedural' programming paradigm, where the focus is on the sequence of instructions to be executed. As we have already discussed in Chapters 13 and 15, in a more elaborate and powerful programming paradigm, computer programs are designed around the concept of 'object' and built in terms of a collection of discrete objects that incorporate both data and behavior. The above-outlined general picture, coupled to a self-guided study of the specific syntax details of Fortran 95, should contain enough information for eager students to proceed independently and write their own first computer programs, or even more elaborate object-based ones, leveraging on the additional discussion on the code examples given in Chapters 13 and 15. For compiling a computer program written in Fortran 95, a freely available resource is the GNU Fortran compiler available for download at https://gcc.gnu.org/wiki/GFortran (accessed on 8 February 2022).

18.2 High-performance and high-throughput computing

The scientific-computing paradigm exemplified in the above discussion is based on the principle that the instructions codified in the program statements are executed by the computer one after another in a predetermined order. This paradigm is traditionally referred to as 'serial computing'. In serial computing, no matter how fast the computer may be in accessing memory locations and executing the instructions, the processing speed and the available 'space' for memory is necessarily finite due to the physical limitations of the involved circuitry. As a result, the performances of a serial computer program, however efficient and optimized its algorithms may be, will ultimately be limited by intrinsic hardware limitations, and in fact the computing power of even a last-generation single computing unit is easily saturated by computational problems addressing increasingly larger systems of scientific interest.

As a result, the increasingly more demanding needs, in terms of computer resources, of the scientific community in the last decades of the last century were very fertile in fostering the development of efficient strategies for addressing problems of increasing complexity on a double front. On the one hand, they channeled an important part of the research towards the development of efficient algorithms and approximate schemes capable of providing accurate results at a low computational cost. For instance, in theoretical and computational chemistry (where, as we have seen in previous chapters, the cost of a calculation typically scales with the square of the system size or more unfavorable power laws) the 'holy grail' has been for a long time the so-called 'linear scaling', i.e., achieving a linear scaling of the computational cost with the size of the system (see, for instance, Ref. [326] for an overview on linear-scaling techniques).

On the other hand, they gave birth to a new computing paradigm (and a brand-new related branch of computer science), broadly referred to as 'parallel computing', where essentially multiple computing units concur simultaneously to the execution of a computer program. While in fact, the main tendency in the market of computing hardware had been for a while that of pushing the performance of individual computing machines (by achieving a shorter 'clock period' in the processors, faster communications, higher miniaturization or by adopting optimization strategies such as 'look ahead and prefetching'), the turn of the 21th century saw a paradigm shift from serial computation to parallel computation based on multicore and multiprocessor machines, where the time and memory barrier to the solution of demanding computational problems are removed by putting together the efforts of multiple computer resources.

In the following, the basic concepts of parallel computing that will form a basis for the discussion of the parallelization of a relativistic density-functional theory program in Section 18.3, will be introduced. An excellent and more extensive tutorial on parallel computing, which has been one of the main sources for the following discussion, is available on the website of the Lawrence Livermore National Laboratory (https://hpc.llnl.gov/documentation/tutorials/introduction-parallel-computing-tutorial).

As already anticipated, the core idea in parallel computing is that of exploiting the simultaneous work of many computing units for the solution of a computational problem. In other words, the computational problem is broken down into computational tasks (some of) which can be independently executed by different computing units at the same time with an evident saving in the overall 'wall-clock' execution time. The computing hardware used for this purpose can be either one computer with multiple processors, or many computers connected by a network, or a combination of the two. This gives rise to three main classes of parallel architectures: the 'shared-memory' one, where all computing units have access to a common memory, the 'distributed-memory' one, where each computing unit has direct access to its own memory but has to resort to network-based communication in order to access to the memory of the other computing units, and a hybrid one resulting for instance from several multicore shared memory–computing units connected by a network.

These hardware concepts map then onto analogous concepts referring to different parallel-programming models. Accordingly, in a shared memory–parallel programming model all parallel processes have the same 'picture' of memory, and in a distributed memory–parallel programming model, each process only sees the local memory and must use communication through a network to access the memory of other processes.

A typical shared memory–parallel architecture is trivially represented by the multicore processors that can be found in modern workstations or laptops, and a common and efficient way of parallelizing a given serial computer program through a shared memory–parallel programming model is that of including in the program appropriate compiler directives through the OpenMP API specification (https://www.openmp.org/).

On the other hand, the spectrum of distributed memory–computing machines can range from so-called computing clusters (resulting from commodity units connected through a network to build a parallel system) to the world's largest and fastest supercomputing machines, of which a list of the top 500 can be accessed at https://www.top500.org/. The parallelization of a computer program based on a distributed memory–programming paradigm is typically more 'invasive' than the parallelization based on a shared-memory paradigm, as here the parallel execution of the program must now involve network communications among the several parallel processes. This typically requires a (sometimes significant) reorganization of the program and the adoption of external so-called 'message-passing' libraries such as those based on the Message Passing Interface MPI (https://www.mpi-forum.org/) standard (see for instance Open MPI https://www.open-mpi.org/).

A schematic illustration of the basic principles of parallel computing based on a distributed-memory architecture is given in Fig. 18.1. Panel a) shows a single computing unit requiring a certain amount of time and a certain amount of memory to perform the job. A first level of parallelization of the program can be obtained by distributing the workload among four network-connected computing units, on each of which the entire data have been replicated (panel b). If the computational problem at hand lends itself to an optimal parallelization of the workload, then when using four computing units the execution time would be reduced of a factor of four, though the memory requirements on each computing unit would remain unchanged with respect to the serial execution. A second level of parallelization can instead be achieved by distributing among the computing units not only the workload, but also the data involved in the execution (panel c), in such a way that each computing unit handles only a portion of the data during the whole execution time. In so doing, if the distribution of the data can be performed in an ideal way, then also the memory requirements would be reduced by a factor of four. Assuming now the same above-mentioned ideal conditions, by doubling the number of concurrent computing units would further halve the computational cost in terms of both time and memory (panel d).

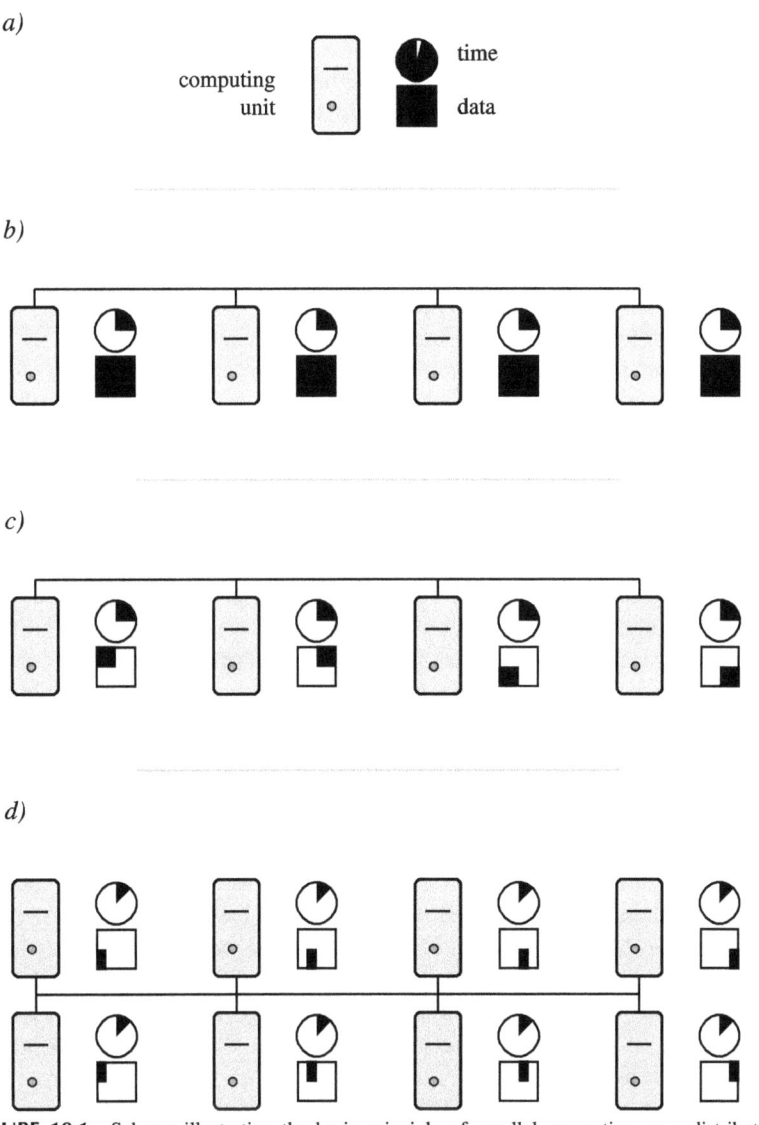

FIGURE 18.1 Scheme illustrating the basic principle of parallel computing on a distributed-memory architecture. Panel a): a single computing unit performs the job in a certain amount of time using a certain amount of memory. Panel b): the workload is distributed among four computing units connected by a network. Panel c): both the workload and the data are distributed among four computing units connected by a network. Panel d) in ideal conditions, doubling the number of concurrent computing units halves both time and memory requirements.

As one might easily guess, the above-mentioned ideal conditions are seldom met in reality. First, not all portions of a computational problem are parallelizable, as some of them might require the results of some others before they can be executed. Furthermore, a full distribution of the involved data might not be possible as the same portion of data might be required by each parallel process and will thus have to be replicated on all the computing units for a correct execution of the local, distributed workload. In the second place, communication between the parallel processes is among the main sources of the so-called 'parallel overhead', i.e., required additional execution time (with respect to useful work) arising from the parallelization of the program. In optimizing the performances of a parallel program, one thus has to take into account a number of factors including the so-called 'granularity', ranging from coarse to fine, which is a measure of the ratio of computation over communication.

The parallel performances are defined in terms of the so-called 'parallel speedup', i.e., the wall-clock time of serial execution over the wall-clock time of parallel execution, and by the so-called 'scalability' of a parallel program, i.e., the scaling of the speedup with the number of utilized computing units. As we shall see in a more detailed way in the next section, the scalability of a parallel program is seldom as linear as that shown in Fig. 18.1, and is usually affected by a number of factors including hardware features, the proneness of the specific problem to parallelization, and the above-mentioned parallel overhead. The parallelization of a computer program thus often involves a profiling of its performances and a fine tuning of the adopted parallelization strategies in order to reduce the adverse factors and achieve the highest possible speedup and scalability.

The above-mentioned architectures, concepts, and tools form the typical domain of so-called high-performance computing (HPC). A related though different parallel computing paradigm is that embodied by the so-called high-throughput computing (HTC). HTC is based on grid infrastructures and has been established in Europe mainly as a support to the high-energy physics transnational community through three subsequent EGEE I, II, and III projects (Enabling Grids for E-sciencE, https://eu-egee-org.web.cern.ch/, accessed on 8 February 2022), later evolved into the European Grid Infrastructure (EGI, https://www.egi.eu/, accessed on 8 February 2022). While in HPC the focus is on tightly coupled parallel processes that benefit the most from low-latency communications between 'local' computer resources gathered in a given site, HTC is based on infrastructures that coordinate (through the use of appropriate so-called 'middleware') a large number of geographically distributed computer resources connected over the public network, and is thus mostly suited for the concurrent execution of a large number of independent or loosely coupled jobs that are individually scheduled on different computer resources across multiple administrative boundaries.

An example of an HTC approach to some of the topics covered in Part II of the book is represented by the assembly of the so-called grid empowered–

molecular simulator GEMS [327–330] for the *a priori* modeling of elementary reactive processes [331–334], which has been successfully used to study the dynamics and compute the reactive probabilities/cross sections and rate coefficients of atom–diatom systems such as H + H_2 [335], Li + FH [334], N + N_2 [336,337] and O + O_2 [73,77]. GEMS is articulated in four blocks (Interaction, Fitting, Dynamics, and Observables) directly related to the stages discussed through Chapters 6 to 9 (i.e., electronic-structure calculations, potential-energy surface fitting, nuclear dynamics, reaction kinetics) featuring a high degree of interoperability thanks to the definition of the common data formats Q5Cost/D5Cost [335,338]. These four blocks are part of a workflow designed to enable the coordinated execution of in-house developed and commercial codes on the distributed platform of the EGI [339] by properly selecting computer resources among the high-performance computing and high-throughput computing available ones.

18.3 Parallelizing a Dirac–Kohn–Sham program

In this section, some of the ideas discussed in the previous section will be illustrated through a real case of parallelization of a computationally demanding scientific problem. In particular, we will address the parallelization of a relativistic density-functional theory program based on the Dirac–Kohn–Sham formalism described in Section 17.2. Based on the results reported in Refs. [300] and [340], to which the reader is referred for further details, we will describe an efficient parallelization strategy based on a distributed memory–parallel programming paradigm and analyze the performances of the resulting parallel program.

As discussed in Section 17.2, full four-component Dirac–Kohn–Sham calculations have an intrinsically higher computational cost, both in terms of computing time and required memory, than analogous nonrelativistic calculations. This arises from the complex representation required for the involved matrices and the intrinsically larger basis sets, descending from the four-component structure of the Dirac–Kohn–Sham equation. Regarding computing time, however, only a larger prefactor in the scaling with the number of atoms or basis set size shows up, not a more unfavorable power law.

As in analogous nonrelativistic calculations, once the overlap (Eq. (17.22)) and one-electron (Eqs. (17.19) and (17.23)) integrals have been evaluated and an initial guess of electron density has been computed, an iterative section is entered involving i) the evaluation of the density-dependent J and K integrals (Eqs. (17.20) and (17.21)) for the assembly of the H_{DKS} (Eq. (17.18)) matrix, and ii) the solution of the eigenvalue problem (Eq. (17.17)) with up-to-date J and K integrals at each cycle. In the program BERTHA, during the SCF procedure the 'bulk' memory allocation is due to the following $2N \times 2N$ ($2N$ being the number of G-spinors employed for the expansion, i.e., the basis-set size) complex matrices: i) the overlap matrix S, ii) the one-electron matrix $\Pi + v$, iii) the Coulomb plus exchange-correlation matrix $J + K$ (once this matrix has been computed, the associated array is conveniently reused to host the whole H_{DKS} matrix to be diagonalized), iv) the matrix of the eigenvectors c, and v) the density matrix D. Accordingly, the memory required for a calculation on a Au_{32} gold cluster with double-ζ quality basis set ($2N = 25216$) amounts at least (additional working space is actually required) to 47.35 GiB using double precision.

In terms of the above-mentioned matrices, the main computational kernels involved in a SCF cycle are:

i. build up the Coulomb plus exchange-correlation $J + K$ matrix (during the first iteration also the one-electron $\Pi + v$ and overlap S matrices are computed and henceforth conveniently stored in memory, as they do not vary from cycle to cycle);

ii. assemble H_{DKS} and solve the eigenvalue problem for a new matrix of the eigenvectors c and density matrix D.

A full-parallel implementation of the program involving the distribution of both the workload and the data, and corresponding to panel c) and d) of Fig. 18.1, would then be based on a scheme whereby all of the required potentially huge complex matrices are, at any time during the execution, apportioned among the parallel processes and the original SCF procedure is replicated – in 'smaller scale' – on all of the processes in a 'symmetric' fashion, so that every process executes the same instructions on subsets of the global matrices. Of course, the distribution scheme for the matrix elements should be such that at each time step data are always distributed among the various processes to obtain the best performance with respect to the actual parallel operations that have to be performed.

Now, as we have discussed above, the SCF procedures involves mainly two computational kernels, the construction of the $J + K$ matrix and the diagonalization of H_{DKS}, and a serious hindrance to the adoption of the above-mentioned distributed-memory scheme is that the data-distribution scheme that better fosters the parallel computation of the $J + K$ matrix is not the same distribution scheme that is optimal for the parallel diagonalization of H_{DKS}. In fact, an optimal parallel construction of the $J + K$ matrix is achieved by cyclically assigning to each process the allocation and computation of blocks of the matrix whose offsets and dimensions depend on the specific structure of the G-spinor basis set. Such a block structure, which we shall refer to as integral-driven distribution (IDD), is naturally dictated by the grouping of G-spinor basis functions in sets characterized by common origin and angular momentum. On the other hand, parallel linear algebra operations are efficiently performed through the ScaLAPACK library routines (http://www.netlib.org/scalapack/, accessed on 8 February 2022), which in turn require a different distribution scheme known as block-cyclic decomposition (BCD). Moreover, the way the local subsets of the global $J + K$ matrix are represented on each of the parallel processes significantly differs in the two distribution schemes. In the BCD scheme, the local subset on one process is represented as a rectangular array whose dimensions are both approximately $2N/P$. In the IDD scheme, instead, the distribution is much less regular with local subsets on each process made up of rectangular blocks of different sizes.

The two steps (i, matrix construction) and (ii, linear algebra) involving the same matrices in two different distribution schemes have thus to be bridged together during the execution by an interprocess communication step where matrix elements are properly exchanged between the parallel processes in order to switch the distribution scheme from IDD (suitable for step i) to BCD (suitable for step ii). In Refs. [300,340], an efficient mapping procedure was devised according to which, once the $J + K$ IDD blocks have been computed on the various processes, the program enters a communication phase where each process, in turn, plays the role of the sender while the other processes act as receivers. The sender broadcasts some preliminary information to all of the other processes and sends properly packed data to each of them; the receivers allocate memory buffers according to the broadcast information, receive the data and unpack them according to the BCD scheme. After completion of step ii), the computed D matrix – which, after its computation, will naturally be distributed among the process according to the BCD scheme – will have to be redistributed among the processes according to the IDD scheme before entering step i) of the next SCF cycle. The algorithm for the backward (BCD to IDD) mapping is essentially the same, in reverse, as the forward one (IDD to BCD), the main difference between the two procedures lying in the pack and unpack phases, where a set of destination indices has to be computed as a function of a set of origin indices.

As mentioned above, such a parallel scheme has the advantage that the flowchart for the iterative section of the SCF procedure (i.e., the analog of the portion highlighted with the light blue (gray in print version) frame in the non-relativistic Hartree–Fock case of Fig. 13.1 in Chapter 13) is identical for all the parallel processes involved and it perfectly superimposes on the original (serial) SCF procedure. The original SCF procedure is in fact replicated on all the parallel processes with the computational effort required by each of them decreasing with the number of processors used as a consequence of the fact that each process works on subsets of the global matrices, at the price of the following additional explicit communications (besides those internal to the ScaLAPACK routines): i) mapping of the $J + K$ matrix from IDD to BCD (also of the $\Pi + v$ and S matrices at the first iteration) and ii) mapping of the D matrix from BCD to IDD.

In order to evaluate the performance of the above-outlined implementation, in Ref. [300] test calculations on a representative set of gold clusters (Au_8, Au_{16}, and Au_{32}) were performed. The corresponding Dirac–Kohn–Sham matrix sizes are 6304 for Au_8 (0.59 GiB), 12 608 for Au_{16} (2.37 GiB), 25 216 for Au_{32} (9.47 GiB). To obtain a quantitative picture of the scaling of the time and memory requirements of the program with the number P of parallel processes (coinciding with the number of utilized computing units), the left panel of Fig. 18.2 reports

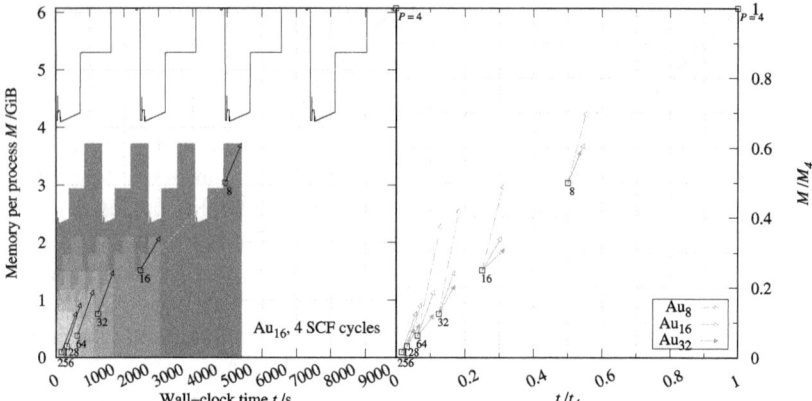

FIGURE 18.2 Left: Memory used by one process as a function of the execution time during 4 SCF cycles of a relativistic Dirac–Kohn–Sham calculation on Au_{16}, using a number of processors P increasing in doubling steps from 4 (larger area) to 256 (smaller area). Arrows are displayed joining the ideal and the actual performance points relative to the $P = 4$ case (see text for discussion). Right: Same plot including Au_8 (up to $P = 64$) and Au_{32}, using scaled coordinates t/t_4 and M/M_4, and retaining only the arrows. The reference values (t_4, M_4) are (1295 s, 1.83 GiB), (9874 s, 6.07 GiB) and (77849 s, 23.3 GiB) for Au_8, Au_{16} and Au_{32}, respectively. Reprinted with permission from Ref. [300]. © 2014 American Chemical Society.

the memory used by one of the parallel processes (which is essentially the same as that used by any other process) as a function of the wall-clock execution time during 4 SCF iterations on Au_{16}.

Seven curves (six of them are color-filled for a better visualization) are displayed, one for each run employing a different number P of processes that, from top to bottom, is 4, 8, 16, 32, 64, 128 and 256. The final point of each curve (indicating the memory used at the end of the corresponding run, which also corresponds to the maximal memory usage) is connected by an arrow to a point on the diagonal of the plot that represents, relative to the $P = 4$ case, the ideal (time, memory) linear performance (which halves at each P doubling step). Thus the length of each arrow is a measure of the efficiency of the corresponding run, both in terms of memory and time requirements. The figure clearly shows that both memory and time requirements are pushed towards the bottom-left area of the plot with increasing number of parallel processes, and thus that the above-outlined approach makes both time and memory scalable with the number of processors used, allowing to virtually overcome at once both the time and the memory barrier associated with Dirac–Kohn–Sham calculations. While the arrows appear to indicate that the memory scaling is somewhat worse than the time scaling (the y component of the arrows is longer than the x component), this is an effect of the included residual (nonscalable) workspace, which is very small in absolute practical terms (few GiB at most) and grows very little with system size.

The right panel of Fig. 18.2 shows the same arrows as the left panel using scaled x and y units, but now includes also the data relative to the Au_{32} and Au_8 systems (only results up to $P = 64$ have been reported in this last case, as higher P runs turned out to be too fast to measure the memory usage). For Au_8, Au_{16} and Au_{32} the reference values (t_4, M_4) are (1295 s, 1.83 GiB), (9874 s, 6.07 GiB) and (77 849 s, 23.3 GiB), respectively. As is evident from the figure, the arrows tend to collapse in going from Au_8 (dashed line empty head) to Au_{16} (solid line empty head) to Au_{32} (solid line filled head) – indicating that better performances are obtained for large, more demanding systems – and the memory usage per process is kept within 2 GiB using 256 processors even for the costly Au_{32} case.

Chapter 19

Virtual reality

"It is nice to know that the computer understands the problem. But I would like to understand it too." This sentence, which is quoted in the opening of an interesting tutorial review by Valle on scientific visualization in quantum chemistry [341], is attributed to the Hungarian-American physicist Eugene Paul Wigner (Nobel Prize winner in Physics in 1963) [342] and may as well serve as the starting point for this chapter. As we have seen throughout the previous chapters, modern molecular sciences make extensive use of computational methods based on an intricate mathematical formalism and involving the management of a huge amount of data. Due to the recent growth of the available computer power, computational campaigns targeting increasingly larger systems result in the production of data sets of increasing size and complexity. Yet, as clearly pointed out in the above-cited Ref. [341], what researchers look for is insight, not numbers.

This chapter is devoted to the transformation of numbers into insight, which is nowadays typically achieved through computer-graphics technology and is the domain of an inherently interdisciplinary branch of computer science referred to as scientific visualization. In particular, Section 19.1 gives an overview on the challenges of scientific visualization with a focus on immersive virtual reality (IVR) environments, while Sections 19.2 and 19.3 illustrate two applications based on different IVR technologies and specifically targeting the topics treated in Parts II and III of the book, i.e., chemical reactions and chemical bonding, respectively.

19.1 Scientific visualization and virtual reality

Chemistry is by itself a deeply visual discipline. Its very language largely consists in graphical diagrams, such as chemical formulae or reaction mechanisms, which can also take the form of three-dimensional models such as the popular ball-and-stick molecular models, with the balls representing atoms and the sticks signifying chemical bonds. On the other hand, as we have learnt through Parts II and III of this book, the mathematical treatment underlying these simplified abstractions is inherently complex, being itself an active research field. Consider an electronic-structure calculation performed through the methodology illustrated in Chapters 12 and 13. Given an input geometry for a given molecular system and a basis-set definition, the results of a calculation formally consist in a potentially huge set of numbers being the collection of the expansion coefficients of the occupied and virtual molecular orbitals, which are in turn related to the spatial distribution of the electrons. Soon after the computation, however, the 'meaning' of this set of numbers in most cases remains opaque to our minds, unless the contained information is converted in shapes and colors through some graphical structures such as, for examples, isodensity surfaces (such as those of Fig. 17.4) or direct volume rendering (as in Fig. 16.2) of the orbitals or of the electron density.

Now, the art of bringing to light the meaning of scientific data by transforming these into graphical structures is the essence of scientific visualization. Scientific visualization aims at complementing our rational activity in the analysis of data with the capability of the human visual system to identify structures, patterns, relations and anomalies in the images, for a quicker and deeper understanding of the visualized data. In an inherently digital era, it will not be surprising that nowadays scientific visualization largely relies on computer-graphics technology, and it will take little effort for the reader to realize that computer graphics has been to a great extent exploited to convey central ideas throughout this book. Computer graphics specifically dedicated to the representation of the molecular world is often referred to as 'molecular graphics'. The interested reader can find interesting accounts on the topic of molecular graphics in Refs. [343] and [344].

In this context, a revolutionary moment has been represented by the advent of IVR technologies, which have opened the unprecedented scenario whereby scientists can immerse themselves in three-dimensional representations of the molecular world. In fact, by means of IVR technology, the real environment in which a user is physically located is either replaced by a three-dimensional environment that is virtual, constructed through computer graphics, or mixed with a virtual-reality environment, or augmented with additional information deriving from virtual-reality content that is updated according to the movements and the position of the user, giving rise to the concepts of virtual-, mixed-, and augmented-reality that are all comprised in the term 'extended reality'.

Scientists have been trying to exploit IVR technologies in molecular sciences since the late 1960s. However, previous generations of hardware and software have been for a long time too limited for an effective usage [345]. For some time, moreover, IVR technology has been the preserve of a few laboratories that could afford expensive, highly specialized hardware to assemble room-sized multiscreen projection theaters, such as the cave automatic virtual environment (CAVE), [346,347]. On the other hand, this scenario is nowadays rapidly changing due to the recent introduction of a new generation of consumer-grade immersive head-mounted displays such as the Oculus Rift (now evolved in Quest2, see https://www.oculus.com/, accessed 10 on February 2022) or HTC Vive (https://www.vive.com, accessed on 10 February 2022) which, thanks to their relatively low cost, will presumably lead in the near future to a widespread usage of IVR technologies in scientific research and education. The reader interested in a detailed account of the use of IVR in chemistry, can find an early review in Ref. [348], while a very recent one is Ref. [349]. Further interesting readings are Ref. [350], specifically targeting the use of IVR in chemical education, Ref. [351], on the relation between IVR and the learning process, and Ref. [352] for a philosophy-of-science perspective on the subject.

There are two main features that make IVR a revolutionary paradigm in doing research and education in Chemistry: i) the fact that users are immersed in a three-dimensional representation of the molecular world in the same way as they are normally immersed in the real physical world, and ii) the fact that users can interact with virtual molecular objects in the same way as they interact with physical objects in their everyday life. Let us examine more closely these two aspects.

In IVR, the immersion in a three-dimensional virtual world allows users to exploit their proprioceptive system to enhance the perception of the represented virtual content, thus fostering spatial deductions about the dimensions, proportions and topology of complex data. This is particularly true for inherently three-dimensional data such as surfaces, volumetric data and vector fields. Think, for instance, of the graphical representations given in Fig. 14.5 of Chapter 14 for the electron-charge redistribution occurring upon formation of NH_4^+ from NH_3 and H^+. In that case, forced to a two-dimensional representation on a book page, only a two-dimensional projection of the three-dimensional data could be shown, with an inevitable loss of information with respect to the original, fully three-dimensional data. In order to reveal some three-dimensional features that would have remained hidden otherwise, the same three-dimensional data were represented twice, i.e., through two-dimensional projections obtained from two points of view, a front view in the left panel and a side view in the right panel. By looking at the two panels of the image, the reader can then in a subsequent, synthetic stage, mentally reassemble the partial information conveyed by the two panels and reconstruct the three-dimensional features of the represented volumetric data. On the contrary, through IVR technology users would share the same environment with a 'floating' three-dimensional representation of the electron-charge redistribution, and they could use their body to move around and inspect the features of three-dimensional objects in exactly the same way that they would naturally do in the real world with real objects. This makes the analysis much more immediate and efficient compared to the on-paper or on-screen visual inspection of two-dimensional projections, often disclosing features that would otherwise remain unnoticed.

Besides the advantages deriving from immersion in a virtual world, however, the true power of IVR derives from the possibility of interacting with the content of that world. In fact, with specific reference to chemistry, this offers the unprecedented perspective of bringing to life, and right under our nose, the microscopic world of atoms and molecules, bridging the gap between that world and our macroscopic one through virtual representations of molecular systems at a human-like space and time scale. This unlocks the unprecedented possibility of interacting with atoms and molecules and watching them respond to the laws of physics and, as a consequence, of learning about their behavior in an incredibly simple and natural way, which would replace or – more fruitfully – complement our understanding of the molecular world deriving from the use of more conventional, century-old 'intermediary' instruments such as mathematical expressions, graphical structures or verbal content. Moreover, the inherent complexity of the mathematical formalism required to model atoms and molecules would remain, so as to say, 'hidden' inside the computer rather than spouting in the form of huge data sets after the calculations. Conversely, the results of a calculation would be provided directly in a most insightful form. Of course, as recently highlighted by Aspuru-Guzik, Lindh and Reiher in Ref. [353], a major challenge in this respect is that of achieving the so-called 'instantaneous computing', i.e., the fact that the computing time for modeling the behavior of a molecular system should be shrunk to the point that the start and the end of the execution can no longer be separated on the time scale of human perception, which is 60 ms for vision and 1 ms for the sense of touch.

Impressive work has been achieved in this direction by the group of Reiher, especially concerning the exploration of potential-energy surfaces through haptic devices [354–356], and by the group of Glowacki, whose work led to the development of a multiuser IVR framework for real-time, interactive, molecular-dynamics simulations [357–359]. In order to provide the reader with a direct feeling of the application of IVR technology to chemistry on themes that have been covered in this book, we shall in the following illustrate two IVR applications developed with contributions from the present author, specifically targeting the exploration of potential-energy surfaces (Section 19.2) and the analysis of chemical bonding (Section 19.3).

19.2 A walk through chemistry: immersive exploration of potential-energy landscapes

The first application, AVATAR (advanced virtual approach to topological analysis of reactivity), was released in 2020 [360]. Based on last-generation head-mounted displays and hand-held controllers, AVATAR takes advantage of IVR for the specific purpose of the immersive exploration of potential-energy surfaces based on the following two key concepts: 1) the reduction of the dimensionality of the potential-energy surface to two process-tailored, physically meaningful internal coordinates, and 2) the analogy between the evolution of a chemical process and a pathway through valleys (potential wells) and mountain passes (saddle points) of the associated potential-energy landscape.

In Section 6.3 of Chapter 6, we explicitly addressed some issues arising when attempting to visually analyze potential-energy surfaces. As discussed therein, the potential-energy surface of a given electronic state (typically the focus is on the electronic ground state) of an N-atom molecular system is an inherently multidimensional function, or hypersurface, depending on the related $3N - 6$ internal coordinates. Such complexity is sometimes oversimplified by reducing the potential-energy surface to a number of 'connected' stationary points making up a one-dimensional diagram, such as that shown in Fig. 9.1, accounting indeed for a large part of the involved chemistry. On the other hand, modeling and interpreting spectroscopic and dynamical features of a molecular system requires the analysis of the potential-energy surface as a function of at least a few degrees of freedom, typically those that are mostly involved in the physical or chemical process of interest.

Given the peculiarities of the human visual system, one of the most informative graphical representations of a potential-energy surface that one can obtain is that resulting from an appropriate reduction of the original $(3N - 6)$-dimensional potential-energy surface to a two-dimensional one. In fact, unless a nongeometric encoding is used for the energy (such as, for instance, color, degree of opacity or even sound), in the three-dimensional space there are only two dimensions available for describing the geometry of the system, since the third dimension has to be used to represent the potential-energy values. As discussed in Section 6.3, a useful approach thus is that of opportunely reducing the dimensionality to two coordinates carefully chosen so as to condense in them the motion relevant to the process of interest. Then, an informative three-dimensional representation can be built where the selected coordinates are used as the latitude and the longitude, and the associated value of the potential-energy surface – opportunely minimized with respect to all other coordinates – is used as the elevation.

We have seen examples of this dimensionality reduction in the context of gas phase–few atom elementary processes through the so-called rectangular and triangular relaxed plots of Figs. 6.4 and 6.5, respectively, and successfully exploited one of these representations for the analysis of the potential-energy surface of the astrochemical reaction $C + CH^+ \rightarrow C_2^+ + H$ in Chapter 10. Another example relating to a different kind of physical process is given by 'circular relaxed plots' obtained by adopting the Cremer–Pople so-called 'ring-puckering' coordinates [361] for the description of ring-puckering motions (which can be probed by spectroscopic techniques) in the context of the conformational analysis of flexible cyclic molecules. In the case of five-term rings, the use of Cremer–Pople coordinates [361–363] allows for a complete and compact description of ring-puckering motions through only two coordinates: a puckering amplitude q, and a pseudorotation angle θ. Useful two-dimensional circular relaxed plots analogous to the already familiar rectangular and triangular relaxed plots can then be obtained by plotting the potential energy as a function of the two polar coordinates q and θ (with $q > 0$ and $0 < \theta < 2\pi$) after minimization of the energy with respect to all of the remaining coordinates.

For illustrative purposes, an example of circular relaxed plot is given in Fig. 19.1 for the 3S-chloro-1,2-dithiolane molecule. In this kind of plot, each couple (q, θ) is associated with a specific conformation. The value of the amplitude q represents a measure of the distances of the five vertices of the ring from the mean plane, and thus relates to a distortion from planarity (for $q = 0$,

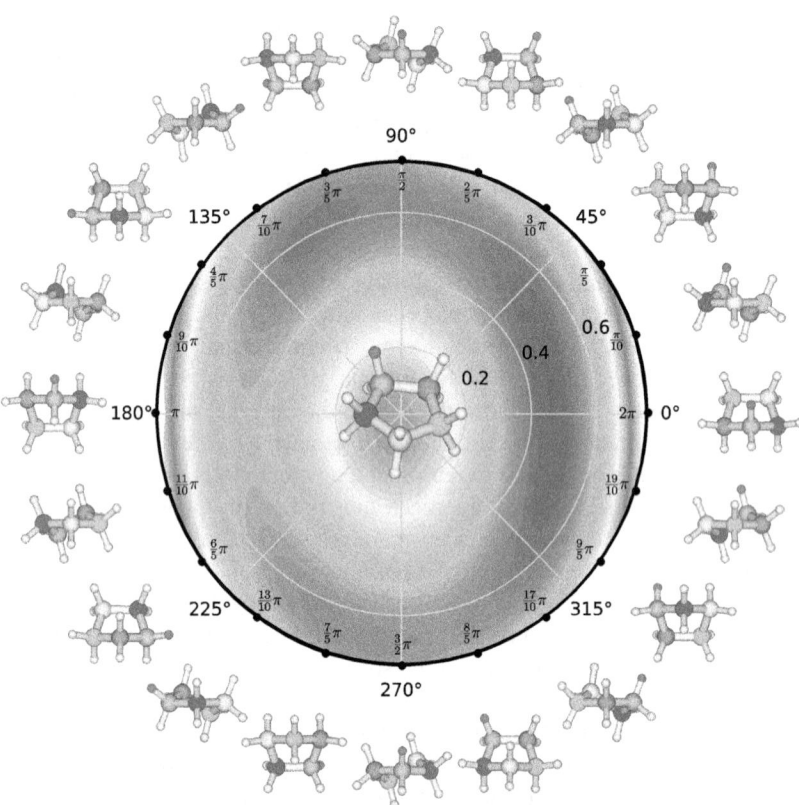

FIGURE 19.1 Two-dimensional potential-energy surface for ring-puckering motions of 3S-chloro-1,2-dithiolane expressed in Cremer–Pople coordinates. For twenty different values of the pseudorotation angle θ, the corresponding conformations are depicted outside the circular plot to clarify the physical meaning of the θ coordinate. Reprinted with permission from Ref. [363]. ©2019 American Chemical Society.

the five vertices lie on the mean plane, and the conformation of the five-term ring is planar). The pseudorotation angle θ relates directly with the possible ring-puckering motions and the description of its meaning in relation to the conformations of the five-term ring is somewhat less straightforward. This can more easily be accessed through the graphical illustration provided in Fig. 19.1, which clarifies the physical meaning of the pseudorotation angle through a representation of each conformation associated to a particular value of θ.

The image in Fig. 19.1, as well as the analogous images reported in Chapters 6 and 10, share many similarities with geographic maps representing the reliefs of a given terrain by means of the same isocontour-line representation. This evokes the suggestive idea of a potential-energy surface as a three-dimensional landscape that can be explored, exactly as a real landscape, simply by walking over it, and suggests a powerful analogy between the evolution of a physical or chemical process and a pathway through valleys (potential wells) and mountain passes (saddle points) of the associated potential energy landscape.

AVATAR is an IVR software specifically designed for the immersive exploration of potential-energy landscapes based on the HTC Vive head-mounted display and available at http://smart.sns.it/avatar (accessed on 10 February 2022). The program was developed using the popular Unity game engine (https://unity.com/, accessed on 10 February 2022), so as to take advantage of the out-of-the-box support for (consumer-grade) head-mounted displays provided by Unity, as well as of the massive amount of dedicated third-party software libraries and graphical assets (many of which are free and/or open source). Among these, the virtual reality toolkit (VRTK, http://vrtk.io/, accessed on 10 February 2022) greatly simplifies the implementation of user interaction in virtual environments, and the Accord.Net framework (http://accord-framework.net, accessed on 10 February 2022), provides a comprehensive set of procedures for linear algebra, geometry, statistics and machine learning.

AVATAR is logically organized in two stages. A first, preparatory stage allows the user to load the data and set various parameters related to the visualization of the potential-energy surface. During this stage, the user interacts with AVATAR through a standard desktop-based user interface. The potential-energy surface, preliminarily sampled on a regular grid defined over the two process coordinates, is provided to AVATAR through the simple and widespread 'xyz' file format with minor additional prescriptions. In particular, AVATAR requires a so-called 'multimodel' xyz file, where each model stores the atomic coordinates associated with a sampled point of the energy surface and, in the comment line, the value of the potential energy as well as the position of the sampled point within the sampling grid.

Then, in a second stage, the immersive exploration takes place. In this stage, the user wears the HTC Vive headset while holding the hand-held controllers, and walks around a virtual world featuring the potential-energy surface as 'terrain'. This stage is graphically portrayed in the two circular panels of Fig. 19.2. In the right panel a user equipped with the HTC Vive head-mounted display and controllers is portrayed in the real physical environment surrounding the researcher. In the left panel, the virtual world seen through the eyes of the user (which completely replaces the real one) is displayed. In particular, with reference to the 'geography' of the map in the rectangular relaxed plot of Fig. 6.4 (Section 6.3, Chapter 6), in the left panel of Fig. 19.2 the user's viewpoint is from the plateau of the reactants (bottom region of the rectangle in Fig. 6.4) and the view is on the valley that has to be crossed in order to reach the plateau of the products (top region of the rectangle in Fig. 6.4).

As shown by the left panel of Fig. 19.2, users find themselves completely immersed in a virtual world in which both the colors and the height of the terrain are functions of the energy values. Moreover, much in an augmented-reality fashion, a ball-and-stick representation of the molecular system under investigation is shown in the scene. This representation floats above the right controller

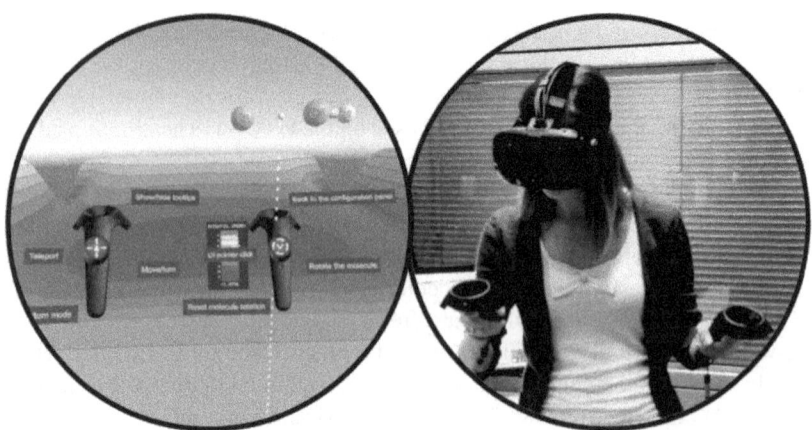

FIGURE 19.2   A researcher tracking the evolution of the $C + CH^+ \rightarrow C_2^+ + H$ reaction by immersive exploration of valleys (potential wells) and mountain passes (saddle points) of the associated potential energy landscape. Reprinted with permission from Ref. [360]. ©2020 Wiley Periodicals, Inc.

at a fixed distance from it, and is centered on the center of mass of the system (indicated by a small white sphere at the beginning of a dotted line joining the sphere to the virtual image of the controller). The relative position of the atoms is updated in real time as the controller is moved over the potential-energy surface, thus allowing the user to have an immediate feeling of the evolution of the system as a pathway on the potential-energy surface is walked along.

An interesting feature of AVATAR is the ability to highlight bond formation and breaking occurring with the evolution of the process. If walking along a reactive pathway, in fact, when the distance between two atoms is lower than a given threshold, a small cylinder appears to represent the bond formation. Then, as the atoms get closer, the radius of the cylinder increases, up to a maximum value.

With respect to user interaction, AVATAR was designed to take the best advantage of the hand-held controllers of the HTC Vive. Broadly speaking, the left controller is used to navigate across the terrain, while the right controller provides useful information on the current energy value and the associated molecular geometry. Although the best way of exploring the energy landscape is by physically moving in the real world (the movement is then mapped into the virtual world thanks to specific sensors tracking the position of the head-mounted display in the physical world), this is only possible for small distances. Therefore, it is also useful to have some sort of additional long-range navigation mechanism. For that purpose, AVATAR provides two common navigation metaphors: 'driving' and 'teleportation'. By touching on the touchpad of the left controller, users can 'slide' over the surface, as if they were driving an invisible car. As an alternative, through a dedicated button on the left controller, the user can enable a teleportation mode displaying a parabolic trajectory and a target that can be moved so as to point in far regions of the landscape and used as the destination of an instantaneous teleportation.

Of course, in addition to the above-mentioned rectangular and circular relaxed plots, any kind of two dimensional–potential energy surface (actually, any generic two-dimensional function) can be loaded and explored through AVATAR, provided that it is made available to the program in the above-mentioned supported xyz file format. Interesting possibilities derive, for instance, from choosing as process coordinates the distances of a breaking and a forming bond, such as in the study of S_N2 reactions, or two dihedral angles, such as in conformational analysis of proteins.

19.3 Chemistry at your fingertips: an immersive laboratory for the analysis of chemical bonding

The second application that will be reviewed in this chapter is based on a rather different IVR environment relying on the more sophisticated technology of cave automatic virtual environments (CAVE) [346,347]. These are a room-sized environment where stereoscopic images are projected onto between three and six walls of the environment. The users physically enter the CAVE with their own bodies wearing three-dimensional glasses and a tracking sensor, and are thus immersed in an extended-reality environment where their bodies and the virtual content generated by the projectors cohabit in the same real environment of the CAVE room.

The possibility of sharing the same 'space' with virtual human-scale, three-dimensional representations of the molecular world suggests the idea that the analysis of the electron-charge redistribution described in Chapters 14 and 15, and shown in action in Chapters 16 and 17, could be performed in a natural and immediate way in the IVR environment, rather than being the result of 'office' work behind a terminal and a keyboard. The CAVE thus becomes the venue where one or a small team of researchers can perform measurements of charge-flow profiles associated with bond formation or electron excitation in a real-time, interactive and collaborative way that resembles closely that of experimentalists in laboratory experiments. This idea is at the heart of the conception and development of the virtual laboratory for chemical bonding described in Refs. [278,364] and based on the Caffeine [365] software, a molecular-graphics program specifically designed to take advantage of IVR technology such as CAVE environments.

The virtual laboratory for the analysis of chemical bonding is conceived as follows. The basic ingredients are: i) one or more researchers equipped with three-dimensional glasses and tracking sensor, ii) a CAVE environment, iii) a three-dimensional representation of an electron-charge redistribution associated to some chemical process of interest (and acting as the 'sample' of an experiment), and iv) an analysis tool capable of performing the CD analysis described

FIGURE 19.3 Immersive session in the virtual laboratory for the analysis of chemical bonding. Two users analyze the coordination bond between nickel and a chelating diphosphine in one of the nickel–dicarbonyl complexes discussed in Section 16.3 of Chapter 16. Researchers interact with the electron-charge redistribution and with the analysis tool (represented by a reference frame for the CD analysis discussed in Section 14.3 of Chapter 14) through a remote-control application for tablet computers. For better clarity, the stereo mode of projectors was temporarily disabled to take this photo. Reprinted with permission from Ref. [278]. ©2018 Wiley Periodicals, Inc.

in Section 14.3 of Chapter 14. These elements are all shown in the portrait of an immersive session in the virtual laboratory provided by Fig. 19.3.

A typical session in the virtual laboratory starts with a preparatory stage where the required files (volumetric data sets of the charge-redistribution and system geometries) are loaded into the Caffeine software from a desktop-based control panel outside the CAVE. Then researchers, equipped with active glasses and a tracking sensor, enter the CAVE room and cooperatively interact with the 3D representations of the loaded volumetric data through a remote control application for tablet computers. Even if, strictly speaking, CAVEs are usually single-user systems (in the sense that the projected images are generated by taking into account the position and the orientation of the head of the user wearing the tracking sensor), small groups of people can participate in the immersive session with satisfactory results as long as they stay close enough to the tracked user and look approximately in the same direction.

The remote-control panel operated through the tablet allows the user to move, rotate and scale the displayed molecular system, as well as to activate and operate the CD analysis tools. Users would typically start by positioning and orientating the representation of the molecular structure and of the related electron-charge redistribution until an optimal point of view of the system under investigation is obtained, and they can physically move themselves within the CAVE around the molecular object so as to analyze its topological features. Then, when ready for CD analysis, they would activate the analysis tool and a special reference frame would appear in the three-dimensional scene, whose origin and z-axis are used as parameters for the CD analysis. This reference frame can be freely moved and rotated so as to be aligned in real time along a desired interaction axis.

Once the analysis tool has been orientated in the desired way, by pressing a dedicated button of the remote control panel the CD function is evaluated for each point along the chosen axis, and the result is plotted on a chart displayed within the three-dimensional scene in an augmented-reality fashion as if it were the outcome of an experimental measurement (the chart and the contained CD function are clearly visible in Fig. 19.3). By default, the chart is always in front of the tracked user and drawn on a semitransparent quad. However, the user can at any time 'pin' the chart in a particular position, switch to a different chart or hide it. Finally, a special tool, referred to as the 'marking plane', is provided to further support the numerical analysis of the CD function. The marking plane is a plane orthogonal to the chosen analysis direction and can be moved along it. Its purpose is to query the value of the CD function at particular z points of interest: once the marking plane has been placed at the desired position, it is sufficient to press a dedicated button on the remote control to insert a corresponding vertical marker on the currently displayed chart and obtain the values of the associated z, $\Delta q(z)$ pair displayed in a table placed below the chart. This process can be repeated, so as to mark the plotted CD function in several points of interest.

Chapter 20

Data-driven chemistry

In the previous chapter, we discussed how increasingly larger data sets resulting from both experiment and calculations require to be analyzed in order to provide scientific insight, and we specifically addressed the issue of converting data into graphical structures to best exploit the capabilities of the brain through our visual system for that purpose. In this chapter, the same issue of gaining knowledge from data will be addressed from the alternative, powerful perspective of artificial intelligence and so called 'machine-learning' techniques.

In particular, in Section 20.1 the emerging paradigm of data-driven science is introduced, and its relation with artificial intelligence and machine learning is discussed. Section 20.2 provides a short introduction to machine-learning techniques and illustrates the basic steps involved in a machine-learning process. In Section 20.3, the integration of machine learning with computational chemistry is discussed by assessing the merits and demerits and highlighting associated open challenges. The discussion in these sections was mainly informed by the nice reviews that can be found in Refs. [366–371], to which the reader is referred for further details.

20.1   A data-driven approach to science

Throughout our illustration of a theoretical and computational approach to chemistry in the previous parts of the book, theory has been the central source from which our discussion, in its several and varied aspects, emanated. Through Chapters 1 to 19, our concern has been essentially to feed the computer with the Schrödinger equation (which embodies the appropriate theory for the treatment of molecular systems) and instruct it to execute a large and intricate set of operations in order to solve that equation and recover from its solutions information on the properties and behavior of a given system. As briefly touched on in Chapter 1, however, theory is the result of the observation of data: humans analyze and rationalize data through their intellectual faculties, and build theoretical models capable of explaining the observed data and make predictions on new ones. But what if we put in the computer this process, too? What if we provided the computer with data and intelligence and let it do the job on its own?

This is the essence of an emerging paradigm of doing science, referred to as 'data-driven science' and sometimes identified as the 'fourth paradigm', unifying the three traditional paradigms centered around theory, experiment, and computation/simulation [372]. With respect to the scheme outlined in Chapter 1,

Chemistry at the Frontier with Physics and Computer Science
https://doi.org/10.1016/B978-0-32-390865-8.00031-3

leading from knowledge (through theory, models, methods, implementation and computation) to data, in data-driven science the process is, in a sense, overturned: leveraging on artificial intelligence and machine-learning algorithms, one goes from data to knowledge [367].

In the above paragraphs, we used jointly the two terms artificial intelligence and machine learning connected to the capability of a computer to learn from data. The two terms, however, are not completely interchangeable and a clarification in this respect might be useful. A first distinction with regard to artificial intelligence is in order. So-called 'general artificial intelligence' refers to the ability of a computer to exhibit broad intelligence and reason across several domains exactly like humans, and is a target not within reach in the near future. In contrast, so-called 'specialty artificial intelligence' refers to the training of a computer program to reason and solve a specific dedicated task. This field has been rapidly advancing over the past two decades, especially thanks to the impressive advances in the development of statistical-learning techniques that gave rise to the category of 'machine learning'. Machine-learning algorithms represent a powerful class of techniques through which specialty artificial intelligence can be implemented and have rapidly become ubiquitous in our contemporary everyday life, being routinely employed in disparate fields ranging from web search, to language translation, self-driving vehicles, and many branches of science.

Early attempts to exploit artificial intelligence in chemistry, especially in the field of computer-assisted molecular design, may be traced back to the pioneering work of Hansch and Fujita [373] and Free and Wilson [374], which established the field of quantitative structure–activity relationship (QSAR) modeling and, contributed to the birth of the so-called 'cheminformatics' in the successive decades. In their work, they used focused data sets consisting of a series of a dozen chemical derivatives to fit equations that could anticipate complex phenotypic effects such as toxicity [368]. Some of the techniques that are nowadays used in data-driven chemistry bear resemblances with these ideas, but others are fundamentally new techniques rooted in the rapidly advancing field of machine learning.

As we shall further discuss in the next section, given a rule-discovery algorithm, through machine-learning techniques a computer has the ability to autonomously determine the 'rules' underlying a given set of data and thus in principle, given enough data, even all physical laws (also potentially some that have somehow escaped our intuition and are currently unknown). A similar scenario has the deep implication that the role itself of the computer has in recent years undergone a radical change. While, in fact, in traditional computational approaches the computer is simply a (however powerful) calculator executing human-driven instructions, with machine learning it gains an unprecedented degree of autonomy that makes it capable of learning rules and building predictive models by assessing portions of data sets [366].

20.2 Machine-learning techniques

A formal definition of learning, in the context of data-driven science, is given in Mitchell's comprehensive book on machine learning [375] (to which the interested reader is referred for a detailed account on the concepts that will be discussed in this section):

> *A computer program is said to learn from experience E with respect to some class of tasks T and performance measure P, if its performance at tasks in T, as measured by P, improves with experience E.*

On a less formal level, the term 'machine learning' refers to a class of algorithms based on powerful statistical-learning techniques and operating on data. The algorithms can be quite generic, thus a few algorithms can handle successfully many different problems. Moreover, the algorithms learn directly from data, thus reducing the need for complex logic and code.

For some aspects, machine-learning techniques share similarities with traditional regression analyses. However, while these aim at tuning the parameters that define a function so as to make this reproduce a set of data points with the lowest error, machine-learning techniques aim at identifying the functions themselves that predict interpolations between data points and thus minimize the prediction error for new data points added later [370]. In other words, whereas in conventional models one starts with clear assumptions about the system that has to be modeled, machine learning focuses on so-called 'universal approximators'. These are models that are in principle able to represent any function with arbitrary accuracy, provided that they are given enough training data and parameters. A popular class of universal approximators, for instance, are the so-called 'neural networks' [376], which mimic the functioning of the brain and fall under the subcategory of machine learning known as 'deep learning'. In particular, in neural networks artificial neurons are arranged in input, output and hidden layers. In the hidden layers, each neuron receives input signals from other neurons, integrates those signals and then uses the result in a straightforward computation. The connections between neurons have weights, and the values of these weights represent the stored knowledge. Learning is thus the process of adjusting the weights so that a given training data is reproduced as accurately as possible [366].

Now, there are three main subcategories of machine-learning techniques, i.e., supervised learning, unsupervised learning, and reinforcement learning. In supervised learning, which is the most mature and widespread approach in physical sciences, the training data consist of sets of input values and associated output values. The goal of the algorithm is to derive a function that, given a set of input values, accurately predicts the associated output values. If the input values are mapped onto a categorical target space, the machine-learning task is a 'classification', if the target space is continuous the task is a 'regression'. The construction of machine-learning force fields (see Section 11.2 of Chapter 11) for molecular-dynamics simulations are an example of regression tasks, while, for instance, a classification task may be used to automatically select appropriate 'model chemistries' (see Fig. 17.1) for a given system [371].

On the contrary, if the training data set consists of only input, then unsupervised learning may be used to identify previously unrecognized trends or patterns in the data set, or to perform a dimensionality reduction or a clustering of the data. Unsupervised learning finds application in computational chemistry in the postprocessing and analysis of molecular simulation data, e.g., in identifying collective variables and reaction pathways [371], or in separating conformers of a molecule from the clustering of data describing exclusively the positions of the atoms [370].

Finally, reinforcement learning combines aspects of supervised and unsupervised learning. In a task of this type, the goal is learning the optimal 'action' on a given 'state' to maximize a specified future reward, and the learning progress is given by a combination of explorative activity and use of the knowledge already gained. An example of reinforcement learning is the application of geometry changes (action) to an unfolded protein (state) with the aim of getting closer to the folded structure with minimum energy (future reward) [371]. Due to their simultaneous features of exploration and training taking place in reinforcement learning tasks, these are being increasingly used for finding molecules with specific properties in large chemical spaces [370].

A machine-learning process, from the preparation of the data to the machine-learning model that is fine tuned, typically consists of the following four stages:

i. preparation of the data;
ii. choice of a model;
iii. training of the model;
iv. testing of the model.

The progression to a good machine-learning model, however, is not necessarily linear, and some of the steps may have to be reiterated as one learns about the problem at hand. The four stages of the above outline workflow are briefly reviewed with further detail in the following paragraphs.

i) From a radical point of view, machine-learning processes could be simply regarded as 'sophisticated parametrizations of data sets' [370]. As a consequence, however sophisticated the architecture of the model may be, an appropriate choice of the reference data is crucial and may deeply affect the effectiveness of the model. Accordingly, the data set should be representative of the problem under study, and should ideally be free of biases or spurious elements, otherwise the model will be incomplete and behave badly for situations that have not been properly represented. Unfortunately, some of these issues often remain unnoticed since data sets that are used to test the model (see later) are usually sampled from the same population as the training set.

ii) Once an appropriate data set has been prepared, a model type (or learner) has to be selected (typically falling in one of the three classes outlined above: supervised, unsupervised, and reinforcement learning). Depending on the type of data and the learning goal, a varied range of learning algorithms (e.g., Bayes classifiers, k-nearest-neighbor, decision trees, kernel methods, artificial and deep neural networks) can be applied, for details on which the reader is referred to the above-cited specialized references. Sometimes it is useful to adopt multiple different models or similar models with different values for their internal parameters, in order to achieve a more robust overall model.

iii) The training process is the stage that bridges together the data set and the learning model. The optimization of the model is guided by a 'loss function' that encodes the following two opposing objectives: i) achieving a good fit to the data, while ii) keeping the parametrization general enough so that the trained model can be applied to data not included in the training set [370]. Many models, however, are not completely autonomous, and require a prior tuning of some internal variables known as 'hyperparameters' which may improve or worsen the learning considerably. The selection of optimal hyperparameters is thus often problematic and the development of automatic optimization procedures is an area of active research [366].

iv) In the final stage, the model is assessed to obtain an estimate of the so-called 'generalization error', relating to how well the model performs on unknown data. For this purpose, a widely adopted strategy is the so-called 'crossvalidation' or 'out-of-sample' testing, whereby one withholds a randomly selected portion of data during the training stage. This withheld data set, known as a test set, is fed into the model once the training is complete. The effectiveness of the training is then evaluated by the extent to which the data of the test set are predicted. It should be stressed, however, that crossvalidation is reliable only when the data samples used for the training stage and the test stage are representative of the whole population, and problems may actually arise if the model is applied to data that are very different from those in the original data set.

20.3 Machine learning in chemistry

The reproducible, closed-form equations of traditional theoretical approaches provide a robust framework for determining the properties and behavior of a given system. However, their applicability is often limited (for instance, as we have seen, by computational reasons). In addressing complex problems (such as those involving massive combinatorial spaces or nonlinear processes), machine learning–based statistical modeling and analysis have emerged as a powerful resource in several scientific fields including chemistry and, in fact, a growing infrastructure of machine-learning tools for generating, testing and refining scientific models is rapidly evolving.

In particular, the integration of machine-learning approaches in theoretical and computational chemistry has been developing rapidly over the past decade, with the number of reported applications growing at an impressive rate. In their review on the integration of machine learning into computational chemistry and material science, Westermayr et al. report a nonexhaustive list of eighteen reviews published in the years 2017–2021, and highlighting progress in various contexts, including catalyst design, the development of force fields and interatomic potentials for ground-state properties and excited states, quantum chemistry, in solving the Schrödinger equation, and on unsupervised learning in atomistic simulation (see [371] and the works cited therein). Additional areas include drug design [368] and the exploration of chemical-reaction networks [377]. A comprehensive report on data-driven chemistry can also be found in a recently published dedicated book edited by Akistu [378].

Over the past decade, the introduction of machine-learning techniques has proven to accelerate scientific progress in guiding chemical synthesis, assisting multidimensional characterization, enhancing theoretical chemistry, and targeting discovery of new compounds [366]. Consider, for instance, the calculation of polyatomic potential-energy surfaces. As we have seen in Chapters 6 and 11, the traditional analytic formulations of either three-atom, reactive potential-energy surfaces (see also Ref. [379]) or many-atom classical force fields rely on preconceived notions on the behavior of the functions to be modeled. In contrast, machine-learning algorithms provide a much more general model class and the resulting machine learning–force fields are able to combine the accuracy of high level–quantum mechanical calculations with the efficiency of classical force fields [380].

Another example of successful integration of machine learning in computational chemistry is in the field of the discovery of new chemicals with desired attributes (e.g., drugs, antivirals, antibiotics, catalysts, battery materials). To provide the reader with a concrete example, we will illustrate with some detail the integration of machine learning in the process of chemical discovery according to the framework described in Ref. [380]. A schematic illustration of the overall process is shown in Fig. 20.1, and may be summarized as follows. Starting from the top-left corner, subsets of relevant chemical-compound space (CCS) are sampled to create data sets of molecular structures. High throughput–quantum mechanical (QM) calculations are subsequently used to construct QM molecular property data sets (top-right corner). Quantum machine-learning (QML) algorithms are then employed to enable interpolation and analysis of QM properties in the CCS (bottom-right corner). QML model analysis is combined with chemical knowledge to extract insights into the CCS (for example, by constructing and analyzing Pareto fronts, as shown in the bottom-left corner of the figure). Finally, the CCS can be further extended and explored with the accumulated knowledge from QML. The main applications of QML in this context up to now have covered CCS of small molecules and ordered extended solids. However, as pictorially summarized in the central panel of the figure, extension of the applicability of QML to biomolecular systems, nanostructures, surfaces, organic-framework materials, supramolecular systems, and even quantum-mechanical model systems should be targeted in the future.

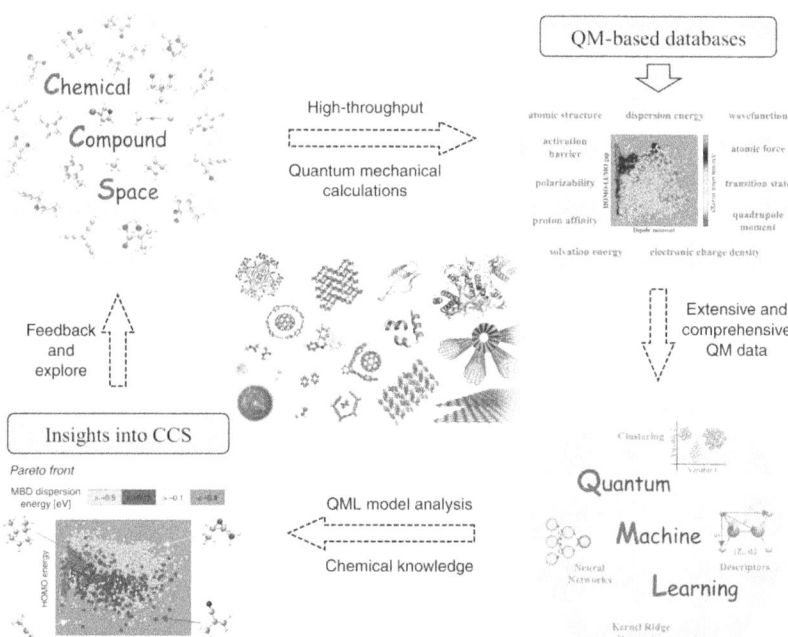

FIGURE 20.1 Schematic illustration of the integration of machine learning in the process of chemical discovery. Reproduced from Ref. [380]. Image licensed under a Creative Commons Attribution 4.0 International License accessible at http://creativecommons.org/licenses/by/4.0/.

On the other hand, a number of major flaws associated with the use of machine-learning approaches have also recently emerged and need to be carefully addressed in the near future. Keith et al., while developing their review [370], launched an anonymous online survey asking the scientific community to express the main concerns on the use of machine-learning models in chemistry. The points raised by the respondents are worth being mentioned here:

i. machine-learning methods are becoming less understood, while they are also more regularly used as black-box tools;
ii. many publications show inadequate technical expertise in machine learning (e.g., inappropriate splitting of training, testing, and validation sets);
iii. it can be difficult to compare different machine-learning methods and know which is the best for a particular application or whether machine-learning should even be used at all;
iv. data quality and context are often missing from machine-learning modeling, and data sets need to be made freely available and clearly explained.

One of the main criticalities of data-based methods is indeed the lack of transparency, and in fact "few examples exist of machine-learning models that have become generally applicable to researchers outside the immediate circle of developers" [371]. As a partial response to this 'reproducibility crisis', Artrith et al. have recently suggested a rigorous 'best practices' protocol to ensure a level of robustness and reproducibility in statistical-learning approaches similar to those of traditional approaches [369]. Moreover, beyond the lack of generality and precision, further issues are the reliance on high-quality data, the inability to derive high-level concepts, and the proneness to artifacts [370]. In this respect, in the same work Keith et al. point out that "the past decade has shown that it has not been enough to just apply existing machine-learning algorithms, but breakthroughs are happening by a handshaking of innovations resulting in novel machine-learning algorithms and architectures driven by the pursuit of novel insights in chemistry while retaining a deep understanding about the underlying physical and chemical principles" [370].

On another front, the establishment of robust data-driven science paradigms also depends on factors that relate to a broader context. On the one hand, an educational challenge has to be faced involving the growing of a new generation of researchers with an inherently interdisciplinary vision, capable of speaking fluently the languages of chemistry, physics and computer science. On the other hand, as pointed out by Butler et al. [366], the accessibility of machine-learning technology relies on factors ("open data, open software and open education") which relate to the emerging demand for a greater openness of science. This leads us straight to the topics covered in the next chapter, where some aspects of the so-called 'open science' will be addressed, especially in their relation with theoretical and computational chemistry.

Chapter 21

Towards open molecular science

In the opening chapter of a book entitled 'Opening Science: The Evolving guide on How the Internet is Changing Research, Collaboration and Scholarly Publishing', Bartling and Friesike briefly recap the history of knowledge creation and dissemination from prehistoric times to current days, rightly highlighting the introduction of the writing system and the invention of the printing press as two important turning points [381]. In particular, the authors discuss how science as we intend it and practice it today is the result of a first scientific revolution occurred in the 17th century, and tightly connected to the definition of the scientific method and the establishing of the scientific-publishing system. Upon such a revolution, knowledge creation underwent a process of professionalization and institutionalization, during which an originally small number of disciplines started to branch into the large number of tiny fields featuring a high degree of specialization that we see today.

Now, as noted by the above authors in their essay, the advent of the Internet at the turn of the 21st century has imposed a dramatic and sudden change in the time and space scale of our communication, representing for the history of knowledge creation and dissemination a twist at least as dramatic as that related to the development of the writing system or of the printing press. As a consequence, our cultural system is now living a profound metamorphosis due to a natural adaptation process (so-called 'legacy gap'), with some aspects rapidly evolving and some others being reluctant to change.

In some respects, the scientific culture has already taken advantage of features enabled by the Web 2.0 and the Internet, certainly on the technological side, and as such it is sometimes termed Science 2.0. However, the most radical changes rooted in the Internet revolution will impact on a much broader social and cultural context, and are yet to come. The awareness of the necessity for such changes, together with a claim for greater openness of science in a world whose communication potential has just been unlocked, have given rise at the beginning of the century to a multifaceted cultural movement referred to as 'open science' (OS).

This chapter attempts to picture this *in fieri* (r)evolution, especially in relation to the scientific community and research practices of the discipline with which this book is concerned. In particular, in Section 21.1 the basic principles of OS are introduced. Section 21.2 focuses on those aspects of OS that are particularly relevant in theoretical and computational chemistry, touching on issues such as the reproducibility of the research, the disclosure of computational source code, and the accessibility of data. Finally, Section 21.3 reports on the growth of several collaborative frameworks, often based on dedicated computing infrastructures, acting as a technological and social venue for OS practices within the scientific community of computational chemists.

21.1 Open-science basics

One of the most basic principles underlying contemporary scientific research is the urge of making public the results of the research activity, so that other researchers can acquire knowledge, and build further knowledge on top of that. The Internet has entered our daily life for at least some twenty years now, and certainly we do not struggle to think that, as noted by Bartling and Friesike, the Internet could provide a powerful venue for such knowledge exchange. Researchers could share in real time the results of their research with potentially anybody. Knowledge would be more fluid and could freely expand throughout the globe, rather than being filtered by institutions and personal networks. Scientists would share the entire research process in more impactful multimedia formats rather than wait for the research to end and make available a 'polished' version of the gained knowledge in the traditional form of an article. In so doing, they would benefit from immediate feedback to calibrate their ongoing work, and other researchers may learn from negative or less significant results that are typically dismissed when assembling a traditional scientific article and shall thus never contribute, no matter how little, to scientific progress.

In 2011, a commentary by Woelfle, Olliaro and Todd [382] published in Nature Chemistry reported on the realization of a similar scenario within a chemical project. The entire research process was since the start exposed on the web, where experts could identify themselves and spontaneously contribute to the research. The take-home message of the short writing was that a similarly 'open' approach to research accelerated the process of discovery. The commentary is entitled 'Open science is a research accelerator', and reports indeed on much of what OS is about.

It might be useful at this point to better frame the concept of OS by examining some definitions that have been recently proposed. According to The Open Science Training Handbook (available at https://book.fosteropenscience.eu/, accessed 15 February 2022), OS is 'the movement to make scientific products and processes accessible to and reusable by all'. Key principles are transparency, reuse, participation, cooperation, accountability and reproducibility of the research, inclusion, fairness, equity, and sharing, and related key practices are opening access to research publications, data sharing, open notebooks, transparency in research evaluation, ensuring the reproducibility of the research, transparency in the research methods, open software and infrastructures, and open education.

In 2018, Vicente-Saez and Martinez-Fuentes performed a systematic literature review in order to forge an integrated definition of the term and arrived at the concise statement that 'Open Science is transparent and accessible knowledge that is shared and developed through collaborative networks' [383].

On another front, the complexity of the phenomenon is analyzed by Fecher and Friesike [384], who describe OS as an 'umbrella term encompassing a multitude of assumptions about the future of knowledge creation and dissemination', and identify five predominant thought schools. The 'public' school focuses on outreach and accessibility of the research. The 'pragmatic' school aims at fostering the collaboration by connecting scholars and improving transparency. The 'infrastructure' school focuses on technological platforms, tools and services for dissemination and collaboration. The 'measurement' school aims at devising alternative metrics for tracking the impact of scholarship. The 'democratic school' claims that scholarly knowledge, including publications and data, should be freely available to all.

This last aspect of OS is perhaps the most mature of all and has already gained enough ground in the academic practices. Exemplified by the label 'open access', it has in fact entered the daily life of scientists due to the mandate adopted by many research institutions, funders or government, requiring that the results of the (often publicly) funded research should be published in open-access form and thus made available to anyone, accessible online free of charge and with no technical obstacles.

Coming back to the commentary by Woelfle, Olliaro and Todd, in the conclusive section the authors wonder:

> The research therefore inevitably proceeded faster than if we had attempted to contact people in our limited professional circle individually, in series. Perhaps this is not surprising, but if it is the case that 'none of us is as smart as all of us' and if we wish to reach scientific goals quickly, why is so much science not practiced this way?

The reason why this is not generally happening is that the above-mentioned urge of making public the results of the research activity almost invariably does not stem only from a disinterested knowledge's sake, but also, and at least equally importantly, responds to an effective and stable way of getting credit for the research, increasing one's own reputation, succeeding in competition, and gaining and exerting varied forms of power. This results from a well-established sociocultural system that has developed around the course established with the above-mentioned first scientific revolution. A whole system has thus now to adapt to a radically new context (ultimately stemming from the impromptu fall of the barriers to global instantaneous communication), for instance ensuring – as wished by Bartling and Friesike in the above-cited essay – that OS practices are appropriately credited by other scientists and by granting authorities, which – as they note – is not yet the case. In this cultural transformation, researchers can choose between playing a passive role in a process driven by other stakeholders of scientific research, or taking an active attitude, and become the leading force in shaping what Bartling and Friesike call 'the second scientific revolution'.

21.2   Open research, open software, open data

As already sketched in the previous section, one of the leading objectives of OS is making transparent the whole research process, in order to improve the reliability, rigor and quality of the research. In this context, a central concept is that of 'reproducibility' of the research experience, which is one of the core principles of the scientific method itself. In simple words, the claim for a scientific finding can be considered seriously only if it is supported by a complete documentation allowing other researchers to achieve the same results by replicating the research experience. Though originally developed in an experimental context, the concept of reproducibility has also transferred over the past decades to the so-called *in silico* experiments, i.e., calculations or simulations performed through computational software, such as those described in this book.

In principle, computational reproducibility could be achieved if enough detail is given on the theoretical methodology and the input data used for the calculations. In this case, other researchers would then be able to write a computer program implementing the same methodology, and thus obtain the same results if using the same input data. However, as pointed out by Gezelter in a viewpoint published in The Journal of Physical Chemistry Letters in 2015 [385] the advances and proliferation of computational approaches over the past decades have now led to a high degree of sophistication in computational software, which makes it in practice impossible to ensure the reproducibility of a computational research without access to the code, data, and the metadata that describes how the data is organized.

Pointing out a general context of 'reproducibility crisis' [386,387], where some scientists are admitting that their own organizations are struggling with reproducing the results of prior publications, Gezelter concludes that "access to the code and data should therefore be an expectation for publication and review in the chemical literature". Similar opinions have been recently put forward by Coudert [388] in 2017 for the adjacent field of computational chemistry of materials, asserting that computational reproducibility needs nowadays an 'opening' of data, of the input and output files, and of the source code of the adopted computational program, and by Walters in 2020 [389], wishing that computational chemistry and cheminformatics journals could establish, as other high-profile journals have already done, policies mandating the release of code accompanying papers that describe computational methods.

On the other hand, in another viewpoint appeared in the same year and journal as Gezelter's one and signed by Krylov, Herbert, Furche, Head-Gordon, Knowles, Lindh, Manby, Pulay, Skylaris, and Werner, ten eminent scientists involved in the development of some of the most widespread computational-chemistry program packages expressed a rather opposing view, advocating the legitimacy of closed-source, commercial software [390]. According to these authors, software is a product and not a scientific finding, "more akin to, say, an NMR spectrometer – a sophisticated instrument – than to the spectra produced by that instrument", and should thus not necessarily be 'included' in a research report. In the view of these scientists, an open-source mandate would in fact also have negative consequences in a competitive context where science funding – as in the United States and elsewhere – is distributed among the best original ideas through merit-based criteria. The authors add, moreover, that commercial software often means greater stability and user support by virtue of the involvement of expert programmers that, explicitly hired for that, are already in a sense addressing the reproducibility issue. They finally note that many commercial codes grant the access to the source code to academic groups for development purposes, defining in such a way a collaborative model that, rather than 'open source', could be termed 'open teamware'.

In a further article appeared on the same journal one year later [391], Jacob noted, however, that the access to commercial software packages as a development platform is not as open as claimed in Ref. [390], as it may involve signing a developer agreement, transferring the intellectual property rights on the new developments, a nondisclosure clause concerning the source code, and sometimes exclusivity, meaning that new source code cannot be contributed to other commercial or open-source packages.

As exemplified by the above discussion, the issue of 'opening the software' is not a trivial one, at least in academic environments such as that of theoretical and computational chemistry. The debate is still ongoing and a general consensus still seems far away. The reader may find an interesting account on this subject, with specific reference to computational chemistry, also in Ref. [392], where the question is examined from an epistemic perspective.

As already mentioned above, a second aspect involved in the reproducibility of the research is the sharing of data. The practice of publishing data as supplemental material to scientific articles to allow for the reproducibility of the research, has now partially found its way among scientists in computational chemistry, a typical example being the attachment, as supplemental material, of the molecular geometries at which some electronic-structure calculations have been performed, provided in the form of xyz files (see later, on data formats). However, as of today, it is often up to the researchers to choose which data should be included, and in which format.

In a broader context of OS, opening up data means that data can be freely accessed, reused, remixed and redistributed, and should be appropriately licensed as such (except for limitations arising in exceptional cases, e.g., when the identity of human subjects has to be protected). Open data should thus adhere to the so-called 'FAIR' principles, i.e., they should be findable, accessible, interoperable and reusable. In other words they should be easily retrievable, accompanied with descriptive metadata, and encoded in format adhering to a well-defined standard, so as to allow complete reusability.

Several databases have been developed in past years with the purpose of exchanging data in a similar way, one of the most popular ones being the protein data bank (PDB, http://www.wwpdb.org/, accessed on 14 February 2022), a repository for structure data and metadata for biological macromolecules, that has enabled reuse of the primary structural data but has also opened up new research lines based on statistical and metaanalysis of the structures [385]. Useful resources for electronic-structure computational chemistry are the Basis Set Exchange portal (available at https://www.basissetexchange.org/), that we have already mentioned in Chapter 13, and the Computational Chemistry Comparison and Benchmark DataBase (https://cccbdb.nist.gov/, accessed on 15 February 2022), containing experimental and computed quantum-mechanical thermochemical data for a selected set of 2069 gas-phase atoms and small molecules. Examples of community-contributed data sets, that we have also already mentioned in Chapter 10, are the KIDA [142] (https://kida.astrochem-tools.org/, accessed 15 January 2022), and the UMIST (University of Manchester Institute of Science and Technology) UDfA [143] (http://udfa.ajmarkwick.net/, accessed 15 January 2022). Another relevant example of data sharing relating to the current global Covid health emergency is given by Ref. [393], which describes freely available data and computational resources that can facilitate and assist the mass spectrometry–based analysis of SARS-CoV-2.

An important aspect for accessibility and reusability of data is the file format, i.e., the way that information is encoded for storage in a computer file. In an OS perspective, open data should be based on so-called open formats, i.e., formats adhering to a freely accessible and duly documented standard. With reference to the theoretical and computational chemistry community, it is worth mentioning here that an attempt at the development of a common data format and library for code interoperability in quantum chemistry, the Q5Cost [338,394], was pursued in a joint endeavor by a European team. As of today, however, no standard data format has been currently established for quantum-chemistry programs and, as a result, these continue to 'speak' each its own peculiar language, sometimes hampering the possibility of combining different features available in different programs in the same research project.

Another relevant initiative in the context of data sharing is the Blue Obelisk, a group of chemists who promote open data, open source and open standards that was initiated in 2005 to associate multiple open-source cheminformatics projects [395].

21.3   Collaborative frameworks

For the purpose of enabling OS practices, in recent years several networks and frameworks providing online virtual environments for collaborative research across continents, time zones and disciplines, have been growing. These connect geographically dispersed researchers to enable them to cooperate seamlessly on their research, sharing research objects as well as ideas and experiences. Examples of large scientific bodies are the Open Knowledge Foundation (OKF) (https://okfn.org/about), a worldwide nonprofit network of people interested in openness and in using advocacy, technology and training to unlock information and enable the sharing of knowledge, and the Open Science Framework (OSF) [396], that provides free and open-source project management support for researchers across the entire research lifecycle leveraging on a flexible repository for storing and archiving research data, protocols, and materials.

Some collaborative frameworks, which are particularly relevant in computational sciences, rely on the sharing of a computing infrastructure. For this purpose, as already discussed in Chapter 18, a particularly suitable model is represented by grid infrastructures, connecting together a large number of geographically distributed computer resources across multiple administrative boundaries, such as the Open Science Grid (OSG) [397,398], a large distributed computational infrastructure in the United States that supports many different high-throughput scientific applications, and partners (federates) with other infrastructures nationally and internationally to form multidomain integrated distributed systems for science, or SHIWA [399], a framework enabling multiworkflow simulations on a mixed grid/cloud infrastructure.

On distributed computing infrastructures, so-called virtual research environments (VRE) can be established, typically focused on a specific discipline, where scientists based in different geographic places can seamlessly access (e.g., through a web browser, a shell terminal, or a specifically developed application) data, software, and computer resources managed by diverse systems in separate administration domains. Examples from the computer-based chemistry community are the OpenMolGrid [400], an open computing grid for molecular science and engineering providing data warehouse for chemical data, software for building QSPR/QSAR models, and molecular-engineering tools for generating compounds with predefined chemical properties or biological activities, or MoSGrid [401], a grid-based science gateway for scientists with all kinds of background and experience levels, supporting applications in the domains of quantum chemistry, molecular dynamics, and docking, and allowing for the usage of well-defined workflows annotated with metadata.

A third example of collaborative frameworks in the context of molecular science is the already-mentioned GEMS [327,330,402], which, as already discussed in Chapter 18, was developed through several initiatives based on the European grid infrastructure, including the D23 METACHEM (https://www.cost.eu/actions/D23/) D37 GRIDCHEM (https://www.cost.eu/actions/D37/) COST actions and the constitution of the COMPCHEM Virtual Organization (VO) and the Chemistry, Molecular and Materials Science and Technologies (CMMST) Virtual Research Community (VRC) within the already cited EGEE III and EGI Inspire EU initiatives.

Laganà, Terstyanszky, and Krüger report in Ref. [403] the description of the proposal SUMO-CHEM – 'Supporting Research in computational and experimental chemistry via Research Infrastructure' submitted to the Horizon 2020 framework call H2020-INFRAIA-2016-2017 (Integrating and opening research infrastructures of European interest) for the creation of a molecular-science European research infrastructure that, leveraging on the above-mentioned experiences, would pioneer "a new way of collaborating between computational and experimental chemists by creating a seamless open environment for joint research, data production and reuse including its transfer in innovation and societal utilization" [18].

The proposal, which encompassed several areas of computational chemistry including electronic structure, nuclear dynamics, the design of smart energy carriers, innovative materials and biomedical processes, and the handling of knowledge for training and education, was submitted within the Horizon 2020 funding program through the European Open Science Cloud for Research pilot (EOSCpilot) project specifically addressing some of the reasons why European research is not yet fully tapping into the potential of data. In the cited article, Laganà, Terstyanszky, and Krüger report that the constitution of an Open Molecular Science (OMS) framework did not find its way among the approved EOSC pilots, and argue that "this is due to a large extent to the insufficient attention paid by the chemistry community to OS in spite of the fact that, in the last 10 years within the activities of the COST actions D23 and D37, of the COMPCHEM VO and of the CMMST VRC an embryo presence of the OMS community among the OS ones has been developed." In spite of that, however, the mentioned EOSC proposal to Horizon 2020 offers a view on the possible directions in which the molecular-science community might move to the end of implementing an open-science cloud. Among these it is worth mentioning the current NFDI4CHEM consortium for research (https://www.nfdi4chem.de/) and the ECTN association (https://ectn.eu/) for education.

Concluding remarks

Here we are at the end of our journey. It might be worth, at this stage, recapping how far we have come.

Taking the two central concepts of chemical reactions and chemical bonding as leading themes, we saw how chemistry stems from different souls of physics – classical mechanics, quantum mechanics, and relativistic theory. In particular, we moved from the analysis of the paradigmatic case of simple atom–diatom reactive processes (exemplified in nature by astrochemical reactions) and showed what physics lies behind the reaction arrow of a chemical equation. We illustrated the basic ideas of both the classical and quantum treatment of these processes, and examined the concrete case of the astrochemical reaction of $C + CH^+$. Then, we looked a little closer to what happens to the electrons when a chemical bond forms, and showed how the physical counterpart of the 'movement' of the electrons in the popular Lewis' diagrams can be found in the redistribution of the electron cloud, how this can be modeled in quantum-mechanical terms, and analyzed through simple computational tools. Armed with these tools, we focused on the concrete case of the metal–carbonyl bond in coordination chemistry and showed how elusive chemical concepts such as σ-donation and π-backdonation can be extracted on a quantitative basis from the electron-charge redistribution, and put in relation to experimental observables. Additionally, the analysis of chemical bonding between superheavy elements and gold fragments gave us the chance to discuss the effects of relativity on the periodic system of the elements.

Both in dealing with chemical reactions and chemical bonding, for a small selection of problems (wavepacket dynamics of collinear reactions, Hartree–Fock electronic-structure calculations, analysis of the electron-charge redistribution) we went the whole way down from theory to computing, hopefully providing a practical guide on how to address a scientific problem from a theoretical and computational perspective. Finally, we discussed the relation between chemistry and computer science, from the safe and well-established field of scientific computing to rapidly evolving and fascinating topics connected with the advent of virtual-reality technologies, with the exploitation of artificial intelligence in chemistry, and with the ongoing cultural changes originating from the impact of global instantaneous communication on knowledge and society.

In the vast landscape of theoretical and computational chemistry sure we went a short way, but hopefully this book provided a map indicating directions and connections to those readers who wish to continue their journey and explore other areas, maybe someday even uncharted ones.

Bibliography

[1] T.S. Kuhn, The Structure of Scientific Revolutions, University of Chicago Press, Chicago, 1962.

[2] A. Messiah, Quantum Mechanics, Dover, New York, 2014.

[3] C. Cohen-Tannoudji, B. Diu, L. Franck, Quantum Mechanics, John Wiley & Sons, New York, 1977.

[4] M. Born, Zur quantenmechanik der stoßvorgänge, Zeitschrift für Physik 37 (1926) 863–867.

[5] E. Schrödinger, An undulatory theory of the mechanics of atoms and molecules, Physical Review 28 (1926) 1049–1070.

[6] M. Born, R. Oppenheimer, Zur Quantentheorie der Molekeln, Annalen der Physik 84 (1927) 457–484.

[7] A.S. Davydov, Quantum Mechanics, Pergamon, Oxford, 1965.

[8] R. Spezia, Filosofia della chimica, Aphex 15 (2017) hal-01575060.

[9] H. Hettema, The Union of Chemistry and Physics. Linkages, Reduction, Theory Nets and Ontology, Springer, Cham, 2017.

[10] K. Gavrlogu, A. Simões, Neither Physics Nor Chemistry. A History of Quantum Chemistry, MIT Press, Cambridge, Massachusetts, 2012.

[11] E. Nagel, The Structure of Science, Hackett Publishing Company, Indianapolis, 1979.

[12] P.A.M. Dirac, Quantum mechanics of many-electron systems, Proceedings of the Royal Society of London. Series A 123 (1929) 714–733.

[13] R.F.W. Bader, C.F. Matta, Atoms in molecules as non-overlapping, bounded, space-filling open quantum systems, Foundations of Chemistry 15 (2013) 253–276.

[14] P.-O. Löwdin, The mathematical definition of a molecule and molecular structure, in: Jean Maruani (Ed.), Molecules in Physics, Chemistry, and Biology: Physical Aspects of Molecular Systems, Springer Netherlands, Dordrecht, 1988, pp. 3–60.

[15] R.F.W. Bader, On the non-existence of parallel universes in chemistry, Foundations of Chemistry 13 (2011) 11–37.

[16] L. Pauling, The Nature of the Chemical Bond and the Structure of Molecules and Crystals: An Introduction to Modern Structural Chemistry, Cornell University Press, Ithaca, 1939.

[17] W. Heisenberg, Physics and Beyond: Encounters and Conversations, Harper & Row, New York, 1972.

[18] A. Laganà, G. Parker, Chemical Reactions, Springer, Cham, 2018.

[19] G.N. Lewis, The atom and the molecule, Journal of the American Chemical Society 38 (1916) 762–785.

[20] N.R. Council, Mathematical Challenges from Theoretical/Computational Chemistry, The National Academies Press, Washington, DC, 1995.

[21] D.R. Hartree, The wave mechanics of an atom with a non-Coulomb central field. Part I. Theory and methods, Mathematical Proceedings of the Cambridge Philosophical Society 24 (1928) 89–110.

[22] J.C. Slater, The theory of complex spectra, Physical Review 34 (1929) 1293–1322.

[23] V. Fock, Näherungsmethode zur lösung des quantenmechanischen mehrkörperproblems, Zeitschrift für Physik 61 (1930) 126–148.

[24] C.C.J. Roothaan, New developments in molecular orbital theory, Reviews of Modern Physics 23 (1951) 69–89.

[25] G.G. Hall, The molecular orbital theory of chemical valency. VIII. A method of calculating ionization potentials, Proceedings of the Royal Society of London A: Mathematical, Physical and Engineering Sciences 205 (1951) 541–552.

[26] S.F. Boys, Electronic wave functions. I. A general method of calculation for the stationary states of any molecular system, Proceedings of the Royal Society of London. Series A, Mathematical and Physical Sciences 200 (1950) 542–554.

[27] P. Casavecchia, Chemical reaction dynamics with molecular beams, Reports on Progress in Physics 63 (2000) 355–414.

[28] J. Hirschfelder, H. Eyring, B. Topley, Reactions involving hydrogen molecules and atoms, Journal of Chemical Physics 4 (1936) 170–177.

[29] G.C. Schatz, A. Kuppermann, Quantum mechanical reactive scattering for three-dimensional atom plus diatom systems. I. Theory, Journal of Chemical Physics 65 (1976) 4642–4667.

[30] G.C. Schatz, A. Kuppermann, Quantum mechanical reactive scattering for three-dimensional atom plus diatom systems. II. Accurate cross sections for $H + H_2$, Journal of Chemical Physics 65 (1976) 4668–4692.

[31] J.M. Bowman, G.C. Schatz, Theoretical studies of polyatomic bimolecular reaction dynamics, Annual Review of Physical Chemistry 46 (1995) 169–196.

[32] S.C. Althorpe, D.C. Clary, Quantum scattering calculations on chemical reactions, Annual Review of Physical Chemistry 54 (2003) 493–529.

[33] M.H. Beck, A. Jäckle, G.A. Worth, H.-D. Meyer, The multiconfiguration time-dependent Hartree (MCTDH) method: a highly efficient algorithm for propagating wavepackets, Physics Reports 324 (2000) 1–105.

[34] E. Herbst, The chemistry of interstellar space, Chemical Society Reviews 30 (2001) 168–176.

[35] C. Chyba, C. Sagan, Endogenous production, exogenous delivery and impact-shock synthesis of organic molecules: an inventory for the origins of life, Nature 355 (1992) 125–132.

[36] T. Helgajer, E. Uggerud, H.J.A. Jensen, Integration of the classical equations of motion on ab initio molecular potential energy surfaces using gradients and hessians: application to translational energy release upon fragmentation, Chemical Physics Letters 173 (1990) 145–150.

[37] W. Chen, W.L. Hase, H.B. Schlegel, Ab initio classical trajectory study of $H_2CO \rightarrow H_2$ + CO dissociation, Chemical Physics Letters 228 (1994) 436–442.

[38] R. Steckler, G.M. Thurman, J.D. Watts, R.J. Bartlett, Ab initio direct dynamics study of OH $+ HCl \rightarrow Cl + H_2O$, Journal of Chemical Physics 106 (1997) 3926–3933.

[39] F. London, Quantenmechanische deutung des vorgangs der aktivierung, Zeitschrift für Elektrochemie und Angewandte Physikalische Chemie 35 (1929) 552–555.

[40] H. Eyring, M. Polanyi, Über einfache gasreaktionen, Zeitschrift für Physikalische Chemie. Abteilung B 12 (1931) 279–311.

[41] S. Sato, On a new method of drawing the potential energy surface, Journal of Chemical Physics 23 (1955) 592–593.

[42] S. Sato, Potential energy surface of the system of three atoms, Journal of Chemical Physics 23 (1955) 2465–2466.

[43] N. Sathyamurthy, Computational fitting of ab initio potential energy surfaces, Computer Physics Reports 3 (1985) 1–69.

[44] D.G. Truhlar, R. Steckler, M.S. Gordon, Potential energy surfaces for polyatomic reaction dynamics, Chemical Reviews 87 (1987) 217–236.

[45] G.C. Schatz, The analytical representation of electronic potential-energy surfaces, Reviews of Modern Physics 61 (1989) 669–688.

[46] G.C. Schatz, Fitting potential energy surfaces, in: A. Laganà, A. Riganelli (Eds.), Reaction and Molecular Dynamics, in: Lecture Notes in Chemistry, vol. 75, Springer Berlin Heidelberg, 2000, pp. 15–32.

[47] A.J.C. Varandas, Intermolecular and Intramolecular Potentials: Topographical Aspects, Calculation, and Functional Representation via a Double Many-Body Expansion Method, John Wiley & Sons, Inc., 2007, pp. 255–338.

[48] T. Hollebeek, T.-S. Ho, H. Rabitz, Constructing multidimensional molecular potential energy surfaces from ab initio data, Annual Review of Physical Chemistry 50 (1999) 537–570.

[49] R. Jaquet, Interpolation and fitting of potential energy surfaces: concepts, recipes and applications, in: A.F. Sax (Ed.), Potential Energy Surfaces, in: Lecture Notes in Chemistry, vol. 71, Springer Berlin Heidelberg, 1999, pp. 97–175.

[50] D. Shepard, A two-dimensional interpolation function for irregularly-spaced data, in: Proceedings of the 1968 23rd ACM National Conference, ACM '68, ACM, New York, NY, USA, 1968, pp. 517–524.

[51] J. Ischtwan, M.A. Collins, Molecular potential energy surfaces by interpolation, Journal of Chemical Physics 100 (1994) 8080–8088.

[52] M.A. Collins, D.H. Zhang, Application of interpolated potential energy surfaces to quantum reactive scattering, Journal of Chemical Physics 111 (1999) 9924–9931.

[53] M.A. Collins, Molecular potential-energy surfaces for chemical reaction dynamics, Theoretical Chemistry Accounts 108 (2002) 313–324.

[54] T.J. Frankcombe, M.A. Collins, G.A. Worth, Converged quantum dynamics with modified Shepard interpolation and Gaussian wave packets, Chemical Physics Letters 489 (2010) 242–247.

[55] T. Ishida, G.C. Schatz, A local interpolation scheme using no derivatives in quantum-chemical calculations, Chemical Physics Letters 314 (1999) 369–375.

[56] G.G. Maisuradze, D.L. Thompson, A.F. Wagner, M. Minkoff, Interpolating moving least-squares methods for fitting potential energy surfaces: detailed analysis of one-dimensional applications, Journal of Chemical Physics 119 (2003) 10002–10014.

[57] A. Aguado, M. Paniagua, A new functional form to obtain analytical potentials of triatomic molecules, Journal of Chemical Physics 96 (1992) 1265–1275.

[58] A. Aguado, C. Tablero, M. Paniagua, Global fit of ab initio potential energy surfaces I. Triatomic systems, Computer Physics Communications 108 (1998) 259–266.

[59] A. Aguado, C. Suárez, M. Paniagua, Accurate global fit of the H_4 potential energy surface, Journal of Chemical Physics 101 (1994) 4004–4010.

[60] C. Tablero, A. Aguado, M. Paniagua, Global fit of ab initio potential energy surfaces: II.2. tetratomic systems A_2B_2 and ABC_2, Computer Physics Communications 140 (2001) 412–417.

[61] A. Aguado, C. Tablero, M. Paniagua, Global fit of ab initio potential energy surfaces: II.1. Tetraatomic systems ABCD, Computer Physics Communications 134 (2001) 97–109.

[62] K.S. Sorbie, J.N. Murrell, Analytical potentials for triatomic molecules from spectroscopic data, Molecular Physics 29 (1975) 1387–1407.

[63] Y. Paukku, K.R. Yang, Z. Varga, D.G. Truhlar, Global ab initio ground-state potential energy surface of N_4, Journal of Chemical Physics 139 (2013) 044309 (8).

[64] Y. Paukku, K.R. Yang, Z. Varga, D.G. Truhlar, Erratum: "Global ab initio ground-state potential energy surface of N_4" [J. Chem. Phys. 139, 044309 (2013)], Journal of Chemical Physics 140 (2014) 019903 (1).

[65] T. Ishida, G.C. Schatz, Automatic potential energy surface generation directly from ab initio calculations using Shepard interpolation: a test calculation for the H_2 + H system, Journal of Chemical Physics 107 (1997) 3558–3568.

[66] M. Majumder, S.A. Ndengue, R. Dawes, Automated construction of potential energy surfaces, Molecular Physics 114 (2016) 1–18.

[67] S. Rampino, Configuration-space sampling in potential energy surface fitting: a space-reduced bond-order grid approach, The Journal of Physical Chemistry A 120 (2016) 4683–4692.

[68] L. Pauling, Atomic radii and interatomic distances in metals, Journal of the American Chemical Society 69 (1947) 542–553.

[69] A. Laganà, A rotating bond order formulation of the atom diatom potential energy surface, Journal of Chemical Physics 95 (1991) 2216–2217.

[70] E. Garcia, A. Laganà, Diatomic potential functions for triatomic scattering, Molecular Physics 56 (1985) 621–627.

[71] E. Garcia, A. Laganà, A new bond-order functional form for triatomic molecules, Molecular Physics 56 (1985) 629–639.

[72] A. Laganà, G. Ochoa de Aspuru, E. Garcia, The largest angle generalization of the rotating bond order potential: three different atom reactions, Journal of Chemical Physics 108 (1998) 3886–3896.

[73] S. Rampino, D. Skouteris, A. Laganà, Microscopic branching processes: the $O + O_2$ reaction and its relaxed potential representations, International Journal of Quantum Chemistry 110 (2010) 358–367.

[74] A. Laganà, S. Crocchianti, N. Faginas Lago, L. Pacifici, G. Ferraro, A nonorthogonal co-ordinate approach to atom-diatom parallel reactive scattering calculations, Collection of Czechoslovak Chemical Communications 68 (2003) 307–330.

[75] S. Rampino, A. Laganà, Bond order uniform grids for quantum reactive scattering, International Journal of Quantum Chemistry 112 (2012) 1818–1828.

[76] P.M. Morse, Diatomic molecules according to the wave mechanics. II. Vibrational levels, Physical Review 34 (1929) 57–64.

[77] S. Rampino, D. Skouteris, A. Laganà, The $O + O_2$ reaction: quantum detailed probabilities and thermal rate coefficients, Theoretical Chemistry Accounts 123 (2009) 249–256.

[78] A. Varandas, A useful triangular plot of triatomic potential energy surfaces, Chemical Physics Letters 138 (1987) 455–461.

[79] L. Verlet, Computer "experiments" on classical fluids. I. Thermodynamical properties of Lennard-Jones molecules, Physical Review 159 (1967) 98–103.

[80] M. Karplus, R.N. Porter, R.D. Sharma, Exchange reactions with activation energy. I. Simple barrier potential for (H, H_2), Journal of Chemical Physics 43 (1965) 3259–3287.

[81] D.E. Manolopoulos, State to state reactive scattering, in: Encyclopedia of Computational Chemistry, John Wiley & Sons, Ltd, 1998, pp. 2699–2708.

[82] R.A. Marcus, On the analytical mechanics of chemical reactions. Quantum mechanics of linear collisions, Journal of Chemical Physics 45 (1966) 4493–4499.

[83] A. Kuppermann, G.C. Schatz, Quantum mechanical reactive scattering: an accurate three-dimensional calculation, Journal of Chemical Physics 62 (1975) 2502–2504.

[84] J. Dai, J.Z.H. Zhang, Time-dependent wave packet approach to state-to-state reactive scattering and application to $H + O_2$ reaction, The Journal of Physical Chemistry 100 (1996) 6898–6903.

[85] D.J. Tannor, D.E. Weeks, Wave packet correlation function formulation of scattering theory: the quantum analog of classical S-matrix theory, Journal of Chemical Physics 98 (1993) 3884–3893.

[86] R.T. Pack, G.A. Parker, Quantum reactive scattering in three dimensions using hyperspherical (APH) coordinates. Theory, Journal of Chemical Physics 87 (1987) 3888–3921.

[87] W.H. Miller, Recent advances in quantum mechanical reactive scattering theory, including comparison of recent experiments with rigorous calculations of state-to-state cross sections for the $H/D + H_2 \rightarrow H_2/HD + H$ reactions, Annual Review of Physical Chemistry 41 (1990) 245–281.

[88] R. Jaquet, Quantum reactive scattering: the time-independent approach. I. Principles and early developments, in: W. Jakubetz (Ed.), Methods in Reaction Dynamics, in: Lecture Notes in Chemistry, vol. 77, Springer Berlin Heidelberg, 2001, pp. 17–82.

[89] R. Jaquet, Quantum reactive scattering: the time independent approach. II Current methods and developments, in: W. Jakubetz (Ed.), Methods in Reaction Dynamics, in: Lecture Notes in Chemistry, vol. 77, Springer Berlin Heidelberg, 2001, pp. 83–126.

[90] N. Balakrishnan, C. Kalyanaraman, N. Sathyamurthy, Time-dependent quantum mechanical approach to reactive scattering and related processes, Physics Reports 280 (1997) 79–144.

[91] S.C. Althorpe, Quantum wavepacket method for state-to-state reactive cross sections, Journal of Chemical Physics 114 (2001) 1601–1616.

[92] G. Balint-Kurti, Wavepacket quantum dynamics, Theoretical Chemistry Accounts 127 (2010) 1–17.

[93] J. Crawford, G.A. Parker, State-to-state three-atom time-dependent reactive scattering in hyperspherical coordinates, Journal of Chemical Physics 138 (2013) 054313 (12).

[94] G. Nyman, H.-G. Yu, Quantum theory of bimolecular chemical reactions, Reports on Progress in Physics 63 (2000) 1001–1059.

[95] J. Mazur, R.J. Rubin, Quantum-mechanical calculation of the probability of an exchange reaction for constrained linear encounters, Journal of Chemical Physics 31 (1959) 1395–1412.

[96] E.A. McCullough, R.E. Wyatt, Dynamics of the collinear H + H$_2$ reaction. II. Energy analysis, Journal of Chemical Physics 54 (1971) 3592–3600.

[97] D. Kosloff, R. Kosloff, A Fourier method solution for the time dependent Schrödinger equation as a tool in molecular dynamics, Journal of Computational Physics 52 (1983) 35–53.

[98] H. Tal-Ezer, R. Kosloff, An accurate and efficient scheme for propagating the time dependent Schrödinger equation, Journal of Chemical Physics 81 (1984) 3967–3971.

[99] E.A. McCullough, R.E. Wyatt, Quantum dynamics of the collinear (H, H$_2$) reaction, Journal of Chemical Physics 51 (1969) 1253–1254.

[100] E.A. McCullough, R.E. Wyatt, Dynamics of the collinear H + H$_2$ reaction. I. Probability density and flux, Journal of Chemical Physics 54 (1971) 3578–3591.

[101] V. Mohan, N. Sathyamurthy, Quantal wavepacket calculations of reactive scattering, Computer Physics Reports 7 (1988) 213–258.

[102] K.C. Kulander, Collision induced dissociation in collinear H + H$_2$: quantum mechanical probabilities using the time-dependent wavepacket approach, Journal of Chemical Physics 69 (1978) 5064–5072.

[103] C. Leforestier, Competition between dissociation and exchange processes in a collinear A + BC collision. I. Exact quantum results, Chemical Physics 87 (1984) 241–261.

[104] M. Feit, J. Fleck, A. Steiger, Solution of the Schrödinger equation by a spectral method, Journal of Computational Physics 47 (1982) 412–433.

[105] M.D. Feit, J.A. Fleck, Solution of the Schrödinger equation by a spectral method II: vibrational energy levels of triatomic molecules, Journal of Chemical Physics 78 (1983) 301–308.

[106] M.D. Feit, J.A. Fleck, Wave packet dynamics and chaos in the Hénon-Heiles system, Journal of Chemical Physics 80 (1984) 2578–2584.

[107] S.K. Gray, G.G. Balint-Kurti, Quantum dynamics with real wave packets, including application to three-dimensional ($j = 0$) D +H$_2$ → HD + H reactive scattering, Journal of Chemical Physics 108 (1998) 950–962.

[108] S.K. Gray, Wave packet dynamics of resonance decay: an iterative equation approach with application to HCO → H+CO, Journal of Chemical Physics 96 (1992) 6543–6554.

[109] B. Liu, Ab initio potential energy surface for linear H$_3$, Journal of Chemical Physics 58 (1973) 1925–1937.

[110] P. Siegbahn, B. Liu, An accurate three-dimensional potential energy surface for H$_3$, Journal of Chemical Physics 68 (1978) 2457–2465.

[111] D.G. Truhlar, C.J. Horowitz, Functional representation of Liu and Siegbahn's accurate ab initio potential energy calculations for H + H$_2$, Journal of Chemical Physics 68 (1978) 2466–2476.

[112] D.G. Truhlar, C.J. Horowitz, Erratum: Functional representation of Liu and Siegbahn's accurate abinitio potential energy calculations for H + H$_2$, Journal of Chemical Physics 71 (1979) 1514.

[113] C.C. Marston, G.G. Balint-Kurti, The Fourier grid Hamiltonian method for bound state eigenvalues and eigenfunctions, Journal of Chemical Physics 91 (1989) 3571–3576.

[114] D.T. Colbert, W.H. Miller, A novel discrete variable representation for quantum mechanical reactive scattering via the S-matrix Kohn method, Journal of Chemical Physics 96 (1992) 1982–1991.

[115] V. Kokoouline, O. Dulieu, R. Kosloff, F. Masnou-Seeuws, Mapped Fourier methods for long-range molecules: application to perturbations in the $Rb_2(0_u^+)$ photoassociation spectrum, Journal of Chemical Physics 110 (1999) 9865–9876.

[116] E. Fattal, R. Baer, R. Kosloff, Phase space approach for optimizing grid representations: the mapped Fourier method, Physical Review E 53 (1996) 1217–1227.

[117] D.G. Truhlar, J.T. Muckerman, Reactive scattering cross sections III: quasiclassical and semiclassical methods, in: R. Bernstein (Ed.), Atom-Molecule Collision Theory, Springer US, 1979, pp. 505–566.

[118] F. Vazart, C. Latouche, D. Skouteris, N. Balucani, V. Barone, Cyanomethanimine isomers in cold interstellar clouds: insights from electronic structure and kinetic calculations, The Astrophysical Journal 810 (2015) 111.

[119] J. Lupi, C. Puzzarini, V. Barone, Methanimine as a key precursor of imines in the interstellar medium: the case of propargylimine, The Astrophysical Journal 903 (2020) L35.

[120] O.K. Rice, H.C. Ramsperger, Theories of unimolecular gas reactions at low pressures, Journal of the American Chemical Society 49 (1927) 1617–1629.

[121] L.S. Kassel, Studies in homogeneous gas reactions. I, The Journal of Physical Chemistry 32 (1927) 225–242.

[122] R.A. Marcus, Unimolecular dissociations and free radical recombination reactions, Journal of Chemical Physics 20 (1952) 359–364.

[123] A. Fernández-Ramos, J.A. Miller, S.J. Klippenstein, D.G. Truhlar, Modeling the kinetics of bimolecular reactions, Chemical Reviews 106 (2006) 4518–4584.

[124] M. Brouard, Reaction Dynamics, Oxford Chemistry Primers, OUP Oxford, 1998.

[125] N.J.B. Green, Introduction, in: N.J. Green (Ed.), Unimolecular Kinetics, in: Comprehensive Chemical Kinetics, vol. 39, Elsevier, 2003, pp. 1–53 (Chapter 1).

[126] V. Barone, Anharmonic vibrational properties by a fully automated second-order perturbative approach, Journal of Chemical Physics 122 (2005) 014108.

[127] J. Bloino, M. Biczysko, V. Barone, General perturbative approach for spectroscopy, thermodynamics, and kinetics: methodological background and benchmark studies, Journal of Chemical Theory and Computation 8 (2012) 1015–1036.

[128] S.E. Stein, B.S. Rabinovitch, Accurate evaluation of internal energy level sums and densities including anharmonic oscillators and hindered rotors, Journal of Chemical Physics 58 (1973) 2438–2445.

[129] T. Beyer, D.F. Swinehart, Algorithm 448: number of multiply-restricted partitions, Communications of the ACM 16 (1973) 379.

[130] F. Wang, D.P. Landau, Efficient, multiple-range random walk algorithm to calculate the density of states, Physical Review Letters 86 (2001) 2050–2053.

[131] C. Zhou, R.N. Bhatt, Understanding and improving the Wang-Landau algorithm, Physical Review E 72 (2005) 025701.

[132] M. Basire, P. Parneix, F. Calvo, Quantum anharmonic densities of states using the Wang-Landau method, Journal of Chemical Physics 129 (2008) 081101 (4).

[133] T.L. Nguyen, J.R. Barker, Sums and densities of fully coupled anharmonic vibrational states: a comparison of three practical methods, The Journal of Physical Chemistry A 114 (2010) 3718–3730.

[134] W.H. Miller, Tunneling corrections to unimolecular rate constants, with application to formaldehyde, Journal of the American Chemical Society 101 (1979) 6810–6814.

[135] S.H. Robertson, M.J. Pilling, L.C. Jitariu, I.H. Hillier, Master equation methods for multiple well systems: application to the 1-,2-pentyl system, Physical Chemistry Chemical Physics 9 (2007) 4085–4097.

[136] A.C. Mansell, D.J. Kahle, D.J. Bellert, Calculating RRKM rate constants from vibrational frequencies and their dynamic interpretation, The Mathematica Journal 19 (2017) 1–20.

[137] S. Nandi, B. Ballotta, S. Rampino, V. Barone, A general user-friendly tool for kinetic calculations of multi-step reactions within the virtual multifrequency spectrometer project, Applied Sciences 10 (2020) 1872.

[138] J.R. Barker, R.E. Weston, Collisional energy transfer probability densities $P(E, J; E', J')$ for monatomics colliding with large molecules, The Journal of Physical Chemistry A 114 (2010) 10619–10633.

[139] D.R. Glowacki, C.-H. Liang, C. Morley, M.J. Pilling, S.H. Robertson, MESMER: an open-source master equation solver for multi-energy well reactions, The Journal of Physical Chemistry A 116 (2012) 9545–9560.

[140] S. Rampino, M. Pastore, E. Garcia, L. Pacifici, A. Laganà, On the temperature dependence of the rate coefficient of formation of C_2^+ from C + CH^+, Monthly Notices of the Royal Astronomical Society 460 (2016) 2368–2375.

[141] L. Pacifici, M. Pastore, E. Garcia, A. Laganà, S. Rampino, A dynamics investigation of the C + CH^+ → C_2^+ + H reaction on an ab initio bond-order like potential, The Journal of Physical Chemistry A 120 (2016) 5125–5135.

[142] V. Wakelam, E. Herbst, J.-C. Loison, I.W.M. Smith, V. Chandrasekaran, B. Pavone, N.G. Adams, M.-C. Bacchus-Montabonel, A. Bergeat, K. Béroff, V.M. Bierbaum, M. Chabot, A. Dalgarno, E.F. van Dishoeck, A. Faure, W.D. Geppert, D. Gerlich, D. Galli, E. Hébrard, F. Hersant, K.M. Hickson, P. Honvault, S.J. Klippenstein, S. Le Picard, G. Nyman, P. Pernot, S. Schlemmer, F. Selsis, I.R. Sims, D. Talbi, J. Tennyson, J. Troe, R. Wester, L. Wiesenfeld, A KInetic Database for Astrochemistry (KIDA), The Astrophysical Journal. Supplement Series 199 (2012) 21 (19).

[143] D. McElroy, C. Walsh, A.J. Markwick, M.A. Cordiner, K. Smith, T.J. Millar, The UMIST database for astrochemistry 2012, Astronomy & Astrophysics 550 (2013) A36.

[144] D.M. Kooij, Über die Zersetzung des Gasförmigen phosphorwasserstoffs, Zeitschrift für Physikalische Chemie. Abteilung B 12 (1893) 155–161.

[145] K. Laidler, A glossary of terms used in chemical kinetics, including reaction dynamics (IUPAC recommendations 1996), Pure and Applied Chemistry 68 (1996) 149–192.

[146] P.M. Solomon, W. Klemperer, The formation of diatomic molecules in interstellar clouds, The Astrophysical Journal 178 (1972) 389–422.

[147] B. Bussery-Honvault, P. Honvault, J.-M. Launay, A study of the C (^1D) + H_2 → CH + H reaction: global potential energy surface and quantum dynamics, Journal of Chemical Physics 115 (2001) 10701–10708.

[148] S.Y. Lin, H. Guo, Case study of a prototypical elementary insertion reaction: C(^1D) + H_2 → CH + H, The Journal of Physical Chemistry A 108 (2004) 10066–10071.

[149] N. Balucani, G. Capozza, L. Cartechini, A. Bergeat, R. Bobbenkamp, P. Casavecchia, F. Javier Aoiz, L. Bañares, P. Honvault, B. Bussery-Honvault, J.-M. Launay, Dynamics of the insertion reaction C(^1D) + H_2: a comparison of crossed molecular beam experiments with quasiclassical trajectory and quantum mechanical scattering calculations, Physical Chemistry Chemical Physics 6 (2004) 4957–4967.

[150] N. Balucani, G. Capozza, E. Segoloni, A. Russo, R. Bobbenkamp, P. Casavecchia, T. Gonzalez-Lezana, E.J. Rackham, L. Bañares, F.J. Aoiz, Dynamics of the C(^1D) + D_2 reaction: a comparison of crossed molecular-beam experiments with quasiclassical trajectory and accurate statistical calculations, Journal of Chemical Physics 122 (2005) 234309 (11).

[151] P. Defazio, C. Petrongolo, B. Bussery-Honvault, P. Honvault, Born-Oppenheimer quantum dynamics of the C(^1D) + H_2 reaction on the CH_2 $\tilde{a}^1A_1$ and $\tilde{b}^1B_1$ surfaces, Journal of Chemical Physics 131 (2009) 114303 (6).

[152] S. Joseph, P.J.S.B. Caridade, A.J.C. Varandas, Quasiclassical trajectory study of the C (^1D) + H_2 reaction and isotopomeric variants: kinetic isotope effect and CD/CH branching ratio, The Journal of Physical Chemistry A 115 (2011) 7882–7890.

[153] Z.S. Sun, C. Zhang, S. Lin, Y. Zheng, Q. Meng, W.B. Bian, Quantum reaction dynamics of the C(^1D) + $H_2(D_2)$ → CH(D) + H(D) on a new potential energy surface, Journal of Chemical Physics 139 (2013) 014306 (6).

[154] Y. Wu, C. Zhang, J. Cao, W. Bian, Quasiclassical trajectory study of the C(^1D) + H_2 → CH + H reaction on a new global ab initio potential energy surface, The Journal of Physical Chemistry A 118 (2014) 4235–4242.

[155] D. Herraez-Aguilar, P.G. Jambrina, M. Menendez, J. Aldegunde, R. Warmbier, F.J. Aoiz, The effect of the reactant internal excitation on the dynamics of the $C^+ + H_2$ reaction, Physical Chemistry Chemical Physics 16 (2014) 24800–24812.

[156] A. Zanchet, G. Godard, N. Bulut, O. Roncero, P. Halvick, C. José, $H_2(v = 0, 1) + C^+ (^2P) \rightarrow H + CH^+$ state-to-state rate constants for chemical pumping models in astrophysical media, The Astrophysical Journal 766 (2013) 80 (8).

[157] T. Stoecklin, P. Halvick, Low temperature quantum rate coefficient of the $H + CH^+$ reaction, Physical Chemistry Chemical Physics 7 (2005) 2446–2452.

[158] R. Warmbier, R. Schneider, Ab initio potential energy surface of CH_2^+ and reaction dynamics of $H + CH^+$, Physical Chemistry Chemical Physics 13 (2011) 10285–10294.

[159] S. Bovino, T. Grassi, F.A. Gianturco, CH^+ destruction by reaction with H: computing quantum rates to model different molecular regions in the interstellar medium, The Journal of Physical Chemistry A 119 (2015) 11973–11982.

[160] M. Boggio-Pasqua, P. Halvick, M.-T. Rayez, J.-C. Rayez, J.-M. Robbe, Ab initio study of the potential energy surfaces for the reaction $C + CH \rightarrow C_2 + H$, The Journal of Physical Chemistry A 102 (1998) 2009–2015.

[161] B.-Y. Tang, M.-D. Chen, K.-L. Han, J.Z.H. Zhang, Time-dependent quantum wave packet study of the C + CH reaction, Journal of Chemical Physics 115 (2001) 731–738.

[162] B.-Y. Tang, M.-D. Chen, K.-L. Han, J.Z.H. Zhang, Time-dependent quantum dynamics study of the C + CH reaction on the $2A'$ surface, The Journal of Physical Chemistry A 105 (2001) 8629–8634.

[163] H. Yang, M. Hankel, Y. Zheng, A.J.C. Varandas, Significant nonadiabatic effects in the C + CH reaction dynamics, Journal of Chemical Physics 135 (2011) 024306 (6).

[164] A.E. Douglas, G. Herzberg, Note on CH^+ in interstellar space and in the laboratory, The Astrophysical Journal 94 (1941) 381.

[165] D. Krankowsky, P. Lämmerzahl, I. Herrwerth, J. Woweries, P. Eberhardt, U. Dolder, U. Herrmann, W. Schulte, J.J. Berthelier, J.M. Illiano, R.R. Hodges, J.H. Hoffman, In situ gas and ion measurements at comet Halley, Nature 321 (1986) 326–329.

[166] M.A. Coplan, K.W. Ogilvie, M.F. A'Hearn, P. Bochsler, J. Geiss, Ion composition and upstream solar wind observations at comet Giacobini-Zinner, Journal of Geophysical Research: Space Physics 92 (1987) 39–46.

[167] G. Winnewisser, The chemistry of interstellar molecules, in: Cosmo- and Geochemistry, in: Topics in Current Chemistry, vol. 99, Springer Berlin Heidelberg, 1981, pp. 39–71.

[168] P. Langevin, Une formule fondamentale de théorie cinétique, Annales de Chimie et de Physique 5 (1905) 245–288.

[169] G. Gioumousis, D.P. Stevenson, Reactions of gaseous molecule ions with gaseous molecules. V. Theory, Journal of Chemical Physics 29 (1958) 294–299.

[170] D.E. Woon, E. Herbst, Quantum chemical predictions of the properties of known and postulated neutral interstellar molecules, The Astrophysical Journal. Supplement Series 185 (2009) 273–288.

[171] V. Wakelam, I. Smith, E. Herbst, J. Troe, W. Geppert, H. Linnartz, K. Öberg, E. Roueff, M. Agúndez, P. Pernot, H. Cuppen, J. Loison, D. Talbi, Reaction networks for interstellar chemical modelling: improvements and challenges, Space Science Reviews 156 (2010) 13–72.

[172] C. Thierfelder, B. Assadollahzadeh, P. Schwerdtfeger, S. Schäfer, R. Schäfer, Relativistic and electron correlation effects in static dipole polarizabilities for the group-14 elements from carbon to element $z = 114$: theory and experiment, Physical Review A 78 (2008) 052506.

[173] S.S. Prasad, W.T. Huntress Jr., A model for gas phase chemistry in interstellar clouds. II - nonequilibrium effects and effects of temperature and activation energies, The Astrophysical Journal 239 (1980) 151–165.

[174] C. Angeli, R. Cimiraglia, J.-P. Malrieu, n-electron valence state perturbation theory: a spinless formulation and an efficient implementation of the strongly contracted and of the partially contracted variants, Journal of Chemical Physics 117 (2002) 9138–9153.

[175] C. Angeli, M. Pastore, R. Cimiraglia, New perspectives in multireference perturbation theory: the n-electron valence state approach, Theoretical Chemistry Accounts 117 (2007) 743–754.

[176] H.-J. Werner, P.J. Knowles, G. Knizia, F.R. Manby, M. Schütz, P. Celani, T. Korona, R. Lindh, A. Mitrushenkov, G. Rauhut, K.R. Shamasundar, T.B. Adler, R.D. Amos, A. Bernhardsson, A. Berning, D.L. Cooper, M.J.O. Deegan, A.J. Dobbyn, F. Eckert, E. Goll, C. Hampel, A. Hesselmann, G. Hetzer, T. Hrenar, G. Jansen, C. Köppl, Y. Liu, A.W. Lloyd, R.A. Mata, A.J. May, S.J. McNicholas, W. Meyer, M.E. Mura, A. Nicklass, D.P. O'Neill, P. Palmieri, K. Pflüger, R. Pitzer, M. Reiher, T. Shiozaki, H. Stoll, A.J. Stone, R. Tarroni, T. Thorsteinsson, M. Wang, A. Wolf, Molpro, version 2010.1, a package of ab initio programs, see http://www.molpro.net, 2010. (Accessed 15 February 2022).

[177] S. Green, P.S. Bagus, B. Liu, A.D. McLean, M. Yoshimine, Calculated potential-energy curves for CH^+, Physical Review A 5 (1972) 1614–1618.

[178] C. Petrongolo, P.J. Bruna, S.D. Peyerimhoff, R.J. Buenker, Theoretical prediction of the potential curves for the lowest-lying states of the C_2^+ molecular ion, Journal of Chemical Physics 74 (1981) 4594–4602.

[179] W.P. Kraemer, B.O. Roos, A CAS SCF CI study of the $^1\Sigma_g^+$ and $^3\Pi_u$ states of the C_2 molecule and the $^4\sigma_g^-$ and $^2\Pi_u$ states of the C_2^+ ion, Chemical Physics 118 (1987) 345–355.

[180] J.D. Watts, R.J. Bartlett, Coupled-cluster calculations on the C_2 molecule and the C_2^+ and c_2^- molecular ions, Journal of Chemical Physics 96 (1992) 6073–6084.

[181] D. Shi, X. Niu, J. Sun, Z. Zhu, Potential energy curves, spectroscopic parameters, and spin-orbit coupling: a theoretical study on 24 λ-S and 54 ω states of C_2^+ cation, The Journal of Physical Chemistry A 117 (2013) 2020–2034.

[182] R. Krishnan, M.J. Frisch, R.A. Whiteside, J.A. Pople, P.v.R. Schleyer, The structure of CCH^+, Journal of Chemical Physics 74 (1981) 4213–4214.

[183] W. Koch, G. Frenking, The low-lying electronic states of protonated C_2, CCH^+, Journal of Chemical Physics 93 (1990) 8021–8028.

[184] K. Huber, G. Herzberg, Molecular Spectra and Molecular Structure. IV Constants of Diatomic Molecules, Springer US, 1979.

[185] J.P. Maier, M. Rösslein, The $\tilde{B}^4\Sigma_u^- - \tilde{X}^4\Sigma_g^-$ electronic spectrum of C_2^+, Journal of Chemical Physics 88 (1988) 4614–4620.

[186] W.L. Hase, R.J. Duchovic, X. Hu, A. Komornicki, K.F. Lim, D.-h. Lu, G.H. Peslherbe, K.N. Swamy, S.R. Van de Linde, A.J.C. Varandas, H. Wang, R.J. Wolf, VENUS96: a general chemical dynamics computer program, Quantum Chemistry Program Exchange Bulletin 16 (1996) 43.

[187] V. Aquilanti, K.C. Mundim, M. Elango, S. Kleijn, T. Kasai, Temperature dependence of chemical and biophysical rate processes: phenomenological approach to deviations from Arrhenius law, Chemical Physics Letters 498 (2010) 209–213.

[188] V. Aquilanti, K.C. Mundim, S. Cavalli, D. De Fazio, A. Aguilar, J.M. Lucas, Exact activation energies and phenomenological description of quantum tunneling for model potential energy surfaces. the $F + H_2$ reaction at low temperature, Chemical Physics 398 (2012) 186–191.

[189] S.T. Banks, D.C. Clary, Reduced dimensionality quantum dynamics of $Cl + CH_4 \rightarrow HCl + CH_3$ on an ab initio potential, Physical Chemistry Chemical Physics 9 (2007) 933–943.

[190] J.M. Bowman, A.F. Wagner, Reduced Dimensionality Theories of Quantum Reactive Scattering: Applications to $Mu + H_2$, $H + H_2$, $O(^3P) + H_2$, D_2 and HD, Springer Netherlands, Dordrecht, 1986, pp. 47–76.

[191] H.-D. Meyer, U. Manthe, L. Cederbaum, The multi-configurational time-dependent Hartree approach, Chemical Physics Letters 165 (1990) 73–78.

[192] H. Meyer, F. Gatti, G. Worth (Eds.), Multidimensional Quantum Dynamics: MCTDH Theory and Applications, Wiley-VCH Verlag GmbH & Co. KGaA, Weinheim, 2009.

[193] D. Skouteris, Time-dependent calculations on systems of chemical interest: dynamical and kinetic approaches, International Journal of Quantum Chemistry 116 (2016) 1618–1622.

[194] H. Wang, M. Thoss, Multilayer formulation of the multiconfiguration time-dependent Hartree theory, Journal of Chemical Physics 119 (2003) 1289–1299.

[195] H. Wang, Multilayer multiconfiguration time-dependent Hartree theory, The Journal of Physical Chemistry A 119 (2015) 7951–7965.

[196] E.J. Heller, Time-dependent approach to semiclassical dynamics, Journal of Chemical Physics 62 (1975) 1544–1555.

[197] I.R. Craig, D.E. Manolopoulos, Chemical reaction rates from ring polymer molecular dynamics, Journal of Chemical Physics 122 (2005) 084106.

[198] D. Chandler, P.G. Wolynes, Exploiting the isomorphism between quantum theory and classical statistical mechanics of polyatomic fluids, Journal of Chemical Physics 74 (1981) 4078–4095.

[199] S. Rampino, Y.V. Suleimanov, Thermal rate coefficients for the astrochemical process C + $CH^+ \rightarrow C2^+$ + H by ring polymer molecular dynamics, The Journal of Physical Chemistry A 120 (2016) 9887–9893.

[200] Y.V. Suleimanov, F.J. Aoiz, H. Guo, Chemical reaction rate coefficients from ring polymer molecular dynamics: theory and practical applications, The Journal of Physical Chemistry A 120 (2016) 8488–8502.

[201] Y. Li, Y.V. Suleimanov, H. Guo, Ring-polymer molecular dynamics rate coefficient calculations for insertion reactions: $X + H_2 \rightarrow HX + H$ (X = N, O), Journal of Physical Chemistry Letters 5 (2014) 700–705.

[202] Y.V. Suleimanov, W.J. Kong, H. Guo, W.H. Green, Ring-polymer molecular dynamics: rate coefficient calculations for energetically symmetric (near thermoneutral) insertion reactions $(X + H_2) \rightarrow HX + H(X = C(^1D), S(^1D))$, Journal of Chemical Physics 141 (2014) 244103.

[203] K.M. Hickson, J.-C. Loison, H. Guo, Y.V. Suleimanov, Ring-polymer molecular dynamics for the prediction of low-temperature rates: an investigation of the $C(^1D) + H_2$ reaction, Journal of Physical Chemistry Letters 6 (2015) 4194–4199.

[204] A.D. MacKerell, D. Bashford, M. Bellott, R.L. Dunbrack, J.D. Evanseck, M.J. Field, S. Fischer, J. Gao, H. Guo, S. Ha, D. Joseph-McCarthy, L. Kuchnir, K. Kuczera, F.T.K. Lau, C. Mattos, S. Michnick, T. Ngo, D.T. Nguyen, B. Prodhom, W.E. Reiher, B. Roux, M. Schlenkrich, J.C. Smith, R. Stote, J. Straub, M. Watanabe, J. Wiórkiewicz-Kuczera, D. Yin, M. Karplus, All-atom empirical potential for molecular modeling and dynamics studies of proteins, The Journal of Physical Chemistry B 102 (1998) 3586–3616.

[205] C.J. Cramer, Essentials of Computational Chemistry. Theories and Models, 2nd edition, John Wiley & Sons, Chichester, 2004.

[206] G.A. Worth, H.-D. Meyer, L.S. Cederbaum, Relaxation of a system with a conical intersection coupled to a bath: a benchmark 24-dimensional wave packet study treating the environment explicitly, Journal of Chemical Physics 109 (1998) 3518–3529.

[207] B.F.E. Curchod, T.J. Martínez, Ab initio nonadiabatic quantum molecular dynamics, Chemical Reviews 118 (2018) 3305–3336.

[208] J.C. Tully, R.K. Preston, Trajectory surface hopping approach to nonadiabatic molecular collisions: the reaction of H^+ with D_2, Journal of Chemical Physics 55 (1971) 562–572.

[209] J.C. Tully, Molecular dynamics with electronic transitions, Journal of Chemical Physics 93 (1990) 1061–1071.

[210] A. Szabo, N.S. Ostlund, Modern Quantum Chemistry. Introduction to Advanced Electronic Structure Theory, Dover, Mineola, 1996.

[211] P. Hohenberg, W. Kohn, Inhomogeneous electron gas, Physical Review 136 (1964) B864–B871.

[212] R.G. Parr, Density-Functional Theory of Atoms and Molecules, Oxford University Press, New York, 1995.

[213] W. Kohn, L.J. Sham, Self-consistent equations including exchange and correlation effects, Physical Review 140 (1965) A1133–A1138.

[214] A.D. Becke, A multicenter numerical integration scheme for polyatomic molecules, Journal of Chemical Physics 88 (1988) 2547–2553.

[215] Konstantin N. Kudin, Gustavo E. Scuseria, Converging self-consistent field equations in quantum chemistry - recent achievements and remaining challenges, ESAIM. Mathematical Modelling and Numerical Analysis 41 (2007) 281–296.

[216] V.R. Saunders, I.H. Hillier, A "level–shifting" method for converging closed shell Hartree-Fock wave functions, International Journal of Quantum Chemistry 7 (1973) 699–705.

[217] E. Clementi, D.R. Davis, Electronic structure of large molecular systems, Journal of Computational Physics 1 (1966) 223–244.

[218] T. Petersson, B. Hellsing, A detailed derivation of Gaussian orbital-based matrix elements in electron structure calculations, European Journal of Physics 31 (2010) 37–46.

[219] S. Obara, A. Saika, Efficient recursive computation of molecular integrals over Cartesian Gaussian functions, Journal of Chemical Physics 84 (1986) 3963–3974.

[220] L.E. McMurchie, E.R. Davidson, One- and two-electron integrals over Cartesian Gaussian functions, Journal of Computational Physics 26 (1978) 218–231.

[221] K.L. Schuchardt, B.T. Didier, T. Elsethagen, L. Sun, V. Gurumoorthi, J. Chase, J. Li, T.L. Windus, Basis set exchange: a community database for computational sciences, Journal of Chemical Information and Modeling 47 (2007) 1045–1052.

[222] B.P. Pritchard, D. Altarawy, B. Didier, T.D. Gibson, T.L. Windus, New basis set exchange: an open, up-to-date resource for the molecular sciences community, Journal of Chemical Information and Modeling 59 (2019) 4814–4820.

[223] R. Ditchfield, W.J. Hehre, J.A. Pople, Self-consistent molecular-orbital methods. IX. An extended Gaussian-type basis for molecular-orbital studies of organic molecules, Journal of Chemical Physics 54 (1971) 724–728.

[224] W.J. Hehre, R. Ditchfield, J.A. Pople, Self-consistent molecular orbital methods. XII. Further extensions of Gaussian-type basis sets for use in molecular orbital studies of organic molecules, Journal of Chemical Physics 56 (1972) 2257–2261.

[225] P.C. Hariharan, J.A. Pople, The influence of polarization functions on molecular orbital hydrogenation energies, Theoretica Chimica Acta 28 (1973) 213–222.

[226] B. Jeziorski, R. Moszynski, K. Szalewicz, Perturbation theory approach to intermolecular potential energy surfaces of van der Waals complexes, Chemical Reviews 94 (1994) 1887–1930.

[227] K. Szalewicz, Symmetry-adapted perturbation theory of intermolecular forces, WIREs Computational Molecular Science 2 (2012) 254–272.

[228] T. Ziegler, A. Rauk, On the calculation of bonding energies by the Hartree Fock Slater method, Theoretica Chimica Acta 46 (2009) 1–10.

[229] L. Zhao, M. von Hopffgarten, D.M. Andrada, G. Frenking, Energy decomposition analysis, WIREs Computational Molecular Science 8 (2018) e1345.

[230] A. Savin, O. Jepsen, J. Flad, O.K. Andersen, H. Preuss, H.G. von Schnering, Electron localization in solid-state structures of the elements: the diamond structure, Angewandte Chemie. International Edition in English 31 (1992) 187–188.

[231] A.D. Becke, K.E. Edgecombe, A simple measure of electron localization in atomic and molecular systems, Journal of Chemical Physics 92 (1990) 5397–5403.

[232] V. Bezugly, P. Wielgus, M. Kohout, F.R. Wagner, Electron localizability indicators ELI–D and ELIA for highly correlated wavefunctions of homonuclear dimers. I. Li_2, Be_2, B_2, and C_2, Journal of Computational Chemistry 31 (2010) 1504–1519.

[233] F. Weinhold, C.R. Landis, Valency and Bonding: A Natural Bond Orbital Donor-Acceptor Perspective, Cambridge University Press, 2005.

[234] S.M. Bachrach, Population analysis and electron densities from quantum mechanics, Reviews in Computational Chemistry 5 (1994) 171–228.

[235] R. Contreras, L.R. Domingo, B. Silvi, Electron Densities: Population Analysis and Beyond, John Wiley & Sons, Ltd, 2017, pp. 1–114.

[236] G. Voronoi, Nouvelles applications des paramètres continus à la théorie des formes quadratiques. Premier mémoire. Sur quelques propriétés des formes quadratiques positives parfaites, Journal für die Reine und Angewandte Mathematik 1908 (1908) 97–102.

[237] Atomic charges from modified Voronoi polyhedra, Journal of Molecular Structure. Theochem 538 (2001) 235–238.

[238] C. Fonseca Guerra, J.-W. Handgraaf, E.J. Baerends, F.M. Bickelhaupt, Voronoi deformation density (VDD) charges: assessment of the Mulliken, Bader, Hirshfeld, Weinhold, and VDD methods for charge analysis, Journal of Computational Chemistry 25 (2004) 189–210.

[239] R.S. Mulliken, Electronic population analysis on LCAO–MO molecular wave functions. I, Journal of Chemical Physics 23 (1955) 1833–1840.

[240] L. Cusachs, P. Politzer, On the problem of defining the charge on an atom in a molecule, Chemical Physics Letters 1 (1968) 529–531.

[241] F.L. Hirshfeld, Bonded-atom fragments for describing molecular charge densities, Theoretica Chimica Acta 44 (1977) 129–138.

[242] P. Politzer, R.R. Harris, Properties of atoms in molecules. I. Proposed definition of the charge on an atom in a molecule, Journal of the American Chemical Society 92 (1970) 6451–6454.

[243] R.F.W. Bader, A quantum theory of molecular structure and its applications, Chemical Reviews 91 (1991) 893–928.

[244] R. Bader, Atoms in Molecules: A Quantum Theory, International Series of Monographs on Chemistry, Clarendon Press, 1994.

[245] P.L.A. Popelier, Atoms in Molecules: An Introduction, Prentice Hall, Harlow, 2000.

[246] R.E. Brown, H. Shull, A configuration interaction study of the four lowest $^1\Sigma^+$ states of the LiH molecule, International Journal of Quantum Chemistry 2 (1968) 663–685.

[247] L. Belpassi, I. Infante, F. Tarantelli, L. Visscher, The chemical bond between Au(I) and the noble gases. comparative study of NgAuF and $NgAu^+$ (Ng = Ar, Kr, Xe) by density functional and coupled cluster methods, Journal of the American Chemical Society 130 (2008) 1048–1060.

[248] L. Sagresti, S. Rampino, Charge-flow profiles along curvilinear paths: a flexible scheme for the analysis of charge displacement upon intermolecular interactions, Molecules 26 (2021) 6409.

[249] G. Bistoni, S. Rampino, F. Tarantelli, L. Belpassi, Charge-displacement analysis via natural orbitals for chemical valence: charge transfer effects in coordination chemistry, Journal of Chemical Physics 142 (2015) 084112.

[250] M. De Santis, S. Rampino, H.M. Quiney, L. Belpassi, L. Storchi, Charge-displacement analysis via natural orbitals for chemical valence in the four-component relativistic framework, Journal of Chemical Theory and Computation 14 (2018) 1286–1296.

[251] S. Potenti, L. Paoloni, S. Nandi, M. Fusé, V. Barone, S. Rampino, Chemical bonding in cuprous complexes with simple nitriles: octet rule and resonance concepts versus quantitative charge-redistribution analysis, Physical Chemistry Chemical Physics 22 (2020) 20238–20247.

[252] W. Li, L. Spada, N. Tasinato, S. Rampino, L. Evangelisti, A. Gualandi, P.G. Cozzi, S. Melandri, V. Barone, C. Puzzarini, Theory meets experiment for noncovalent complexes: the puzzling case of pnicogen interactions, Angewandte Chemie. International Edition 57 (2018) 13853–13857.

[253] D.A. Obenchain, L. Spada, S. Alessandrini, S. Rampino, S. Herbers, N. Tasinato, M. Mendolicchio, P. Kraus, J. Gauss, C. Puzzarini, J.-U. Grabow, V. Barone, Unveiling the sulfur-sulfur bridge: accurate structural and energetic characterization of a homochalcogen intermolecular bond, Angewandte Chemie. International Edition 57 (2018) 15822–15826.

[254] A. Patti, S. Pedotti, G. Mazzeo, G. Longhi, S. Abbate, L. Paoloni, J. Bloino, S. Rampino, V. Barone, Ferrocenes with simple chiral substituents: an in-depth theoretical and experimental VCD and ECD study, Physical Chemistry Chemical Physics 21 (2019) 9419–9432.

[255] M. De Santis, S. Rampino, L. Storchi, L. Belpassi, F. Tarantelli, The chemical bond and s-d hybridization in coinage metal(I) cyanides, Inorganic Chemistry 58 (2019) 11716–11729.

[256] M. Mitoraj, A. Michalak, Natural orbitals for chemical valence as descriptors of chemical bonding in transition metal complexes, Journal of Molecular Modeling 13 (2007) 347–355.

[257] A. Michalak, M. Mitoraj, T. Ziegler, Bond orbitals from chemical valence theory, The Journal of Physical Chemistry A 112 (2008) 1933–1939.

[258] P. Löwdin, On the non-orthogonality problem connected with the use of atomic wave functions in the theory of molecules and crystals, Journal of Chemical Physics 18 (1950) 365–375.

[259] I. Mayer, On Löwdin's method of symmetric orthogonalization, International Journal of Quantum Chemistry 90 (2002) 63–65.

[260] B.C. Carlson, J.M. Keller, Orthogonalization procedures and the localization of Wannier functions, Physical Review 105 (1957) 102–103.

[261] R.F. Nalewajski, A.M. Köster, K. Jug, Chemical valence from the two-particle density matrix, Theoretica Chimica Acta 85 (1993) 463–484.

[262] R.F. Nalewajski, J. Mrozek, Modified valence indices from the two-particle density matrix, International Journal of Quantum Chemistry 51 (1994) 187–200.

[263] R.F. Nalewajski, J. Mrozek, A. Michalak, Two-electron valence indices from the Kohn-Sham orbitals, International Journal of Quantum Chemistry 61 (1997) 589–601.

[264] M. Radoń, On the properties of natural orbitals for chemical valence, Theoretical Chemistry Accounts 120 (2008) 337–339.

[265] C.J. Calzado, J.-P. Malrieu, J. Cabrero, R. Caballol, Excitation energy dedicated molecular orbitals. Method and applications to magnetic systems, The Journal of Physical Chemistry A 104 (2000) 11636–11643.

[266] M. Mandado, N. Ramos-Berdullas, Analyzing the electric response of molecular conductors using "electron deformation" orbitals and occupied-virtual electron transfer, Journal of Computational Chemistry 35 (2014) 1261–1269.

[267] G. Booch, Object-oriented development, IEEE Transactions on Software Engineering SE-12 (1986) 211–221.

[268] M.G. Gray, R.M. Roberts, P.F. Dubois, Object-based programming in Fortran 90, Computers in Physics 11 (1997) 355–361.

[269] L. Cardelli, P. Wegner, On understanding types, data abstraction and polymorphism, ACM Computing Surveys 17 (1985) 471–522.

[270] A.M. Gorelik, Object-oriented programming in modern Fortran, Programming and Computer Software 30 (2004) 173–179.

[271] J.R. Cary, S.G. Shasharina, J.C. Cummings, J.V. Reynders, P.J. Hinker, Comparison of C++ and Fortran 90 for object-oriented scientific programming, Computer Physics Communications 105 (1997) 20–36.

[272] M. Metcalf, J. Reid, The F Programming Language, Oxford University Press, Inc., New York, NY, USA, 1996.

[273] M.J. Frisch, G.W. Trucks, H.B. Schlegel, G.E. Scuseria, M.A. Robb, J.R. Cheeseman, G. Scalmani, V. Barone, G.A. Petersson, H. Nakatsuji, X. Li, M. Caricato, A.V. Marenich, J. Bloino, B.G. Janesko, R. Gomperts, B. Mennucci, H.P. Hratchian, J.V. Ortiz, A.F. Izmaylov, J.L. Sonnenberg, D. Williams-Young, F. Ding, F. Lipparini, F. Egidi, J. Goings, B. Peng, A. Petrone, T. Henderson, D. Ranasinghe, V.G. Zakrzewski, J. Gao, N. Rega, G. Zheng, W. Liang, M. Hada, M. Ehara, K. Toyota, R. Fukuda, J. Hasegawa, M. Ishida, T. Nakajima, Y. Honda, O. Kitao, H. Nakai, T. Vreven, K. Throssell, J.A. Montgomery Jr., J.E. Peralta, F. Ogliaro, M.J. Bearpark, J.J. Heyd, E.N. Brothers, K.N. Kudin, V.N. Staroverov, T.A. Keith, R. Kobayashi, J. Normand, K. Raghavachari, A.P. Rendell, J.C. Burant, S.S. Iyengar, J. Tomasi, M. Cossi, J.M. Millam, M. Klene, C. Adamo, R. Cammi, J.W. Ochterski, R.L. Martin, K. Morokuma, O. Farkas, J.B. Foresman, D.J. Fox, Gaussian 16 Revision A.03, Gaussian Inc., Wallingford CT, 2016.

[274] S. Rampino, CUBES: a library and a program suite for manipulating orbitals and densities, VIRT&L-COMM 7 (2015) 6.

[275] M.J.S. Dewar, Bulletin de la Société Chimique de France (1951) C71–C79.

[276] J. Chatt, L.A. Duncanson, 586. Olefin co-ordination compounds. Part III. Infra-red spectra and structure: attempted preparation of acetylene complexes, Journal of the Chemical Society (1953) 2939–2947.

[277] J. Jover, N. Fey, The computational road to better catalysts, Chemistry - An Asian Journal 9 (2014) 1714–1723.

[278] A. Salvadori, M. Fusè, G. Mancini, S. Rampino, V. Barone, Diving into chemical bonding: an immersive analysis of the electron charge rearrangement through virtual reality, Journal of Computational Chemistry 39 (2018) 2607–2617.

[279] G. Bistoni, S. Rampino, N. Scafuri, G. Ciancaleoni, D. Zuccaccia, L. Belpassi, F. Tarantelli, How π back-donation quantitatively controls the CO stretching response in classical and non-classical metal carbonyl complexes, Chemical Science 7 (2016) 1174–1184.

[280] H. Willner, J. Schaebs, G. Hwang, F. Mistry, R. Jones, J. Trotter, F. Aubke, Bis(carbonyl)gold(I) undecafluorodiantimonate(V), [Au(CO)$_2$][Sb$_2$F$_{11}$]: synthesis, vibrational, and carbon-13 NMR study and the molecular structure of bis(acetonitrile)gold(I) hexafluoroantimonate(V), [Au(NCCH$_3$)$_2$][SbF$_6$], Journal of the American Chemical Society 114 (1992) 8972–8980.

[281] B. Liang, L. Andrews, Reactions of laser-ablated Ag and Au atoms with carbon monoxide: matrix infrared spectra and density functional calculations on Ag(CO)$_n$ ($n = 2, 3$), Au(CO)$_n$ ($n = 1, 2$) and M(CO)$_n^+$ ($n = 1$-4; M = Ag, Au), The Journal of Physical Chemistry A 104 (2000) 9156–9164.

[282] C. Dash, P. Kroll, M. Yousufuddin, H.V.R. Dias, Isolable, gold carbonyl complexes supported by N-heterocyclic carbenes, Chemical Communications 47 (2011) 4478–4480.

[283] S. Martínez-Salvador, J. Forniés, A. Martín, B. Menjón, [Au(CF$_3$) (CO)]: a gold carbonyl compound stabilized by a trifluoromethyl group, Angewandte Chemie. International Edition 50 (2011) 6571–6574.

[284] D.B. Dell'Amico, F. Calderazzo, P. Robino, A. Segre, Halogenocarbonyl complexes of gold, Journal of the Chemical Society. Dalton Transactions (1991) 3017–3020.

[285] M. Joost, L. Estévez, S. Mallet-Ladeira, K. Miqueu, A. Amgoune, D. Bourissou, Enhanced π-backdonation from Gold(I): isolation of original carbonyl and carbene complexes, Angewandte Chemie. International Edition 53 (2014) 14512–14516.

[286] A.J. Lupinetti, S.H. Strauss, G. Frenking, Nonclassical Metal Carbonyls, John Wiley & Sons, Ltd, 2001, pp. 1–112.

[287] P.C. Kamer, P.W. van Leeuwen, Phosphorus (III) Ligands in Homogeneous Catalysis: Design and Synthesis, John Wiley & Sons, 2012.

[288] M. Fusè, I. Rimoldi, E. Cesarotti, S. Rampino, V. Barone, On the relation between carbonyl stretching frequencies and the donor power of chelating diphosphines in nickel dicarbonyl complexes, Physical Chemistry Chemical Physics 19 (2017) 9028–9038.

[289] M. Fusè, I. Rimoldi, G. Facchetti, S. Rampino, V. Barone, Exploiting coordination geometry to selectively predict the σ-donor and π-acceptor abilities of ligands: a back-and-forth journey between electronic properties and spectroscopy, Chemical Communications 54 (2018) 2397–2400.

[290] P. Pyykkö, J.P. Desclaux, Relativity and the periodic system of elements, Accounts of Chemical Research 12 (1979) 276–281.

[291] T. Saue, Relativistic Hamiltonians for chemistry: a primer, ChemPhysChem 12 (2011) 3077–3094.

[292] P.A.M. Dirac, The quantum theory of the electron, Proceedings of the Royal Society of London. Series A 117 (1928) 610–624.

[293] H.E. White, Pictorial representations of the Dirac electron cloud for hydrogen-like atoms, Physical Review 38 (1931) 513–520.

[294] H.M. Quiney, H. Skaane, I.P. Grant, Ab Initio Relativistic Quantum Chemistry: Four-Components Good, Two-Components Bad!, Advances in Quantum Chemistry, vol. 32, Academic Press, 1998, pp. 1–49.

[295] I.P. Grant, H.M. Quiney, Application of relativistic theories and quantum electrodynamics to chemical problems, International Journal of Quantum Chemistry 80 (2000) 283–297.

[296] H.M. Quiney, H. Skaane, I.P. Grant, Relativistic calculation of electromagnetic interactions in molecules, Journal of Physics. B, Atomic, Molecular and Optical Physics 30 (1997) L829.

[297] H.M. Quiney, P. Belanzoni, Relativistic density functional theory using Gaussian basis sets, Journal of Chemical Physics 117 (2002) 5550–5563.

[298] I.P. Grant, Relativistic Quantum Theory of Atoms and Molecules: Theory and Computation, Springer Series on Atomic, Optical, and Plasma Physics, Springer, Berlin, 2007.

[299] L. Belpassi, L. Storchi, H.M. Quiney, F. Tarantelli, Recent advances and perspectives in four-component Dirac-Kohn-Sham calculations, Physical Chemistry Chemical Physics 13 (2011) 12368–12394.

[300] S. Rampino, L. Belpassi, F. Tarantelli, L. Storchi, Full parallel implementation of an all-electron four-component Dirac-Kohn-Sham program, Journal of Chemical Theory and Computation 10 (2014) 3766–3776.

[301] B. Swirles, The relativistic self-consistent field, Proceedings of the Royal Society of London. Series A, Mathematical and Physical Sciences 152 (1935) 625–649.

[302] A.K. Rajagopal, J. Callaway, Inhomogeneous electron gas, Physical Review B 7 (1973) 1912–1919.

[303] G. Vignale, W. Kohn, Current-dependent exchange-correlation potential for dynamical linear response theory, Physical Review Letters 77 (1996) 2037–2040.

[304] S. Kümmel, L. Kronik, Orbital-dependent density functionals: theory and applications, Reviews of Modern Physics 80 (2008) 3–60.

[305] E.R. Scerri, The Periodic Table. A Very Short Introduction, Oxford University Press, Oxford, 2011.

[306] Y.T. Oganessian, F.S. Abdullin, P.D. Bailey, D.E. Benker, M.E. Bennett, S.N. Dmitriev, J.G. Ezold, J.H. Hamilton, R.A. Henderson, M.G. Itkis, Y.V. Lobanov, A.N. Mezentsev, K.J. Moody, S.L. Nelson, A.N. Polyakov, C.E. Porter, A.V. Ramayya, F.D. Riley, J.B. Roberto, M.A. Ryabinin, K.P. Rykaczewski, R.N. Sagaidak, D.A. Shaughnessy, I.V. Shirokovsky, M.A. Stoyer, V.G. Subbotin, R. Sudowe, A.M. Sukhov, Y.S. Tsyganov, V.K. Utyonkov, A.A. Voinov, G.K. Vostokin, P.A. Wilk, Synthesis of a new element with atomic number $Z = 117$, Physical Review Letters 104 (2010) 142502.

[307] Y.T. Oganessian, V.K. Utyonkov, Y.V. Lobanov, F.S. Abdullin, A.N. Polyakov, R.N. Sagaidak, I.V. Shirokovsky, Y.S. Tsyganov, A.A. Voinov, G.G. Gulbekian, S.L. Bogomolov, B.N. Gikal, A.N. Mezentsev, S. Iliev, V.G. Subbotin, A.M. Sukhov, K. Subotic, V.I. Zagrebaev, G.K. Vostokin, M.G. Itkis, K.J. Moody, J.B. Patin, D.A. Shaughnessy, M.A. Stoyer, N.J. Stoyer, P.A. Wilk, J.M. Kenneally, J.H. Landrum, J.F. Wild, R.W. Lougheed, Synthesis of the isotopes of elements 118 and 116 in the ^{249}Cf and ^{245}Cm+^{48}Ca fusion reactions, Physical Review C 74 (2006) 044602.

[308] S.-G. Wang, W. Schwarz, Icon of chemistry: the periodic system of chemical elements in the new century, Angewandte Chemie. International Edition 48 (2009) 3404–3415.

[309] B.F. Thornton, S.C. Burdette, The ends of elements, Nature Chemistry 5 (2013) 350–352.

[310] K.S. Pitzer, Are elements 112, 114, and 118 relatively inert gases?, Journal of Chemical Physics 63 (1975) 1032–1033.

[311] N. Gaston, I. Opahle, H. Gäggeler, P. Schwerdtfeger, Is eka-mercury (element 112) a group 12 metal?, Angewandte Chemie. International Edition 46 (2007) 1663–1666.

[312] R. Eichler, N.V. Aksenov, A.V. Belozerov, G.A. Bozhikov, V.I. Chepigin, S.N. Dmitriev, R. Dressler, H.W. Gäggeler, V.A. Gorshkov, F. Haenssler, M.G. Itkis, A. Laube, V.Y. Lebedev, O.N. Malyshev, Y.T. Oganessian, O.V. Petrushkin, D. Piguet, P. Rasmussen, S.V. Shishkin, A.V. Shutov, A.I. Svirikhin, E.E. Tereshatov, G.K. Vostokin, M. Wegrzecki, A.V. Yeremin, Chemical characterization of element 112, Nature 447 (2007) 72–75.

[313] R. Eichler, N. Aksenov, A. Belozerov, G. Bozhikov, V. Chepigin, S. Dmitriev, R. Dressler, H. Gäggeler, A. Gorshkov, M. Itkis, F. Haenssler, A. Laube, V. Lebedev, O. Malyshev, Y. Oganessian, O. Petrushkin, D. Piguet, A. Popeko, P. Rasmussen, S. Shishkin, A. Serov, A. Shutov, A. Svirikhin, E. Tereshatov, G. Vostokin, M. Wegrzecki, A. Yeremin, Thermochemical and physical properties of element 112, Angewandte Chemie. International Edition 47 (2008) 3262–3266.

[314] R. Eichler, N.V. Aksenov, Y.V. Albin, A.V. Belozerov, G.A. Bozhikov, V.I. Chepigin, S.N. Dmitriev, R. Dressler, H.W. Gäggeler, V.A. Gorshkov, G.S. Henderson, A. Johnsen, J.M. Kenneally, V.Y. Lebedev, V.Y. Lebedev, O. Malyshev, K. Moody, Y. Oganessian, O. Petrushkin, D. Piguet, A. Popeko, P. Rasmussen, S. Shishkin, A. Serov, A. Shutov, M. Stoyer,

N. Stoyer, A. Svirikhin, E. Tereshatov, G. Vostokin, M. Wegrzecki, P. Wilk, D. Wittwer, A. Yeremin, Indication for a volatile element 114, Radiochimica Acta 98 (2010) 133–139.

[315] P. Schwerdtfeger, One flerovium atom at a time, Nature Chemistry 5 (2013) 636.

[316] A. Yakushev, J.M. Gates, A. Türler, M. Schädel, C.E. Düllmann, D. Ackermann, L.-L. Andersson, M. Block, W. Brüchle, J. Dvorak, K. Eberhardt, H.G. Essel, J. Even, U. Forsberg, A. Gorshkov, R. Graeger, K.E. Gregorich, W. Hartmann, R.-D. Herzberg, F.P. Heßberger, D. Hild, A. Hübner, E. Jäger, J. Khuyagbaatar, B. Kindler, J.V. Kratz, J. Krier, N. Kurz, B. Lommel, L.J. Niewisch, H. Nitsche, J.P. Omtvedt, E. Parr, Z. Qin, D. Rudolph, J. Runke, B. Schausten, E. Schimpf, A. Semchenkov, J. Steiner, P. Thörle-Pospiech, J. Uusitalo, M. Wegrzecki, N. Wiehl, Superheavy element Flerovium (element 114) is a volatile metal, Inorganic Chemistry 53 (2014) 1624–1629.

[317] C.S. Nash, Atomic and molecular properties of elements 112, 114, and 118, The Journal of Physical Chemistry A 109 (2005) 3493–3500.

[318] D. Hofmann, Synthesis of superheavy elements by cold fusion, Radiochimica Acta 99 (2011) 405–428.

[319] Y.T. Oganessian, Synthesis of the heaviest elements in ^{48}Ca-induced reactions, Radiochimica Acta 99 (2011) 429–439.

[320] M. Schädel, Chemistry of superheavy elements, Angewandte Chemie. International Edition 45 (2006) 368–401.

[321] M. Schädel, Chemistry of superheavy elements, Radiochimica Acta 100 (2012) 579–604.

[322] A. Türler, V. Pershina, Advances in the production and chemistry of the heaviest elements, Chemical Reviews 113 (2013) 1237–1312.

[323] R. Eichler, First foot prints of chemistry on the shore of the island of superheavy elements, Journal of Physics. Conference Series 420 (2013) 012003.

[324] I. Zvára, The Inorganic Radiochemistry of Heavy Elements – Methods for Studying Gaseous Compounds, Springer, Dordrecht, 2008.

[325] S. Rampino, L. Storchi, L. Belpassi, Gold–superheavy-element interaction in diatomics and cluster adducts: a combined four-component Dirac-Kohn-Sham/charge-displacement study, Journal of Chemical Physics 143 (2015) 024307.

[326] R. Zalesny, M.G. Papadopoulos, P.G. Mezey, J. Leszczynski (Eds.), Linear-Scaling Techniques in Computational Chemistry and Physics, Springer, Dordrecht, Heidelberg, London, New York, 2011.

[327] A. Laganà, A. Costantini, O. Gervasi, N. Faginas Lago, C. Manuali, S. Rampino, COMPCHEM: progress towards GEMS a grid empowered molecular simulator and beyond, Journal of Grid Computing 8 (2010) 571–586.

[328] S. Rampino, Workflows and data models for atom diatom quantum reactive scattering calculations on the Grid, Ph.D. thesis, Università degli Studi di Perugia, 2011.

[329] C. Manuali, A. Laganà, S. Rampino, GriF: a grid framework for a web service approach to reactive scattering, Computer Physics Communications 181 (2010) 1179–1185.

[330] S. Rampino, N. Faginas Lago, A. Laganà, F. Huarte-Larrañaga, An extension of the grid empowered molecular simulator to quantum reactive scattering, Journal of Computational Chemistry 33 (2012) 708–714.

[331] S. Rampino, D. Skouteris, A. Laganà, E. Garcia, A comparison of the isotope effect for the N + N$_2$ reaction calculated on two potential energy surfaces, in: S. Misra, O. Gervasi, B. Murgante, E. Stankova, V. Korkhov, C. Torre, A.M.A. Rocha, D. Taniar, B.O. Apduhan, E. Tarantino (Eds.), Computational Science and Its Applications – ICCSA 2008, in: Lecture Notes in Computer Science, vol. 5072, Springer Berlin Heidelberg, 2008, pp. 1081–1093.

[332] A. Laganà, N. Faginas Lago, S. Rampino, F. Huarte Larrañaga, E. García, Thermal rate coefficients in collinear versus bent transition state reactions: the N + N$_2$ case study, Physica Scripta 78 (2008) 058116.

[333] S. Rampino, F. Pirani, E. Garcia, A. Laganà, A study of the impact of long range interactions on the reactivity of N + N$_2$ using the Grid Empowered Molecular Simulator GEMS, International Journal of Web and Grid Services 6 (2010) 196–212.

[334] A. Laganà, S. Rampino, A grid empowered virtual versus real experiment for the barrierless Li + FH → LiF + H reaction, in: Computational Science and Its Applications – ICCSA 2014, in: Lecture Notes in Computer Science, vol. 8579, Springer International Publishing, 2014, pp. 571–584.

[335] S. Rampino, A. Monari, E. Rossi, S. Evangelisti, A. Laganà, A priori modeling of chemical reactions on computational grid platforms: workflows and data models, Chemical Physics 398 (2012) 192–198.

[336] S. Rampino, D. Skouteris, A. Laganà, E. García, A. Saracibar, A comparison of the quantum state-specific efficiency of N + N_2 reaction computed on different potential energy surfaces, Physical Chemistry Chemical Physics 11 (2009) 1752–1757.

[337] S. Rampino, E. Garcia, F. Pirani, A. Laganà, Accurate quantum dynamics on grid platforms: some effects of long range interactions on the reactivity of N + N_2, in: D. Taniar, O. Gervasi, B. Murgante, E. Pardede, B.O. Apduhan (Eds.), Computational Science and Its Applications – ICCSA 2010, in: Lecture Notes in Computer Science, vol. 6019, Springer Berlin Heidelberg, 2010, pp. 1–12.

[338] E. Rossi, S. Evangelisti, A. Laganà, A. Monari, S. Rampino, M. Verdicchio, K.K. Baldridge, G.L. Bendazzoli, S. Borini, R. Cimiraglia, C. Angeli, P. Kallay, H.P. Lüthi, K. Ruud, J. Sanchez-Marin, A. Scemama, P.G. Szalay, A. Tajti, Code interoperability and standard data formats in quantum chemistry and quantum dynamics: the Q5/D5Cost data model, Journal of Computational Chemistry 35 (2014) 611–621.

[339] The European grid infrastructure, http://www.egi.eu/. (Accessed 22 December 2021).

[340] L. Storchi, S. Rampino, L. Belpassi, F. Tarantelli, H.M. Quiney, Efficient parallel all-electron four-component Dirac–Kohn–Sham program using a distributed matrix approach II, Journal of Chemical Theory and Computation 9 (2013) 5356–5364.

[341] M. Valle, Visualization: a cognition amplifier, International Journal of Quantum Chemistry 113 (2013) 2040–2052.

[342] H.M. Nussenzveig, Diffraction Effects in Semiclassical Scattering, Cambridge University Press, Cambridge (UK), 1992.

[343] J.L. Atwood, L.J. Barbour, Molecular graphics: from science to art, Crystal Growth & Design 3 (2003) 3–8.

[344] X. Martinez, M. Krone, N. Alharbi, A.S. Rose, R.S. Laramee, S. O'Donoghue, M. Baaden, M. Chavent, Molecular graphics: bridging structural biologists and computer scientists, Structure 27 (2019) 1617–1623.

[345] A. van Dam, A.D. Forsberg, D.H. Laidlaw, J.J. LaViola, R.M. Simpson, Immersive VR for scientific visualization: a progress report, IEEE Computer Graphics and Applications 20 (2000) 26–52.

[346] C. Cruz-Neira, D.J. Sandin, T.A. DeFanti, R.V. Kenyon, J.C. Hart, The CAVE: audio visual experience automatic virtual environment, Communications of the ACM 35 (1992) 64–72.

[347] C. Cruz-Neira, D.J. Sandin, T.A. DeFanti, Surround-screen projection-based virtual reality: the design and implementation of the CAVE, in: Proceedings of the 20th Annual Conference on Computer Graphics and Interactive Techniques, SIGGRAPH '93, ACM, New York, NY, USA, 1993, pp. 135–142.

[348] W.-D. Ihlenfeldt, Virtual reality in chemistry, Molecular Modeling Annual 3 (1997) 386–402.

[349] A. Fombona-Pascual, J. Fombona, E. Vázquez-Cano, VR in chemistry, a review of scientific research on advanced atomic/molecular visualization, Chemistry Education Research and Practice (2022), https://doi.org/10.1039/D1RP00317H, published online.

[350] Z.A. Jiménez, Teaching and Learning Chemistry via Augmented and Immersive Virtual Reality, 2019, pp. 31–52, Chapter 3.

[351] B. Chavez, S. Bayona, Virtual reality in the learning process, in: Á. Rocha, H. Adeli, L.P. Reis, S. Costanzo (Eds.), Trends and Advances in Information Systems and Technologies, Springer International Publishing, Cham, 2018, pp. 1345–1356.

[352] K. Mainzer, Computational models and virtual reality: new perspectives of research in chemistry, HYLE - An International Journal for the Philosophy of Chemistry 5 (1999) 117–126.

[353] A. Aspuru-Guzik, R. Lindh, M. Reiher, The matter simulation (r)evolution, ACS Central Science 4 (2018) 144–152.

[354] K.H. Marti, M. Reiher, Haptic quantum chemistry, Journal of Computational Chemistry 30 (2009) 2010–2020.

[355] M.P. Haag, K.H. Marti, M. Reiher, Generation of potential energy surfaces in high dimensions and their haptic exploration, ChemPhysChem 12 (2011) 3204–3213.

[356] M.P. Haag, M. Reiher, Studying chemical reactivity in a virtual environment, Faraday Discussions 169 (2014) 89–118.

[357] M. O'Connor, H.M. Deeks, E. Dawn, O. Metatla, A. Roudaut, M. Sutton, L.M. Thomas, B.R. Glowacki, R. Sage, P. Tew, M. Wonnacott, P. Bates, A.J. Mulholland, D.R. Glowacki, Sampling molecular conformations and dynamics in a multiuser virtual reality framework, Science Advances 4 (2018) eaat2731.

[358] S. Amabilino, L.A. Bratholm, S.J. Bennie, A.C. Vaucher, M. Reiher, D.R. Glowacki, Training neural nets to learn reactive potential energy surfaces using interactive quantum chemistry in virtual reality, Journal of Chemical Physics 123 (2019) 4486–4499.

[359] M.B. O'Connor, S.J. Bennie, H.M. Deeks, A. Jamieson-Binnie, A.J. Jones, R.J. Shannon, R. Walters, T.J. Mitchell, A.J. Mulholland, D.R. Glowacki, Interactive molecular dynamics in virtual reality from quantum chemistry to drug binding: an open-source multi-person framework, Journal of Chemical Physics 150 (2019) 220901.

[360] M. Martino, A. Salvadori, F. Lazzari, L. Paoloni, S. Nandi, G. Mancini, V. Barone, S. Rampino, Chemical promenades: exploring potential-energy surfaces with immersive virtual reality, Journal of Computational Chemistry 41 (2020) 1310–1323.

[361] D. Cremer, J.A. Pople, General definition of ring puckering coordinates, Journal of the American Chemical Society 97 (1975) 1354–1358.

[362] D. Cremer, Calculation of puckered rings with analytical gradients, Journal of Physical Chemistry 94 (1990) 5502–5509.

[363] L. Paoloni, S. Rampino, V. Barone, Potential-energy surfaces for ring-puckering motions of flexible cyclic molecules through Cremer-Pople coordinates: computation, analysis, and fitting, Journal of Chemical Theory and Computation 15 (2019) 4280–4294.

[364] J. Lupi, M. Martino, A. Salvadori, S. Rampino, G. Mancini, V. Barone, Virtual reality tools for advanced modeling, AIP Conference Proceedings 2145 (2019) 020001.

[365] A. Salvadori, G. Del Frate, M. Pagliai, G. Mancini, V. Barone, Immersive virtual reality in computational chemistry: applications to the analysis of QM and MM data, International Journal of Quantum Chemistry 116 (2016) 1731–1746.

[366] K.T. Butler, D.W. Davies, H. Cartwright, O. Isayev, A. Walsh, Machine learning for molecular and materials science, Nature 559 (2018) 547–555.

[367] J. Gasteiger, Chemistry in times of artificial intelligence, ChemPhysChem 21 (2020) 2233–2242.

[368] N. Brown, P. Ertl, R. Lewis, T. Luksch, D. Reker, N. Schneider, Artificial intelligence in chemistry and drug design, Journal of Computer-Aided Molecular Design 34 (2020) 709–715.

[369] N. Artrith, K.T. Butler, F.-X. Coudert, S. Han, O. Isayev, A. Jain, A. Walsh, Best practices in machine learning for chemistry, Nature Chemistry 13 (2021) 505–508.

[370] J.A. Keith, V. Vassilev-Galindo, B. Cheng, S. Chmiela, M. Gastegger, K.-R. Müller, A. Tkatchenko, Combining machine learning and computational chemistry for predictive insights into chemical systems, Chemical Reviews 121 (2021) 9816–9872.

[371] J. Westermayr, M. Gastegger, K.T. Schütt, R.J. Maurer, Perspective on integrating machine learning into computational chemistry and materials science, Journal of Chemical Physics 154 (2021) 230903.

[372] A. Agrawal, A. Choudhary, Perspective: materials informatics and big data: realization of the "fourth paradigm" of science in materials science, APL Materials 4 (2016) 053208.

[373] C. Hansch, T. Fujita, p-σ-π analysis. A method for the correlation of biological activity and chemical structure, Journal of the American Chemical Society 86 (1964) 1616–1626.

[374] S.M. Free, J.W. Wilson, A mathematical contribution to structure-activity studies, Journal of Medicinal Chemistry 7 (1964) 395–399.

[375] T.M. Mitchell, Machine Learning, Mc Graw-Hill, Cambridge, 1997.

[376] M. Leshno, V.Y. Lin, A. Pinkus, S. Schocken, Multilayer feedforward networks with a non-polynomial activation function can approximate any function, Neural Networks 6 (1993) 861–867.

[377] J.P. Unsleber, M. Reiher, The exploration of chemical reaction networks, Annual Review of Physical Chemistry 71 (2020) 121–142.

[378] T. Akitsu (Ed.), Computational and Data-Driven Chemistry Using Artificial Intelligence, Elsevier, Amsterdam, Oxford, Cambridge (US), 2022.

[379] D. Licari, S. Rampino, V. Barone, Machine learning of potential-energy surfaces within a bond-order sampling scheme, in: S. Misra, O. Gervasi, B. Murgante, E. Stankova, V. Korkhov, C. Torre, A.M.A. Rocha, D. Taniar, B.O. Apduhan, E. Tarantino (Eds.), Computational Science and Its Applications – ICCSA 2019, in: Lecture Notes in Computer Science, vol. 11624, Springer International Publishing, Cham, 2019, pp. 388–400.

[380] A. Tkatchenko, Machine learning for chemical discovery, Nature Communications 11 (2020) 4125.

[381] S. Bartling, S. Friesike, Towards another scientific revolution, in: S. Bartling, S. Friesike (Eds.), Opening Science: The Evolving Guide on How the Internet Is Changing Research, Collaboration and Scholarly Publishing, Springer International Publishing, Cham, 2014, pp. 3–15.

[382] M. Woelfle, P. Olliaro, M.H. Todd, Open science is a research accelerator, Nature Chemistry 3 (2011) 745–748.

[383] R. Vicente-Saez, C. Martinez-Fuentes, Open science now: a systematic literature review for an integrated definition, Journal of Business Research 88 (2018) 428–436.

[384] B. Fecher, S. Friesike, Open science: one term, five schools of thought, in: S. Bartling, S. Friesike (Eds.), Opening Science: The Evolving Guide on How the Internet Is Changing Research, Collaboration and Scholarly Publishing, Springer International Publishing, Cham, 2014, pp. 17–47.

[385] J.D. Gezelter, Open source and open data should be standard practices, Journal of Physical Chemistry Letters 6 (2015) 1168–1169.

[386] J.P.A. Ioannidis, Why most published research findings are false, PLoS Medicine 2 (2005) e124.

[387] F. Prinz, T. Schlange, K. Asadullah, Believe it or not: how much can we rely on published data on potential drug targets?, Nature Reviews Drug Discovery 10 (2011) 712.

[388] F.-X. Coudert, Reproducible research in computational chemistry of materials, Chemistry of Materials 29 (2017) 2615–2617.

[389] W.P. Walters, Code sharing in the open science era, Journal of Chemical Information and Modeling 60 (2020) 4417–4420.

[390] A.I. Krylov, J.M. Herbert, F. Furche, M. Head-Gordon, P.J. Knowles, R. Lindh, F.R. Manby, P. Pulay, C.-K. Skylaris, H.-J. Werner, What is the price of open-source software?, Journal of Physical Chemistry Letters 6 (2015) 2751–2754.

[391] C.R. Jacob, How open is commercial scientific software?, Journal of Physical Chemistry Letters 7 (2016) 351–353.

[392] A. Hocquet, F. Wieber, Epistemic issues in computational reproducibility: software as the elephant in the room, European Journal for Philosophy of Science 11 (2021) 38.

[393] W. Bittremieux, C. Adams, K. Laukens, P.C. Dorrestein, N. Bandeira, Open science resources for the mass spectrometry-based analysis of SARS-CoV-2, Journal of Proteome Research 20 (2021) 1464–1475.

[394] S. Rampino, A. Monari, S. Evangelisti, E. Rossi, K. Ruud, A. Laganà, A priori modeling of chemical reactions on a grid based virtual laboratory, in: M. Bubak, M. Turala, K. Wiatr (Eds.), Proceedings of the Cracow Grid Workshop 2009 (CGW'09), Cracow, Poland, 2010, pp. 164–171.

[395] N.M. O'Boyle, R. Guha, E.L. Willighagen, S.E. Adams, J. Alvarsson, J.-C. Bradley, I.V. Filippov, R.M. Hanson, M.D. Hanwell, G.R. Hutchison, C.A. James, N. Jeliazkova, A.S. Lang, K.M. Langner, D.C. Lonie, D.M. Lowe, J. Pansanel, D. Pavlov, O. Spjuth, C. Steinbeck, A.L. Tenderholt, K.J. Theisen, P. Murray-Rust, Open data, open source and open standards in chemistry: the blue obelisk five years on, Journal of Cheminformatics 3 (2011) 37.

[396] I. Sullivan, A. DeHaven, D. Mellor, Open and reproducible research on open science framework, Current Protocols Essential Laboratory Techniques 18 (2019) e32.

[397] M. Altunay, P. Avery, K. Blackburn, B. Bockelman, M. Ernst, D. Fraser, R. Quick, R.W. Gardner, S. Goasguen, T. Levshina, M. Livny, J. McGee, D. Olson, R. Pordes, M. Potekhin, A. Rana, A. Roy, C. Sehgal, I. Sfiligoi, F. Würthwein, A science driven production cyberinfrastructure - the open science grid, Journal of Grid Computing 9 (2011) 201–218.

[398] R. Pordes, D. Petravick, B. Kramer, D. Olson, M. Livny, A. Roy, P. Avery, K. Blackburn, T. Wenaus, F. Würthwein, I. Foster, R. Gardner, M. Wilde, A. Blatecky, J. McGee, R. Quick, The open science grid, Journal of Physics. Conference Series 78 (2007) 012057.

[399] P. Kacsuk, G. Terstyánszky, Á. Balaskó, K. Karóczkai, Z. Farkas, Executing multi-workflow simulations on a mixed grid/cloud infrastructure using the SHIWA and SCI-BUS technology, in: E.H. D'Hollander, J.J. Dongarra, I.T. Foster, L. Grandinetti, G.R. Joubert (Eds.), Transition of HPC Towards Exascale Computing, in: Advances in Parallel Computing, vol. 24, IOS Press, 2013, pp. 141–160.

[400] S. Sild, U. Maran, A. Lomaka, M. Karelson, Open computing grid for molecular science and engineering, Journal of Chemical Information and Modeling 46 (2006) 953–959.

[401] J. Krüger, R. Grunzke, S. Gesing, S. Breuers, A. Brinkmann, L. de la Garza, O. Kohlbacher, M. Kruse, W.E. Nagel, L. Packschies, R. Müller-Pfefferkorn, P. Schäfer, C. Schärfe, T. Steinke, T. Schlemmer, K.D. Warzecha, A. Zink, S. Herres-Pawlis, The MoSGrid science gateway - a complete solution for molecular simulations, Journal of Chemical Theory and Computation 10 (2014) 2232–2245.

[402] S. Rampino, L. Storchi, A. Laganà, Automated simulation of gas-phase reactions on distributed and cloud computing infrastructures, in: Computational Science and Its Applications – ICCSA 2017, Springer International Publishing, Cham, 2017, pp. 60–73.

[403] A. Laganà, G. Terstyanszky, J. Krüger, Open molecular science for the open science cloud, in: O. Gervasi, B. Murgante, S. Misra, G. Borruso, C.M. Torre, A.M.A. Rocha, D. Taniar, B.O. Apduhan, E. Stankova, A. Cuzzocrea (Eds.), Computational Science and Its Applications – ICCSA 2017, Springer International Publishing, Cham, 2017, pp. 29–43.

Index